THE OFFSHORING OF ENGINEERING

Facts, Unknowns, and Potential Implications

Committee on the Offshoring of Engineering

NATIONAL ACADEMY OF ENGINEERING
OF THE NATIONAL ACADEMIES

THE NATIONAL ACADEMIES PRESS
Washington, D.C.
www.nap.edu

THE NATIONAL ACADEMIES PRESS 500 Fifth Street, N.W. Washington, DC 20001

NOTICE: The project that is the subject of this report was approved by the Governing Board of the National Research Council, whose members are drawn from the councils of the National Academy of Science, the National Academy of Engineering, and the Institute of Medicine. The members of the committee responsible for the report were chosen for their special competences and with regard for appropriate balance.

Support for this project was provided by the National Science Foundation (Award No. EEC-0434993), the United Engineering Foundation, a generous gift from NAE member Gordon Bell, and the National Academy of Engineering Fund. The views presented in this report are those of the National Academy of Engineering Committee on the Offshoring of Engineering and are not necessarily those of the funding organizations.

Library of Congress Cataloging-in-Publication Data

National Academy of Engineering. Committee on the Offshoring of Engineering.
The offshoring of engineering : facts, unknowns, and potential implications / Committee on the Offshoring of Engineering, National Academy of Engineering.
p. cm.
"February 2008."
Includes bibliographical references.
ISBN 978-0-309-11483-7 (pbk.) — ISBN 978-0-309-11484-4 (pdf) 1. Engineering—United States—Management. 2. Offshore outsourcing—United States. I. Title.
TA23.N275 2008
620.0068'4—dc22
2008013220

Copies of this report are available from the National Academies Press, 500 Fifth Street, N.W., Lockbox 285, Washington, DC 20055; (888) 624-8373 or (202) 334-3313 (in the Washington metropolitan area); online at *http://www.nap.edu.*

Printed in the United States of America

THE NATIONAL ACADEMIES

Advisers to the Nation on Science, Engineering, and Medicine

The **National Academy of Sciences** is a private, nonprofit, self-perpetuating society of distinguished scholars engaged in scientific and engineering research, dedicated to the furtherance of science and technology and to their use for the general welfare. Upon the authority of the charter granted to it by the Congress in 1863, the Academy has a mandate that requires it to advise the federal government on scientific and technical matters. Dr. Ralph J. Cicerone is president of the National Academy of Sciences.

The **National Academy of Engineering** was established in 1964, under the charter of the National Academy of Sciences, as a parallel organization of outstanding engineers. It is autonomous in its administration and in the selection of its members, sharing with the National Academy of Sciences the responsibility for advising the federal government. The National Academy of Engineering also sponsors engineering programs aimed at meeting national needs, encourages education and research, and recognizes the superior achievements of engineers. Dr. Charles M. Vest is president of the National Academy of Engineering.

The **Institute of Medicine** was established in 1970 by the National Academy of Sciences to secure the services of eminent members of appropriate professions in the examination of policy matters pertaining to the health of the public. The Institute acts under the responsibility given to the National Academy of Sciences by its congressional charter to be an adviser to the federal government and, upon its own initiative, to identify issues of medical care, research, and education. Dr. Harvey V. Fineberg is president of the Institute of Medicine.

The **National Research Council** was organized by the National Academy of Sciences in 1916 to associate the broad community of science and technology with the Academy's purposes of furthering knowledge and advising the federal government. Functioning in accordance with general policies determined by the Academy, the Council has become the principal operating agency of both the National Academy of Sciences and the National Academy of Engineering in providing services to the government, the public, and the scientific and engineering communities. The Council is administered jointly by both Academies and the Institute of Medicine. Dr. Ralph J. Cicerone and Dr. Charles M. Vest are chair and vice chair, respectively, of the National Research Council.

www.national-academies.org

COMMITTEE ON THE OFFSHORING OF ENGINEERING

WILLIAM J. SPENCER *(chair)*, Chairman Emeritus, SEMATECH
LINDA M. ABRIOLA, Dean of Engineering, Tufts University
PETER R. BRIDENBAUGH, Retired Executive Vice President, Automotive, Aluminum Company of America
STEPHEN W. DREW, Science Partners LLC
SAMUEL C. FLORMAN, Chairman, Kreisler Borg Florman General Construction Company
SUSAN L. GRAHAM, Pehong Chen Distinguished Professor, Computer Science Division-EECS, University of California, Berkeley
LORI G. KLETZER, Professor of Economics and Department Chair, University of California, Santa Cruz
ANNE L. STEVENS, Chairman, President, and CEO, Carpenter Technology Corporation
GEORGE J. TAMARO, Partner, Mueser Rutledge Consulting Engineers
MARIE C. THURSBY, Hal and John Smith Chair in Entrepreneurship, DuPree College of Management, Georgia Institute of Technology

Project Staff

THOMAS ARRISON, Study Director, Policy and Global Affairs, National Academies
CAROL R. ARENBERG, Senior Editor, National Academy of Engineering
PENELOPE J. GIBBS, Senior Program Associate, Program Office, National Academy of Engineering
ROBERT P. MORGAN, Consultant
PROCTOR P. REID, Director, Program Office, National Academy of Engineering

Preface

In recent years, the offshoring of high-skill service jobs previously performed in the United States has attracted a great deal of media attention and sparked a spirited policy debate. The decline in U.S. manufacturing jobs relative to the total workforce is a decades-long trend driven by the expansion of international trade in goods and increases in manufacturing productivity. Several important changes in the business environment in the late 1990s facilitated the emergence and rapid growth of services offshoring, including the offshoring of activities with significant engineering content. These changes include advances in information technology, an increase in the demand for certain types of technical skills, and the emergence of appropriately skilled, low-wage workforces in India, China, and elsewhere.

Criticism of offshoring and the presumed "hollowing out" of the U.S. engineering workforce are reminiscent of the debates of 20 years ago about U.S. standing in international trade and manufacturing industries. A number of groups and prominent individuals have long argued that offshoring hurts U.S. workers and the U.S. economy. Others counter that offshoring is a benign trend that enables U.S.-based companies and entrepreneurs to develop and market innovations more quickly and cost effectively.

Several reports and statements by U.S. science and engineering organizations—including the National Academies report *Rising Above the Gathering Storm* (NAS/NAE/IOM, 2007)—that have been published concurrently with the offshoring debate have argued that long-term U.S. leadership in science and engineering is at risk. Almost all of them express a central concern that if U.S. companies increasingly move R&D offshore to China, India, and other locations that provide high value in terms of science and engineering human resources, America's ability to innovate and sustain economic growth would be seriously undermined, leading to a long-term decline. As the present report goes to press in mid-2008, in the midst of a presidential election campaign and a slowdown in the U.S. economy, the globalization of engineering work remains in the news and is still being hotly debated (Shirouzu, 2008; Valcourt, 2008).

Throughout the debate about the costs and benefits of offshoring for the U.S. economy and U.S. workers, arguments on both sides have been bolstered by a variety of anecdotes and statistics. Surprisingly, however, little is definitively known about the effects of offshoring on overall services or on specific engineering subfields in particular industries. We do know, despite the paucity of definitive data, that we are in the midst of important global shifts in how and where engineering is being practiced and that these shifts will have major long-term effects on the U.S. engineering enterprise, including engineering education, practice, and management.

In January 2006, Wm. A. Wulf, then president of the National Academy of Engineering (NAE), appointed an ad hoc committee of experts to organize, conduct, and plan a public workshop on engineering offshoring and prepare a summary report of the proceedings. The committee met in Washington, D.C., in April 2006 to plan the workshop and other fact-finding activities and to evaluate proposals for commissioned papers on engineering offshoring in specific industry sectors to be presented at the workshop. Approximately 100 participants were invited to attend the two-day event in October 2006 at the facilities of the National Academies in Washington, D.C. Following the meeting, the committee developed its summary report.

This volume includes the committee's summary and findings, the commissioned papers, and several edited presentations from the workshop. Taken together, these documents provide a snapshot of the current state of knowledge about engineering offshoring in six major industrial sectors, identify gaps in knowledge and future areas for research, and suggest implications for the U.S. engineering enterprise, including educational institutions, industry, government, engineering societies, and individual engineers.

On behalf of NAE, I thank the committee chair, William J. Spencer, and the committee members for their considerable efforts on this project. I also want to thank Thomas Arrison, the study director, who managed the project; Proctor P. Reid, director of the NAE Program Office, who provided oversight and was actively involved in the workshop and the completion of the report; Penelope Gibbs and Nathan Kahl from the NAE Program Office who provided critical administrative and logistical support; Carol Arenberg, NAE senior editor, who was instrumental in preparing the report for publication; and Robert P. Morgan, former NAE Fellow, who prepared an extensive background paper for the committee and assisted the NAE Council and NAE Program Office in the development of the project.

I also extend the committee's thanks to the authors of the commissioned papers, workshop attendees, and others who contributed to the project. Finally, I would like to express my appreciation to the National Science Foundation and the United Engineering Foundation for their generous support.

Charles M. Vest
President
National Academy of Engineering

REFERENCES

NAS/NAE/IOM. 2007. Rising Above the Gathering Storm. Washington, D.C.: The National Academies Press. Available online at *http://www.nap.edu/catalog.php?record_id=11463*.

Shirouzu, N. 2008. Engineering Jobs Become Car Makers' New Export. *Wall Street Journal*, February 7, p. A13.

Valcourt, J. 2008. Chrysler Begins Overhaul in Engineering. *Wall Street Journal*, February 19, p. A13.

Acknowledgments

This report was reviewed in draft form by individuals chosen for their diverse perspectives and technical expertise, in accordance with procedures approved by NAE. The purpose of the independent review is to provide candid and critical comments that will assist NAE in making its published report as sound as possible and to ensure that the report meets institutional standards for objectivity, evidence, and responsiveness to the study charge. The review comments and draft manuscript remain confidential to protect the integrity of the deliberative process. We wish to thank the following individuals for their review of this report:

Cristina H. Amon, University of Toronto
Erich Bloch, Washington Advisory Group
David Cheney, SRI International
Ron Hira, Rochester Institute of Technology
Louis Martin-Vega, North Carolina State University
Paul S. Peercy, University of Wisconsin-Madison
Hal Salzman, Urban Institute
Anna Lee Saxenian, University of California, Berkeley
Adrian Zaccaria, Bechtel Group Inc.

Although the reviewers listed above have provided many constructive comments and suggestions, they were not asked to endorse the conclusions or recommendations, nor did they see the final draft of the report before its release. The review of this report was overseen by George Hornberger, University of Virginia. Appointed by NAE, he was responsible for making certain that an independent examination of this report was carried out in accordance with institutional procedures and that all review comments were carefully considered. Responsibility for the final content of this report rests entirely with the authoring committee and NAE.

Contents

PART II: COMMISSIONED PAPERS AND WORKSHOP PRESENTATIONS

Commissioned Papers

Workshop Presentations

Appendixes

Executive Summary

Spurred in part by a decades-long decline in manufacturing employment, the implications of globalization for the United States are a source of considerable debate. The emergence of "offshoring"—the transfer of work from the United States to affiliated and unaffiliated entities abroad—has raised additional concerns about the impacts of globalization. Among the occupations subject to offshoring are highly paid professions, including engineering, that are essential to U.S. technological progress, economic growth, and national security.

The National Academy of Engineering (NAE) recognizes that offshoring raises significant challenges not only for engineers themselves, but also for industry, educational institutions, government, and professional societies. Many engineering tasks can now be performed anywhere in the world by qualified professionals with access to appropriate connectivity. To sustain and strengthen U.S. engineering capabilities in this new environment, the United States may need to consider new approaches to education, career development, management, and policy, and make changes where appropriate.

NAE launched this project in 2006 with support from the National Science Foundation, United Engineering Foundation, and internal NAE funds. In the preliminary discussions, it became clear that developing policy recommendations would not be possible based on available data and information in the literature. Therefore, a major goal of this study is to assess the knowledge base and identify gaps, data needs, and areas for future study. The focus of the project was a public workshop featuring the discussion of commissioned papers on the offshoring of engineering in six industry sectors—software development, semiconductors, personal computer (PC) manufacturing, automobiles, construction engineering and services, and pharmaceuticals—and presentations by experts on engineering education and management, the engineering workforce, and the engineering profession. The study committee is aware that not all industries or aspects of engineering were included.

OFFSHORING OF ENGINEERING: TRENDS AND IMPACTS

FINDING 1. *The offshoring of engineering, an inevitable aspect of globalization, has significantly impacted the U.S. engineering enterprise. However, the effects of globalization and offshoring have been uneven, and disparities among industry sectors and engineering sectors are likely to continue.*

One area of rapid increase in offshoring has been in information-technology (IT)-related industries, such as software development, semiconductors, and PC manufacturing. Today both established U.S.-based firms and start-ups are locating at least some engineering work in India or China. In fact, this offshoring is now taken for granted, and reportedly is even required, by some venture capitalists (Hira, 2005). Employment and exports in the Indian software-services industry have grown at annual rates of 30 to 40 percent over the past decade. In the semiconductor industry, 18 of the top 20 U.S.-based companies have opened design centers in India, nine of them since 2004. In the PC industry, much of the product design and engineering work is done by original design manufacturers based mainly in Taiwan; manufacturing is increasingly being done in China.

In the automotive and construction engineering and services industries, engineering activity has long been internationalized. However, in the automotive industry today, engineering workforces are increasingly being configured to develop global platforms, rather than to work on products targeting local markets. Construction engineering and services firms that operate globally have always required engineering help in the countries where projects are located. Today, overseas engineers are increasingly performing tasks related to U.S. projects. In general, offshoring of less complex engineering work is increasing in both the automotive and construction industries.

Finally, offshoring of research and development (R&D) to developing and emerging economies such as China and India is increasing rapidly in pharmaceuticals and some other industries. More than half of more than 200 U.S.- and Europe-based companies that responded to a recent survey anticipate that their technical workforces in China, India, and other parts of Asia will increase in the next three years (Thursby and Thursby, 2006).

FINDING 2. *More and better data on offshoring and other issues discussed in this report, such as the effects on the engineering workforce and engineering education, are necessary for discerning overall trends. As has been pointed out in other recent reports, better U.S. and international statistics on trade in services and employment would give us a much better grasp of basic trends.*

With the emergence of offshoring, a growing portion of the U.S. workforce, including engineers and many other services professionals, have become subject to international competition. For the United States to adopt policies that support continued economic vitality and ensure that the United States remains a premier location for engineering work, policy makers must have a good understanding of changes in comparative salaries, education levels, language skills, productivity and other trends, and the causes of those trends.

Unfortunately, current published estimates and projections on offshoring of engineering include significant uncertainties. McKinsey Global Institute (2005), for example, estimates that more than half of engineering jobs in the industries it analyzed could be performed anywhere in the world. However, it would be wrong to conclude that half of the 1.5 to 2 million U.S. engineers are in danger of losing their jobs in the next few years. Indeed, the U.S. engineering workforce is expected to grow by 13 percent between 2004 and 2014 (CPST, 2006), a substantial increase although smaller than the expected increase in the workforce as a whole. In addition, there are limits to how quickly India and China can improve the quality and increase the quantity of their engineering graduates.

Significant data gaps have prevented policy makers and the public from getting an accurate read on trade in services and offshoring (GAO, 2005a,b; NAPA, 2006; Sturgeon, 2006; etc.), and it may be some time before the most glaring deficiencies are addressed. One difficulty is that offshoring within companies is difficult to track through trade statistics. Another difficulty is that companies are reluctant to make information about their offshoring practices public. Thus industry-specific analyses will continue to be important sources of information but can only provide a snapshot of a rapidly changing phenomenon.

FINDING 3. *Offshoring appears to have contributed to the competitive advantage of U.S.-based firms in a variety of industries, and the negative impacts of offshoring on U.S. engineering appear to have been relatively modest to date. However, the negative effects have been much more severe in some industry sectors and for some jobs than others.*

Global disaggregation, a long-standing aspect of business models in several U.S. industries, has enabled U.S.-based companies in the semiconductor and PC industries to establish and retain global leadership. The key to long-term success for companies that offshore engineering activities is protecting the interface with customers and the resulting information flow, which feeds into product definition, high-level design, and sophisticated engineering tasks.

Cutting costs was the initial motivation for offshoring of services, including engineering, especially in IT-related industries. However, a major factor in the offshoring of R&D facilities to emerging economies, such as China, is the desire to establish a full-spectrum presence in a rapidly growing market. On the flip side, there has been significant "onshoring" of R&D and other engineering work in some industries as multinational companies based in Europe and Asia establish or acquire operations in the United States. Even some companies based in India and China are investing in R&D in the United States, mainly through acquisitions (see Cooney, this volume).

Although the inadequacy of available data makes it difficult to measure the negative impacts of offshoring on engineering jobs and salaries, we can say that the negative impacts have not been evenly distributed. It is logical to infer that, when certain types of routine engineering tasks are sourced in India or China, the U.S. engineers who performed that work lose their jobs. Even though new jobs may be created for U.S. engineers who perform higher level tasks and those who can move to other sectors, those new jobs do not replace the jobs that were lost. The negative individual and social impacts of mass layoffs in general, not necessarily in engineering, are described by Uchitelle (2006).

IMPLICATIONS FOR ENGINEERING EDUCATION

FINDING 4. *Engineering education at the undergraduate and graduate levels has been a major source of strength for the U.S. engineering enterprise. Even today, engineers* *educated in the United States remain among the best trained and most flexible in the world. At a time when other nations are making significant efforts to upgrade their engineering education capabilities, the United States will be challenged to sustain engineering education as a national asset.*

It was clear from the workshop discussions that participants from both industry and academia consider U.S. engineering education a valuable asset. It is also clear that other countries and regions, most prominently China and India, are working hard to upgrade their engineering education capabilities. In addition, large numbers of students from China and India continue to come to the United States for graduate engineering education.

Workshop participants repeatedly stressed that U.S. engineers will need better management and communications skills and that engineers who master the principles of business and management will be rewarded with leadership positions. The same needs have been stressed in reports and statements by professional societies and reports from the NAE Engineer of 2020 Project (NAE, 2004, 2005).

***FINDING 5.** Although individual engineers must ultimately take responsibility for their own careers, industry, government, universities, professional societies, and other groups with a stake in the U.S. engineering enterprise should consider supporting programs and other approaches to helping engineers manage their careers, renew and update their skills, and sustain their capacity to innovate, create, and compete.*

A continuing theme in the workshop discussions was the effect of offshoring on engineers whose jobs are vulnerable, even though their wages may be increasing. For example, in the semiconductor industry, wages are increasing, but very slowly (see Brown and Linden, this volume). The environment for engineering work has changed significantly as organizations grow and shrink and jobs are gained and lost. Some engineers who are proactive in keeping their skills up to date and are able to take advantage of the trend toward more frequent job and career shifts are adapting well. But many workshop participants called for renewed efforts on the part of all stakeholders in U.S. engineering—educators, government, professional societies, and employers—to address the needs of mid-career engineers who need help developing new skills and abilities for a constantly changing job market.

In addition to educational approaches to ameliorating the effects of offshoring, many have called for direct assistance to engineers and other service workers whose jobs are displaced. Approaches that have been discussed include (1) expanding eligibility for Trade Adjustment Assistance to include engineers and other service-industry workers and (2) providing some form of wage insurance to help displaced workers who are forced to take lower paying jobs.

***FINDING 6.** Over the past several decades, engineering has become less attractive to U.S. students as a field of study and as a career compared to some other professions. Although it is widely assumed that globalization and offshoring are contributing to this relative decline in popularity, it is impossible to know how important globalization is compared to other factors. A great deal more needs to be understood about the relationship between offshoring and the attractiveness of engineering as a career.*

Concerns were raised repeatedly about whether offshoring is negatively affecting the public perception of engineering and whether this perception has led (and will lead) to fewer talented U.S. students choosing to pursue careers in engineering. We do not have enough data at this point either to support or allay these concerns. We do know, however, that over the past several decades, the relative popularity of engineering as a major has declined in comparison with other fields that have experienced strong long-term growth. The committee believes that this issue should be thoroughly investigated.

IMPLICATIONS FOR POLICY

***FINDING 7.** For the United States, attracting and retaining world-class engineering activities in an increasingly competitive global environment will require that core U.S. strengths be sustained. Perhaps the most critical task in doing so will be to avoid complacency.*

Workshop participants pointed out the strengths of the United States and argued that the biggest risk to future success is complacency. Public and private efforts to tackle large-scale problems, for example in energy and transportation, could lead to the creation of entirely new industries and would go a long way toward creating new opportunities for engineers.

***FINDING 8.** Plausible scenarios have been developed showing that offshoring either helps, is neutral, or hurts engineering in the United States. Only continued discussions and further studies will lead to a thorough understanding of the potential benefits and costs of offshoring.*

Offshoring in general, and offshoring of engineering in particular, has both costs and benefits, although we cannot paint a clear picture of these based on available data. Nevertheless, the workshop did provide a basis for making general statements about the costs and benefits so far.

On the benefit side, offshoring appears to be adding to the competitiveness and profitability of the U.S.-based companies that manage it effectively. In addition, it has long been assumed that globalization and trade in services will ultimately yield net benefits for the U.S. economy. If offshoring is like other forms of trade in this respect, it too should deliver net economic benefits. However, some questions have been raised about whether this will be the case.

Offshoring is proving to be a boon to several emerging economies, particularly India and China, and long-term U.S. interests will be served by these countries and other developing economies becoming integrated into the global economy and raising their standards of living. Inevitably, this will also lead to improved engineering capabilities in these countries relative to the United States. If America maintains its engineering capability, and if the emerging global networks are open to participation by Americans and American organizations, this might then be a "win-win" situation, because U.S. companies would also benefit directly through expanded markets for their products.

But what of the possible downsides? It has been argued that offshoring and other forms of trade can be harmful to the U.S. economy and U.S. national interests. For example, even if offshoring brings short-term economic benefits to the United States in the form of gains to companies and consumers, it could eventually undermine America's ability to innovate.

In addition, some prominent economists are concerned that the distributional impacts of offshoring on engineers and other service-sector workers in the United States will pose serious challenges to freer trade. They argue that offshoring could lead to the degradation of overall engineering capability in the United States. Thus, even if the U.S. engineering enterprise and economy as a whole are better off with offshoring, those who are most vulnerable to competition might suffer severe hardships. The question is how we should address these distributional issues.

FINDING 9. *As the debate about offshoring continues, it will be important to determine whether current U.S. policies, including immigration policies, provide artificial advantages or incentives for offshoring.*

Although a detailed examination of immigration policies is beyond the scope of this study, immigration issues are closely related to offshoring. The immigration of scientists and engineers, the training of foreign students, and the overall openness of the United States to foreign talent have clearly been a boon to U.S. engineering activities and the U.S. economy. But some argue that the current H-1B and L-1 visa programs facilitate offshoring. Policies that, in effect, subsidize or provide artificial incentives for the offshoring of engineering, they say, are just as counterproductive and market-distorting as artificial barriers or penalties for offshoring would be. Future studies should investigate the interactions between immigration policies and offshoring, particularly in engineering.

FINDING 10. *Security concerns related to the offshoring of engineering have been raised, specifically for the information technology and construction industries.*

Finally, national security concerns have been raised that offshoring in the construction engineering and services industry might lead to detailed plans and other information about U.S. buildings and infrastructure, as well as geospatial data, falling into the wrong hands. Relevant professional societies are already working to ensure that sensitive information can be protected within the existing legal framework.

Concerns have also been raised about whether the globalization of software development could pose a serious threat to national security. For example, accidental defects or maliciously placed code might compromise the security of Department of Defense networks. The Defense Science Board is examining those concerns.

REFERENCES

CPST (Commission on Professionals in Science and Technology). 2006. STEM Employment Forecasts and Distributions among Employment Sector: STEM Workforce Data Project: Report No. 7. Washington, D.C.: CPST. Available online at *www.cpst.org*.

GAO (Government Accountability Office). 2005a. International Trade: U.S. and India Data on Offshoring Show Significant Differences. GAO-06-116. Washington, D.C.: GAO.

GAO. 2005b. Offshoring of Services: An Overview of the Issues. GAO-06-5. Washington, D.C.: GAO. Available online at *www.gao.gov/new.items/d065.pdf*.

Hira, R. 2005. Impacts and trends of offshoring engineering tasks and jobs. The Bridge 35(3): 22–27.

McKinsey Global Institute. 2005. The Emerging Global Labor Market. New York: McKinsey & Company. Available online at *www.mckinsey.com/mgi/publications/emerginggloballabormarket/index.asp*.

NAE (National Academy of Engineering). 2004. The Engineer of 2020: Visions of Engineering in the New Century. Washington, D.C.: The National Academies Press. Available online at *http://newton.nap.edu/catalog/10999.html*.

NAE. 2005. Educating the Engineer of 2020: Adapting Engineering Education to the New Century. Washington, D.C.: The National Academies Press. Available online at *http://newton.nap.edu/catalog/11338.html*.

NAPA (National Academy of Public Administration). 2006. Off-shoring: An Elusive Phenomenon. Washington, D.C.: NAPA.

Sturgeon, T.J. 2006. Services Offshoring Working Group Final Report. Cambridge, Mass.: MIT Industrial Performance Center. Available online at *http://web.mit.edu/ipc/publications/pdf/06-006.pdf*.

Thursby, J., and M. Thursby. 2006. Here or There? A Survey on the Factors in Multinational R&D Location. Washington, D.C.: The National Academies Press. Available online at *http://www.nap.edu/catalog/11675.html*.

Uchitelle, L. 2006. The Disposable American. New York: Alfred A. Knopf.

Part I

Consensus Report

1

Introduction

The phenomenon of "offshoring"—the transfer of work previously performed in the United States to affiliated and unaffiliated entities abroad—suddenly emerged as a major issue in the U.S. political debate a few years ago. At the time, employment and wages were recovering slowly from a recession precipitated by the "dot-com bust" and the 9/11 attacks. Particularly during 2003–2004, news reports of companies simultaneously cutting staff in the United States and launching extensive new operations in lower wage economies abroad attracted attention, and criticism, from many quarters.

Particular concerns were raised about the transfer of work in engineering and information technology (IT). These jobs not only had high skill requirements; they also commanded higher than average wages. In addition, the emergence of offshoring coincided with high levels of unemployment in some engineering specialties, such as electrical and computer engineering. The widely held assumption that U.S. engineering and high-technology jobs were invulnerable to international competition was suddenly called into question.

With the subsequent economic recovery and lower unemployment rates among engineers and other affected groups, fewer headlines referred to offshoring. However, this important aspect of the global economy is not well understood, especially how it fits into the broader context of globalization.

Clearly, business infrastructure, particularly in IT-related businesses, has developed to the point that many service jobs are now "tradable." These include customer-service functions, such as call centers, tax preparation, and accounting, and a variety of IT-related jobs (e.g., database administration). Over time, we might expect the kinds of tasks that can be offshored to increase. The availability of significant numbers of appropriately skilled overseas workers who are willing to work for salaries significantly lower than prevailing U.S. salaries provides an incentive for companies to achieve cost savings by offshoring. Even if wages for the most accessible and skilled of these workers are bid up to levels near those of wages in developed countries, we can expect the supply of workers to increase over time as other individuals, firms, and countries seek out, or begin to provide, the training and connectivity they need to participate in a global service economy.

A number of individual scholars and organizations are investigating the offshoring phenomenon, and several useful studies and analyses have recently been published. Nevertheless, significant gaps in knowledge remain. In fact, there are formidable barriers to compiling a reasonably complete picture of current and likely future conditions. For example, existing categories in official statistics of production, trade, and the labor force reflect past, rather than present (or future), business structures and economic activities.

In addition, much of the information about the microeconomic trends in individual companies and whole industries, which is necessary to construct a complete picture of offshoring, is considered proprietary. This is largely the result of controversies that arose in 2003–2004, when companies that engaged in offshoring were heavily criticized in the media. Since that time, these companies have been careful about releasing information that might open them to heightened scrutiny or criticism (see Dobbs, 2004).

THE GOALS AND PROCESSES OF THIS STUDY

Offshoring raises basic questions for the engineering profession and enterprise in the United States that must be answered before rational decisions can be made about policies (e.g., the debate over H-1B visas) or strategies to address the consequences. For example, we need to determine which fields of engineering and what types of engineering work (e.g., research and development [R&D], R&D management, design, manufacturing, marketing, customer support, and so forth) are being offshored and why. We need to know if the rationale for offshoring in engineering differs from industry to industry, and if so, how. We need to know if the rationale varies over time. What makes some industries more susceptible to offshoring than others (e.g., government regulation, intellectual property laws, and other factors)? How do the effects of offshoring compare/interact with the effects of other factors, such as increased automation, improved technology, or reorganization? What impact do these factors have on the number and composition of engineering jobs in different sectors? How do patterns of engineering offshoring compare with patterns of "onshoring" (bringing in engineering jobs from other countries through direct foreign investment)? What is the relationship between offshoring and the immigration of skilled workers, both temporary and permanent? How much do foreign companies rely on engineering services performed in the United States? Can we characterize differences in performance between engineering service-sector jobs performed abroad and those performed at home? Has offshoring impacted our security? Many, many more questions could be added to this list.

The National Academy of Engineering (NAE) launched this study to help fill in some of the information gaps. Because the engineering enterprise is a pillar of U.S. national and homeland security, economic vitality, and innovation, this study will be of great interest to many people outside the engineering community. The primary goal of the study is to improve our understanding of the scope, composition, and motivation for offshoring and to consider the implications for the future of U.S. engineering practice, labor markets, education, and research. The specific statement of task for the committee is provided in Box 1-1.

For several reasons, distinctions are made in the papers and analysis between U.S.-based companies and companies based elsewhere. First, the industry-focused papers show that U.S.-based companies have tended to undertake offshoring earlier and more extensively than firms based elsewhere. Second, although firms based outside the United States employ a significant and growing share of the overall U.S. workforce, including U.S. engineers, the majority of U.S. engineers are still employed by U.S.-based companies, and the actions of U.S.-based companies still have a disproportionate impact on

BOX 1-1
Project Statement of Task

National Academy of Engineering
Committee on the Offshoring of Engineering

Statement of Task

The National Academy of Engineering will form an ad hoc committee to organize and conduct a public workshop on the issue of offshoring of U.S.-based jobs having significant engineering content.

Workshop presentations and commissioned papers will present what is known about offshoring from a broad perspective and in specific industries, such as information technology, construction and civil engineering, automobiles, and pharmaceuticals. The workshop will bring together analysts from government statistical agencies (e.g., National Science Foundation, Bureau of the Census, and others); experts from engineering professional societies, industry, foundations, and academia; and leaders in engineering education who have collected data and can offer insights and observations.

Based on the workshop, the committee will prepare a report aimed at improving understanding of the scope, composition, motivation, and outlook for offshoring, and on considering the implications for the future of U.S. engineering practice, labor markets, education, and research. The questions to be addressed include:

(1) What do we definitively know about the current status and trends regarding offshoring of work with significant engineering content, including the extent, motivation, types of work subject to offshoring, industry-specific characteristics, and future prospects?

(2) What are the key areas where data is lacking, and how might information gaps be filled?

(3) Given what we currently know, are there actions or options that engineering educators, professional societies, industry leaders, policy makers, and the engineering community at large should consider to strengthen the U.S. engineering enterprise in the face of offshoring and the continuing globalization of the engineering enterprise?

U.S. engineering. Although the interests of U.S. engineers and the engineering enterprise are not exactly the same as those of U.S.-based companies, the location of corporate headquarters still matters in important ways.

Clearly, NAE's underlying interest is in the long-term health and prosperity of the engineering enterprise in the United States. The engineering enterprise includes engineering professionals, the organizations that employ them, the institutions that educate and train them, the government entities that support and rely on engineering, and the societies and associations that serve the engineering profession.

NAE President Wm. A. Wulf appointed an ad hoc steering committee composed of eight NAE members representing a range of engineering fields and two additional experts to oversee the drafting of the commissioned papers, develop the agenda for the public workshop, and prepare the final report. The papers provide an overview of offshoring in specific industries—software, personal computer manufacturing, automobiles, semiconductors, construction engineering and services, and pharmaceuticals. Taken together, these six industries account for a significant share of U.S. engineering activity. In all of the selected sectors, significant research on globalization and U.S. competitiveness has been done in recent years. However, some important industries that also employ engineers were not included, such as financial services, transportation/logistics, aerospace, and others. The papers can be found in Part 2 of this report.

The committee met face to face in April 2006 and held regular teleconferences throughout the project. The public workshop was held in October 2006. Following the workshop, the steering committee prepared a summary report, including findings, and provided suggestions to the authors of the commissioned papers, who then revised their work. In addition, several experts who made presentations at the workshop were invited to convert their presentations into brief papers (see Part 2). By its nature, this project does not constitute a comprehensive examination of all industries or all aspects of engineering.

Following the workshop, the steering committee developed this report, which includes an overview of the current state of knowledge based on available contextual materials (Chapter 2) and summaries of the insights from the workshop (Chapters 3 and 4). Chapter 4 also includes the committee's findings and conclusions, restatements of outstanding questions and issues, and suggestions for next steps by government and the private sector.

In the course of organizing the workshop and preparing the summary, the committee reviewed some recent analyses of offshoring, as well as articles that have appeared in the business and general press. Because the offshoring of engineering is a complex, controversial phenomenon that is changing rapidly, the conclusions of scholars and analysts on all sides of the issues were questioned and their ideas debated.

Some of the examinations of offshoring the committee found most useful have been called into question because they were produced by organizations affiliated with companies or associations with financial or other interests in offshoring. The committee kept these affiliations in mind in preparing the report. However, because the report does not include policy recommendations, and because one of the key findings is that more data are needed on offshoring, the committee chose not to continually raise questions about sources that have not been challenged on substantive grounds. In addition, the NAE Program Office commissioned an overview paper to review statistical and other sources (Morgan, 2006). Finally, although a variety of sources is referenced in the summary, the primary bases for the committee's findings are the industry-focused commissioned papers and the workshop discussions.

REFERENCES

Dobbs, L. 2004. A Home Advantage for U.S. Corporations. Commentary, August 27. Available online at *http://www.cnn.com/2004/US/08/27/home.advantage/index.html*.

Morgan, R.P. 2006. The Impact of Offshoring on the Engineering Profession. Background paper prepared for the National Academy of Engineering. Available online at *http://www.nae.edu/nae/engecocom.nsf/weblinks/PGIS-6WHU3R/$file/Morgan%20Paper.pdf*.

2

Offshoring and Engineering: The Knowledge Base and Issues

Engineering has been defined as "the application of scientific and mathematical principles to practical ends, such as the design, manufacture, and operation of efficient and economical structures, machines, processes, and systems . . . (and) . . . the profession of or the work performed by an engineer" (Pickett et al., 2000). The National Academy of Engineering (NAE) identifies engineering as a key factor in our economic well-being, health, and quality of life (NAE, 2004). The overall importance of engineering is apparent in NAE's list of "Great Engineering Achievements of the 20th Century," which includes electrification, water supply and purification, the automobile, and the Internet.[1] Table 2-1 provides an overview of the engineering profession in terms of demographics (e.g., gender, ethnicity, and proportion of foreign born) and other indicators (e.g., number of engineers and average salaries).

In spite of the benefits of engineering to society, the profession is still "under-examined, under-scrutinized, and poorly understood" (Morgan, 2006). In fact, the available data are not sufficiently detailed to provide a clear understanding of the boundaries, composition, and dynamics of engineering. One difficulty is that engineering is a "porous profession," that is, a significant percentage of the individuals who receive engineering degrees ultimately pursue careers in non-engineering or non-technical fields. At the same time, some individuals who do not have engineering degrees hold jobs with "engineer" in the title.

Thus it is important to keep in mind that engineers are not a homogeneous group, and a study of the offshoring of engineering requires taking into consideration the wide range of engineering capabilities and tasks, both within and between industries and locations. These differences are considered in the commissioned papers where data are available. Another difficulty is that engineering is divided into disciplines (Table 2-2), only some of which require licensing or certification to practice.

UNCERTAINTIES ABOUT THE FUTURE

Today the engineering profession in the United States faces many challenges and uncertainties. One long-term concern is whether engineering will continue to attract sufficient numbers of young people, particularly U.S. citizens, to enter the profession. The overall number of engineering bachelor's degrees granted in the United States, which had been dropping, has gone up in recent years but appears to have reached a peak (Heckel, 2006).[2] Figure 2-1 shows the long-term trend. It is important to note that, although the number of engineering bachelor's degrees has declined somewhat over the past 20 years and the number of engineering and computer science bachelor's degrees combined has increased by about 20 percent, the total number of bachelor's degrees increased by more than 40 percent. Thus overall technical degrees have been less popular than other, nontechnical majors.

A number of reasons have been put forward to explain the long-term decline in interest among U.S. students in engineering, a trend that predates the emergence of offshoring. The reasons include slower salary growth than in other occupations that have less difficult academic requirements

[1]See *http://www.nationalacademies.org/greatachievements/index.html*.

[2]Analysts are fairly certain that the number of engineering degrees has reached a peak because the overall number of degrees has also peaked, reflecting the decrease in the number of 18 to 24 year-olds.

TABLE 2-1 A Snapshot of Metrics and Trends in U.S. Engineering

Metric	Data	Trends/Comments
Total U.S. workforce (2003)	138 million	37% increase since 1983
Total science, technology, engineering, math (STEM) workforce (2003)	7.5 million	70% increase since 1983
Total engineering workforce (2003)	2 million	25% increase since 1983; about 1.4 % of the total workforce, compared with roughly 1.6 % in 1983
Proportion of engineering workforce (2003) that is		
Female	10%	Up from 6% in 1983
African American	3%	
Hispanic	7%	
Asian	10%	
Proportion of the engineering workforce that is foreign-born (2002)	16%	Increase of 2% from 1994.
Average annual salary for engineers (2005)	$63,526	Represents 1.8 times the average salary of the entire U.S. workforce
Engineering degrees awarded in the United States (2004)		
Bachelor's	64,675	Down from 72,670 in 1983; the bachelor's number has tended to fluctuate
Master's	33,872	Up from 18,886 in 1983, reflecting a fairly steady increase
Doctorates	5,776	Up from 2,781 in 1983, this figure has also increased steadily
Projected increase in the engineering workforce between 2004 and 2014	13%	Note that this is a projection, not a certainty. The 13% projected increase in engineering is roughly the same as that projected for the overall U.S. workforce

Note: This presentation is meant to provide a broad overview and therefore does not delve into the subtleties involved in measuring the engineering workforce. Abt Associates (2004) provides a good discussion of the various issues and uncertainties. Perhaps most important, these figures for the engineering workforce DO NOT include "mathematical and computer science professions," which means that the population of interest to this study is somewhat larger than is reflected in the chart.

Source: Adapted from Commission on Professionals in Science and Technology, 2004–2007. Drawn from various tables and charts.

(e.g., business and finance); negative stereotypes of engineers; and, possibly, the perception that offshoring and other aspects of globalization portend a decline in engineering in the United States. All of these factors combined could raise significant barriers to students choosing to major in engineering.

Unfortunately, data to counter these perceptions are difficult to come by. Data on salaries, for instance, are ambiguous. On the one hand, starting salaries for new engineers with bachelor's degrees are significantly higher than starting salaries in many other fields (NAE, 2007). On the other hand, salaries for Ph.D. holders in engineering are lower than, and have not grown as quickly as, salaries of other professionals, such as doctors and lawyers (Freeman, 2005a). Thus students might be justified in believing that the extra work and effort required to earn an advanced degree in engineering might not be as well rewarded financially as advanced degrees in other fields.

A related concern is the increasing reliance of the U.S. engineering enterprise on students from abroad, particularly at the graduate level. Much more than half of engineering doctorates and roughly 40 percent of engineering master's degrees from U.S. institutions are awarded to foreign nationals (Heckel, 2006). Traditionally, many of these graduates have remained in the United States to build their careers and have contributed substantially to U.S.-based innovation (COSEPUP, 2005).

With the number of U.S. citizens entering engineering programs perhaps in decline (perhaps a cyclical decline, but perhaps a longer term trend), a drop in the number of foreign students entering these programs, or a decrease in the number of foreign engineers who stay in the United States after earning degrees, could affect the future overall size and capability of the U.S. engineering workforce. In 2003, 26 percent of engineering degree holders in the United States were foreign-born (22 percent of bachelor's degree holders, 38 percent of master's degree holders, and 51 percent of doctoral degree holders).

Despite the stringent U.S. immigration policies since the 9/11 attacks, current data on foreign enrollments and "stay rates" indicate that the United States is still attracting foreign students who pursue degrees in engineering and launch

TABLE 2-2 Engineering Workforce by Discipline and Other Relevant Occupations, 2006

Discipline	Number of Engineers
Aerospace	87,000
Agricultural	3,000
Biomedical	14,000
Chemical	29,000
Civil	237,000
Computer hardware	74,000
Electrical and electronics	280,000
Environmental	51,000
Industrial, including health/safety	223,000
Marine engineers/naval architects	8,000
Materials	21,000
Mechanical	218,000
Mining/geological	7,000
Nuclear	15,000
Petroleum	15,000
Engineering managers	184,000
Other	156,000
Total	1,622,000
Other Relevant Occupations	**Number Employed**
Computer Scientists and Systems Analysts	678,000
Computer Software Engineers	802,000
Total Engineering and Other Relevant	3,102,000

Notes: Rounded to the nearest thousand. The total for engineers is somewhat lower than that contained in Table 2-1, reflecting different years and methods of compilation.
Source: Bureau of Labor Statistics. May 2006 National Occupational Employment and Wage Estimates. Accessed November 1, 2007. Available online at *http://www.bls.gov/oes/current/oes_nat.htm*.

their careers here (Council of Graduate Schools, 2006). For example, in 2003, one-year stay rates were estimated at 71 percent, five-year stay rates at 67 percent, and ten-year stay rates at 58 percent for foreign students (temporary visa holders) who received science and engineering doctoral degrees from U.S. institutions (Finn, 2005). However, some analysts believe that a growing number of U.S.-educated foreign scientists and engineers are returning to their home countries after graduation (Heenan, 2005; Newman, 2006).

The attractiveness of engineering as a profession in the United States depends on it being considered a satisfying, stable, well compensated career, relative to other professions. However, the current picture and outlook appear to be mixed (Morgan, 2006). In the early years of this decade, unemployment in electrical engineering and fields related to information-technology (IT) industries reached historic highs (Harrison, 2005). The factors contributing to the rise in unemployment included the bursting of the dot-com bubble, rapid changes in technology, and increasing globalization, perhaps including offshoring. Although the unemployment rate for electrical engineers dropped back to its normal (in historic terms) low level between 2003 and 2005, this might reflect slow growth or even shrinkage in the profession, rather than a true recovery.[3]

High levels of unemployment and slow salary growth from 2002 to 2004 and longer term changes in engineering work have raised persistent concerns about the future of the profession. For example, Jones and Oberst (2003) described engineering employment as becoming "more volatile with each decade," as careers characterized by upward mobility and advancement are replaced by work patterns that require numerous lateral job shifts. They ascribe the changes to the "commoditization" of engineering work, that is, the breaking down of jobs into highly specific tasks that can be performed by employees, outsourced to contractors, or sent offshore. At the same time, Sperling and others believe that more and more demands are being made of engineers in terms of responsibilities and skills (Sperling, 2006). One can infer from both of these analyses that lifelong learning may well become more important, both for the profession as a whole and for individual engineers.

The important points to keep in mind in this introductory summary are (1) engineering, like other professions and other job categories, is changing; and (2) technological advances and globalization are two of the forces driving this change. Analyses of the industry-specific studies (provided in Part 2 and summarized in Chapter 3) indicate that engineers are being affected by these changes in different ways, depending on engineering discipline, age, access to continuing education, and educational background.

With improvements in the economy, job prospects, and salary growth in 2006 and 2007, engineers today are feeling more upbeat about their careers, more secure in their jobs, and more inclined to recommend engineering as a career choice than they were just a few years ago (Bokorney, 2006). Although these cyclical improvements in employment prospects are encouraging, they may not relieve apprehensions about long-term trends, including offshoring, and their potential implications and risks.

In his description of the relatively new field of networking, Rappaport (this volume) touches on several of the trends and perceptions that underlie anxieties about the future of U.S. engineering. Networking is a field that combines hardware and software aspects of computing and telecommunications. As U.S.-based corporate research has declined in recent years, firms based elsewhere are increasing their activity. Research based at U.S. universities remains strong, but top university graduate programs are increasingly reliant on students from abroad.

[3]The Occupational Employment Statistics produced by the Bureau of Labor Statistics cannot be used to compare employment levels in some employment categories, such as electrical engineering, over time, because the survey and statistical techniques used to produce a "snapshot" of employment levels at a particular time have changed over time. Thus results are not always comparable.

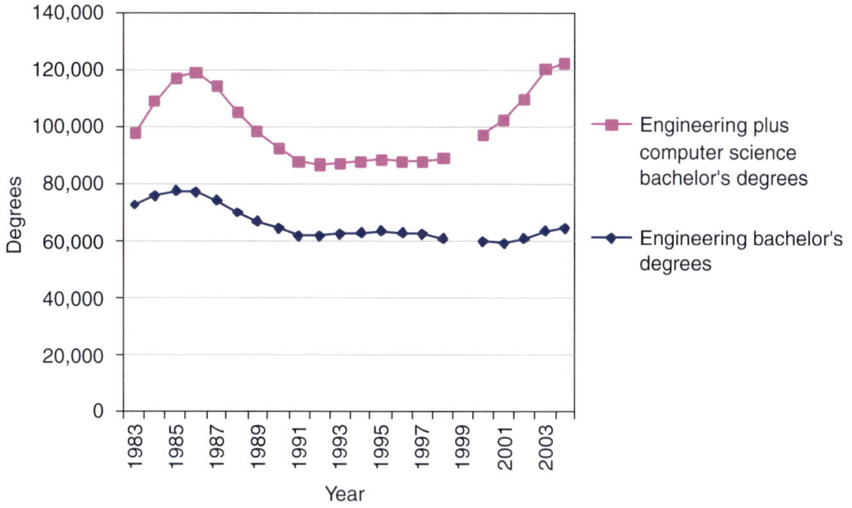

FIGURE 2-1 Bachelor's degrees in engineering and computer science, 1983–2004. Source: National Science Foundation/Division of Science Resources Statistics; data from Department of Education/National Center for Education Statistics; Integrated Postsecondary Data System Completions Survey.

THE INSTITUTIONAL AND HISTORICAL CONTEXT OF OFFSHORING

The NAE Committee on the Offshoring of Engineering defined "globalization" as the broad, long-standing process whereby national economies and business activities are becoming increasingly integrated and interdependent, mainly through expanded trade, capital flows, and foreign direct investment. "Offshoring" was defined as a more recent phenomenon whereby work is being relocated and diffused across national borders, enabled by advances in communications technology and changes in management practices. A wide range of services work is being offshored, but this workshop and report focus only on engineering.

Ideally, the committee would define offshoring of engineering as engineering work transferred from the United States to other locations, both by outsourcing the work to other organizations and by establishing or expanding subsidiary operations in the offshore destination. In practice, there are several difficulties with this definition. First, based on existing data, it is difficult to track the expansion of overseas jobs and the contraction of U.S.-based jobs in a way that establishes a relationship between them. Second, the expansion in overseas engineering work by firms with extensive U.S. engineering operations is not necessarily accompanied by a corresponding contraction in U.S. engineering activity; in addition, the jobs being created overseas may be qualitatively different from those that might be cut in the United States. Even with much better data, it would be very difficult to tell if offshoring is taking place, as described in the "ideal" definition given above. In the industry-focused papers (Part 2) and elsewhere in the report, expansion of overseas engineering work, both through outsourcing and subsidiaries, is considered evidence of offshoring.

Other factors related to offshoring included in this study are specific business practices (e.g., the international diffusion of corporate research and development [R&D]) that preceded the recent wave of offshoring but have taken new directions since it began, the movement of engineering work as a result of the relocation of manufacturing activities, and "onshoring" (engineering work being moved to the United States from abroad).

One important topic discussed in several of the papers but not a focus of the committee's summary is the management of offshoring by onshore firms, including effective practices and barriers to success. Clearly, companies in a variety of industries perceive benefits from offshoring. However, it should not be inferred that offshoring is an easy, frictionless process. The Boeing 787 is a recent example of the complications that can arise (Lunsford, 2007). A growing body of literature on the management of offshoring and multinational product-development teams describes barriers to offshoring in an organizational context and ways to overcome them (see, for example, Carmel and Tjia, 2005).

Offshore Sourcing of Engineering Work: India as an Example

In the context of engineering, the definition of offshoring encompasses several distinct phenomena and business practices that have emerged over the past several decades in particular industries. With rapid changes in technology and markets, these phenomena and business practices, which have somewhat different motivations and destination countries or regions, sometimes overlap and blend into the broader trend of the globalization of innovation.

The business models and infrastructure for a wide range of services offshoring, including business-process offshoring, emerged mostly in the software industry, principally in India. As a context for the discussion of offshoring in specific industries in Chapter 3, we briefly review the historical development of services offshoring and India's role in that development.

From the time of India's independence until the early 1990s, the Indian economy was highly regulated and controlled by the government (Dossani and Kenney, this volume). Indian international trade and investment were based on a protectionist, import-substitution philosophy. At the same time, a focus of public policy in India was investing in science and engineering research and higher education, which included the founding and expansion of Indian institutes of technology (IITs) (Murali, 2003). However, the IITs served a relatively small portion of the population, and many graduates continued to go overseas for graduate training. When the Japanese, South Korean, and other Asian economies underwent rapid economic growth fueled by manufacturing for the global market, India was largely cut off from the global economy. Nevertheless, its pool of skilled, English-speaking workers continued to grow.

During the 1970s and 1980s, India developed a small software industry focused on its domestic market (Aspray et al., 2006). The international Indian software industry began with Tata Consultancy Services, a pioneering firm that provided Indian programmers to work at customer sites in the United States. As this kind of activity increased during the 1980s, the Indian government became aware of the value of the software industry and adopted several preferential policies (e.g., exempting export revenue from taxation) that encouraged growth and kept the industry focused on the international market. Cultural, technological, and business factors came together during the late 1980s and 1990s to accelerate the growth of India's software industry.

Cultural factors included the tendency of educated Indians to become proficient in English. Because of this, India, along with Israel and Ireland, became a destination for the early offshoring of software work for U.S. multinational companies. All three countries offered low labor costs and skilled, English-speaking programmers. Another cultural factor was the presence of Indian-born engineers who had been educated and had worked in the United States (Saxenian, 2006). More than one-quarter of U.S. engineering and technology firms launched between 1995 and 2005 had at least one key founder who was foreign-born, with the largest number from India (Wadhwa et al., 2007).

As India's software industry grew and its global orientation became more prominent, Indian expatriates actively contributed to the development of new Indian-based companies and the operations of U.S.-based IT companies in India. As a result, the Indian government adopted policies to support the software industry, such as raising the standards for physical infrastructure and opening the economy to global trade. Indian expatriates have increasingly focused their efforts on developing entrepreneurial ventures that combine U.S.-based financing and market acumen with India-based engineering implementation.

Technological factors were also important to offshoring of IT-related work to India. The widespread adoption by the computer industry of the Unix workstation standard and the C programming language in the 1980s enabled the modularization of programming. This made it possible for independent software vendors to use standardized tools to develop programs for a wide range of operating systems and applications. During the 1990s, PCs with X86 microprocessors and Windows operating systems replaced RISC/Unix workstations in programming, and the Internet "provided a platform for networked development of software and software installation, hosting, and maintenance" (Dossani and Kenney, this volume). The availability of widely used word processing, spreadsheets, computer-aided design, and drafting software combined with the Internet to enable remote, distributed approaches to technical work. The point is not that these changes gave India unique advantages, but that technological advances made it possible to undertake a wide range of IT-related work in widely dispersed locations at the same time that the development of India's institutions and human-resource base made it an attractive location.

Business factors, which have led to the development of new business models in global service industries, also contributed to the offshoring of engineering and other services work to India. For example, Indian companies and the Indian affiliates of multinational corporations were well positioned to undertake much of the necessary software coding and maintenance work in response to the Y2K crisis in the late 1990s (Sturgeon, 2006). This led to the upgrading and expansion of the business infrastructure, which, in turn, led to the expansion of IT-related business-process offshoring.

The contracting of outside firms to manage data-processing functions has a long history. Large multinational consulting companies prominent in this line of business, such as Accenture, EDS, and IBM Global Services, had also been doing Y2K-related work. Many business-process operations required custom-software development, which overlapped with the skills offered by Indian organizations and individual programmers.

As the costs of telecommunications fell and the demand for skilled IT labor in the United States rose during the dot-com boom, India-based activities serving markets in developed countries increased in scale and in scope. Call centers, accounting, finance, human resources, and other business functions became targets for reengineering and offshoring. Indeed, importing services from India has been a key element in the IT-enabled restructuring of services work that some analysts predict will fuel U.S. productivity growth in the coming years (Mann, 2003).

The prospects for growth and development in this type of offshoring are explored in later chapters. For now, it is important to note that "engineering-services outsourcing" is considered by the Indian IT industry as an area for significant growth (NASSCOM, 2006). At the same time, India is aware that it faces significant challenges in sustaining economic growth and becoming a location for increasingly sophisticated engineering work. For example, increasing the capacity and quality of Indian higher education remains an essential, but difficult, task (Agarwal, 2006).

In addition to the offshoring of services, a great deal of overseas engineering involves engineering of manufactured components incorporated into goods sold by U.S.-based companies, and even entire products. In the United States as of 2004, about 40 percent of engineering employment was in the manufacturing sector, even though manufacturing constituted only about 20 percent of the U.S. GDP (BLS, 2005). Semiconductor manufacturing (Brown and Linden, this volume) and PC manufacturing (Dedrick and Kraemer, this volume) are perhaps the best examples (see Chapter 3 for more detail).

In the semiconductor industry, for example, "fabless" companies (mainly based in the United States) contract their manufacturing to "foundry" companies (such as TSMC and UMC, based in Taiwan). In the PC manufacturing industry, much of the detailed engineering of PCs sold by U.S. companies is done in Taiwan, but manufacturing is increasingly concentrated in China. Although this regional specialization in electronics innovation may not fit into the definition of offshoring used by most analysts, the value chains of both industries have been disaggregated over a number of years, a harbinger, perhaps, of offshoring-enabled shifts in business models for many other industries.

Globalization of R&D and Engineering

Foreign direct investment in R&D by multinational companies is a long-standing practice (Mansfield et al., 1979). Several of the papers in this volume describe how some aspects of innovation have historically been internationalized in certain industries, such as automobiles, construction engineering, and pharmaceuticals. In the 1930s, 7 percent of R&D by the largest U.S. and European firms was done outside of their home countries (Cantwell, 1998). From 1965 to 1995, foreign direct investment in R&D increased as multinational business increased. A survey of 32 large multinational companies based in the United States, Europe, and Japan revealed that in 1995 they performed 25.8 percent of their R&D abroad, which partly reflects the strong tendency for Europe-based companies to perform R&D abroad (Kuemmerle, 1999). In 2004, 15 percent of the R&D of U.S.-based multinationals was performed by foreign affiliates (Yorgason, 2007).

In the late 1980s and early 1990s, a large number of investments by Japanese and other foreign companies in R&D in the United States led some to question whether such investment was good or bad for the U.S. research enterprise (NAE, 1996). Concerns were raised that companies based outside the United States might "cherry pick" the results of publicly supported research through acquisitions and incremental investments in university research.

Many academic studies of overseas R&D by multinationals appeared in the 1990s, particularly on the motivations for investment. In a summary of the literature, Kuemmerle (1999) distinguishes between "home-base-exploiting R&D," in which investing companies want to exploit their existing technological capabilities in the foreign country where they are performing R&D, and "home-base-augmenting R&D," in which investing companies try to access unique assets in the foreign country by performing R&D there.

For a long time, overseas R&D was largely limited to multinational companies based in the developed world that were establishing or acquiring R&D facilities in other developed countries (Kuemmerle, 1999). In the late 1990s, however, global companies such as Motorola began to establish R&D centers in China and other emerging economies (GUIRR, 1998). Since then, the trend toward R&D investments in emerging economies such as India, China, and Russia has continued (UNCTAD, 2005a,b).

In contrast to other kinds of offshoring described above, some research suggests that the primary motivation for R&D investments in emerging economies is not cost reduction (Thursby and Thursby, 2006). A recent survey of R&D facility-location decisions by multinationals showed that they were influenced by a variety of factors. Interviewees cited the growth potential of the market in the destination country and the quality of R&D personnel as the most important attractors, indicating that both home-base-exploiting and home-base-augmenting motives came into play.

Thus the globalization of R&D is a complex, rapidly changing phenomenon, and the trend of global companies locating R&D facilities in emerging economies is relatively recent. Academic and policy research on these trends is ongoing.[4]

TRENDS AND PROSPECTS

The phenomenon of offshoring is important not only for engineering, but also for all economic activity in the United States and around the world. In this section, we review some trends and developments in offshoring that have been identified in the recent literature and were discussed at the workshop.

Economics

The National Academy of Public Administration (NAPA) (2006) points out the difficulty of assessing the impacts of

[4]Three ongoing National Academies projects worth mentioning in this connection are an examination of the globalization of innovation by the Board on Science, Technology, and Public Policy (STEP), an examination of the innovation systems of India and China, also by STEP, and a study of the changing ecosystem for information technology R&D by the Computer Science and Telecommunications Board.

offshoring based on international trade and domestic labor markets. Several recent economic analyses have been undertaken to determine the impacts of offshoring on the U.S. economy and the labor force and to project future trends. One point of consensus in these analyses is that available data are not comprehensive or specific enough to determine how many U.S. jobs have been lost as a result of offshoring, the scale of indirect effects on employment that would create new jobs in the United States, and the effects of offshoring on economic growth and incomes. NAPA (2006) cites estimates of annual job losses attributable to offshoring of 15,000 to 192,000. Although this is a large range, even the larger number, 192,000, is small compared with typical quarterly job losses and gains of seven or eight million in the U.S. economy. So, although the statistics do not show evidence of massive U.S. job losses attributable to offshoring in the short term, this does not mean that important longer term shifts will not become apparent in the future.

Data on trade in services are used to measure the actual flow of offshoring work between the United States and major offshoring destinations such as India. However, as the Government Accountability Office (GAO) points out, the Indian figure for exports to the United States is 20 times the U.S. figure for imports from India (see Figure 2-2) (GAO, 2005a). The GAO report lists differences in the way Indian and U.S. data are compiled that could account for this discrepancy. For example, transactions between affiliated entities are not counted in the U.S. data. So, for example, if Accenture is working on an IT consulting project for a U.S. customer, and if Accenture's operation in India does work under that contract, the work would not be counted as services trade in the U.S. data but would be counted in the Indian data. Another possible source of underreporting of U.S. imports of services might be that many transactions fall below the reporting threshold of the survey or analysis.

Sturgeon (2006) analyzes the limitations of available trade and workforce data and develops a detailed program for addressing the inadequacies in current data. A GAO report that covers similar ground also notes the lack of data in some areas and catalogs potential costs and benefits to the U.S. economy of offshoring (GAO, 2005b).

Some economists argue that the United States will enjoy a significant benefit from offshoring (Mann, 2003). According to one estimate, gains from services offshoring accounted for about 10 percent of U.S. productivity growth from 1992 to 2000 (Amiti and Wei, 2006). Others argue that the United States could suffer a net economic loss in the long term if innovative U.S. industries are undermined by offshoring (Gomory and Baumol, 2001). Freeman (2005a) predicts that the globalization of scientific and engineering talent, of which offshoring is one important aspect, is likely to erode the comparative U.S. advantage in high-technology industries. Given uncertainties in the underlying data and differences in the assumptions of these and other economists, debates over the actual and potential impacts of offshoring are likely to continue.

In a more recent development, analysts have questioned whether U.S. manufacturing output is overstated (Mandel, 2007). If it is, the overstatement could lead to a corresponding overstatement of U.S. productivity growth, meaning that U.S. economic performance in recent years might not be as strong as statistics suggest. If this is true, it would weaken support for the argument that the balance of benefits and

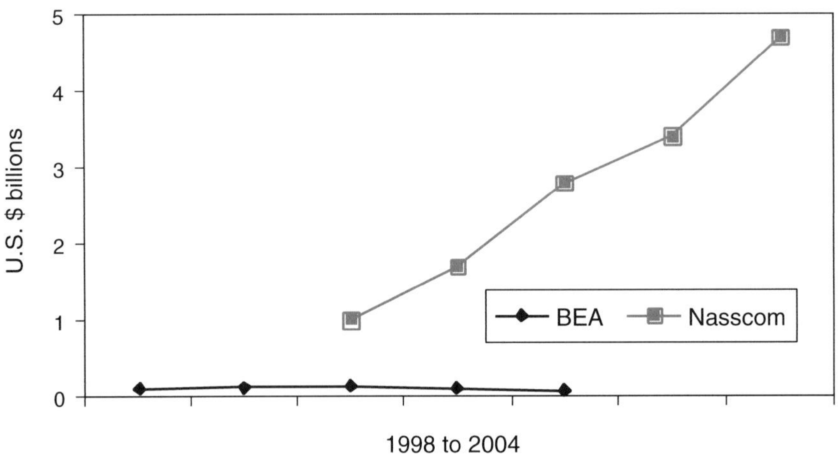

FIGURE 2-2 U.S. software imports from India according to U.S. and Indian statistics, 1998–2004, in billions of dollars. Source: Dossani and Kenney (this volume), based on data from the Bureau of Economic Analysis, U.S. Department of Commerce, and National Association of Software and Services Companies (NASSCOM) of India.

costs to the United States from globalization has been overwhelmingly positive.

Regardless of whether there are long-term net gains or losses for the U.S. economy as a whole, offshoring raises distributional issues, such as possible exacerbations of income inequality and the costs of job displacement, that are borne disproportionately by particular individuals in certain job categories and regions. Possible ways of addressing the distributional issues, such as extending Trade Adjustment Assistance to people who lose their jobs as a result of international trade, and providing wage insurance, are discussed in Chapter 4.

Other analyses have attempted to predict how offshoring might evolve in the future. For example, Jensen and Kletzer (2006) find that the number of U.S. workers engaged in potentially tradable services industries (i.e., workers whose jobs may be vulnerable to offshoring) is higher than the number of workers in manufacturing industries who are vulnerable to potential trade-related job losses.

Several consulting firms that advise companies on offshoring decisions have also developed estimates and projections of future trends. One analysis, by McKinsey Global Institute (2005), argues that, although the supply of young, college-educated workers employable in offshored services work, including engineering, will continue to expand, the supply is not inexhaustible. They cite several reasons for this. First, the rate at which India and other developing economies can expand their higher education infrastructure is limited. Second, only a fraction of potential workers in the pool of young, college-educated workers in China, India, and other emerging economies is suitable for employment by global companies. Most of the potential labor pool is disqualified because of a lack of language skills, a lack of practical skills due to deficiencies in the educational systems of some countries, or a poor cultural fit (e.g., attitudes toward teamwork and flexible working hours). The implication is that wages for the best qualified workers in destination countries will be extremely competitive, thus reducing the cost advantage of offshoring.

One "big picture question" related to offshoring concerns the long-term impacts of economic volatility. Some have argued that the U.S. economy can tolerate a high level of volatility (or flexibility) in labor and other markets because of its openness to trade and, therefore, can innovate and grow more quickly than other developed economies (Brown et al., 2006). Others point out that in the 1990s large emerging economies in India, China, and Russia approximately doubled the global supply of labor, thus decreasing returns to labor and increasing returns to capital (Freeman, 2005a,b). Rapid globalization that increases real and perceived job insecurity for a large portion of the U.S. workforce, they say, may sow the seeds of its own destruction by fueling voter demands for protection from international competition (Anderson and Gascon, 2007).

Politics and Policies

Trade policy is a perennial issue in U.S. politics, and the policy debates over offshoring represent a continuation of that tradition (e.g., Dorgan, 2006; Mankiw and Swagel, 2006). In recent policy debates, there are clear linkages between offshoring and other aspects of globalization, such as immigration. Even before widespread offshoring, some services workers, particularly IT professionals, had raised concerns about job dislocations and slow wage growth brought on by the availability of skilled immigrants holding H-1B and L-1 visas.

Some prominent analyses explicitly link offshoring with immigration (Hira and Hira, 2005). On the one hand, they say, offshoring can act as a substitute for immigration; by performing work overseas, U.S.-based companies have less need to hire immigrants. On the other hand, immigrant engineers with U.S. corporate experience are a valuable resource for companies that want to launch or expand their offshoring activities. Thus policies must be carefully considered because they can have both positive and negative consequences. For example, policies that attract more skilled immigrants to study science and engineering in U.S. graduate schools could not only increase the supply of talent, but also suppress wages, thereby reducing the incentive for U.S. citizens to pursue science and engineering degrees.

As can be seen from the discussion above, there are numerous gaps in the state of knowledge about broad issues raised by offshoring of engineering. To supplement the existing knowledge base, offshoring in six specific industries was explored in commissioned papers and workshop discussions. The results are summarized in Chapter 3.

REFERENCES

Abt Associates. 2004. Engineers in the United States: An Overview of the Profession. Cambridge, Mass.: Abt Associates. Available online at *www.abtassociates.com/reports/NSF_EWP_report2.pdf*.

Agarwal, P. 2006. Higher Education in India: The Need for Change. Indian Council for Research on International Economic Relations, New Delhi, India. Working Paper 180. June.

Amiti, M., and S.-J. Wei. 2006. Service Offshoring and Productivity: Evidence from the United States. Working Paper 11926. Cambridge, Mass.: National Bureau of Economic Research.

Anderson, R.G., and C.S. Gascon. 2007. The Perils of Globalization: Offshoring and Economic Insecurity of the American Worker. Working Paper 2007-004A. St. Louis, Mo.: Federal Reserve Bank of St. Louis.

Aspray, W., F. Mayadas, and M.Y. Vardi, eds. 2006. Globalization and Offshoring of Software: A Report of the ACM Job Migration Task Force. New York: Association for Computing Machinery. Available online at *http://www.acm.org/globalizationreport/*.

BLS (Bureau of Labor Statistics). 2005. Occupational Outlook Handbook, 2006–2007 Edition. Washington, D.C.: U.S. Government Printing Office.

Bokorney, J. 2006. Engineers optimistic about industry comeback. EE Evaluation Engineering (April). Available online at *http://www.highbeam.com/doc/1G1-144966005.html*.

Brown, C., J. Haltiwanger, and J. Lane. 2006. Economic Turbulence: Is a Volatile Economy Good for America? Chicago, Ill.: University of Chicago Press.

Cantwell, J. 1998. The Globalization of Technology: What Remains of the Product-Cycle Model? Pp. 263–288 in The Dynamic Firm, edited by A. Chandler, P. Hagstrom, and O. Solvell. New York: Oxford University Press.

Carmel, E., and P. Tjia. 2005. Offshoring Information Technology: Sourcing and Outsourcing to a Global Workforce. Cambridge, U.K.: Cambridge University Press.

COSEPUP (Committee on Science, Engineering, and Public Policy). 2005. Policy Implications of International Graduate Students and Postdoctoral Students in the United States. Washington, D.C.: The National Academies Press. Available online at *http://newton.nap.edu/catalog/11289.html*.

Council of Graduate Schools. 2006. Findings from the 2006 CGS Annual Admissions Survey, Phase III: Admissions and Enrollment. October. Washington, D.C.: Council of Graduate Schools. Available online at *http://www.cgsnet.org/portals/0/pdf/R_intlenrl06_III.pdf*.

Dorgan, B. 2006. Take This Job and Ship It: How Corporate Greed and Brain-Dead Politics Are Selling Out America. New York: Thomas Dunne Books.

Finn, M.G. 2005. Stay Rates of Foreign Doctorate Recipients from U.S. Universities, 2003. Report prepared for the Division of Science Resources Studies, National Science Foundation. Available online at *http://orise.orau.gov/sep/files/stayrate05.pdf*.

Freeman, R.B. 2005a. Does Globalization of the Scientific/Engineering Workforce Threaten U.S. Economic Leadership? NBER Working Paper 11457. June. Cambridge, Mass.: NBER. Available online at *http://papers.nber.org/papers/w11457.pdf*.

Freeman, R.B. 2005b. The Great Doubling: Labor in the New Global Economy. Usery Lecture in Labor Policy, Georgia State University, Atlanta. April.

GAO (Government Accountability Office). 2005a. International Trade: U.S. and India Data on Offshoring Show Significant Differences. GAO-06-116. Washington, D.C.: GAO.

GAO. 2005b. Offshoring of Services: An Overview of the Issues. GAO-06-5. Washington, D.C.: GAO. Available online at *www.gao.gov/new.items/d065.pdf*.

Gomory, R., and W.J. Baumol. 2001. Global Trade and Conflicting National Interests. Cambridge, Mass.: MIT Press.

GUIRR (Government-University-Industry Research Roundtable). 1998. National Science and Technology Strategies in a Global Context: Report of an International Symposium. Washington, D.C.: National Academy Press.

Harrison, R. 2005. Employment data paints a disturbing picture. IEEE-USA Today's Engineer Online. September. Available online at *http://www.todaysengineer.org/2005/Sep/pulse.asp*.

Heckel, R.W. 2006. Historical Trends and Near-Term Predictions of Statistics on Degrees, Enrollments and Research Expenditures for Engineering Education in the United States. Available online at *http://engtrends.com/IEE/0806A.pdf*.

Heenan, D. 2005. Flight Capital. Mountain View, Calif.: Davies-Black.

Hira, R., and A. Hira. 2005. Outsourcing America: What's Behind Our National Crisis and How We Can Reclaim American Jobs. New York: AMACOM. Available online at *http://www.amazon.com/gp/product/0814408680/qid=1133982951/sr=8-1/ref=pd_bbs_1/002-9456996-1835242?n=507846&s=books&v=glance*.

Jensen, J.B., and L.G. Kletzer. 2006. Tradable Services: Understanding the Scope and Impact of Services Offshoring. Pp. 75–116 in Offshoring White-Collar Work, edited by L. Brainard and S.M. Collins. Washington, D.C.: The Brookings Institution.

Jones, R.C., and B.S. Oberst. 2003. Are current engineering graduates being treated as commodities? European Journal of Engineering Education 28(3): 395–402.

Kuemmerle, W. 1999. Foreign direct investment in industrial research in the pharmaceutical and electronics industries: results from a survey of multinational firms. Research Policy 28(2-3): 179–193.

Lunsford, J.L. 2007. Globally outsourced 787 now mess for Boeing Co. Wall Street Journal, December 7. P. A1.

Mandel, M. 2007. The Real Cost of Offshoring. BusinessWeek, June 18. Available online at *http://www.businessweek.com/magazine/content/07_25/b4039001.htm?chan=top+news_top+news+index_businessweek+exclusives*.

Mankiw, G.N., and P.L. Swagel. 2006. The Politics and Economics of Offshore Outsourcing. July. NBER Working Paper No. 12398. Cambridge, Mass.: NBER.

Mann, C.L. 2003. Globalization of IT Services and White Collar Jobs: The Next Wave of Productivity Growth. International Economics Policy Brief PB03-11. Washington, D.C. Available online at *http://www.iie.com/publications/pb/pb03-11.pdf*.

Mansfield, E., D.J. Teece, and A. Romeo. 1979. Overseas research and development by U.S.-based firms. Economica 46(182): 187–196.

McKinsey Global Institute. 2005. The Emerging Global Labor Market. New York: McKinsey & Company. Available online at *www.mckinsey.com/mgi/publications/emerginggloballabormarket/index.asp*.

Morgan, R.P. 2006. The Impact of Offshoring on the Engineering Profession. Background paper prepared for the National Academy of Engineering. Available online at *http://www.nae.edu/nae/engecocom.nsf/weblinks/PGIS-6WHU3R/$file/Morgan%20Paper.pdf*.

Murali, K. 2003. The IIT story: issues and concerns. Frontline 20(3). Available online at *http://www.flonnet.com/fl2003/stories/20030214007506500.htm*.

NAE (National Academy of Engineering). 1996. Foreign Participation in U.S. Research and Development: Asset or Liability? Washington, D.C.: National Academy Press. Available online at *http://newton.nap.edu/catalog/4922.html*.

NAE. 2004. The Engineer of 2020: Visions of Engineering in the New Century. Washington, D.C.: The National Academies Press. Available online at *http://newton.nap.edu/catalog/10999.html*.

NAE. 2007. *Engineer Girl!* Website. Available online at *http://www.engineergirl.org/CMS/Careers/3845.aspx*.

NAPA (National Academy of Public Administration). 2006. Off-shoring: An Elusive Phenomenon. Washington, D.C.: NAPA. Available online at *http://www.bea.gov/papers/pdf/NAPAOff-ShoringJan06.pdf*.

NASSCOM (National Association of Software and Service Companies). 2006. Globalization of Engineering Services: The Next Frontier for India. Washington, D.C.: NASSCOM. Available online at *http://www.nasscom.in/Nasscom/templates/NormalPage.aspx?id=49790*.

Newman, R.J. 2006. Can America keep up? U.S. News and World Report, March 19. Available online at *http://www.usnews.com/usnews/biztech/articles/060327/27global.htm*.

Pickett, J.P., et al., eds. 2000. American Heritage Dictionary of the English Language, 4th ed. Boston: Houghton Mifflin Company. Available online at *http://www.bartleby.com/61/*.

Saxenian, A. 2006. International Mobility of Engineers and the Rise of Entrepreneurship in the Periphery. Research Paper 2006/142. United Nations University World Institute for Development Economics Research. Available online at *http://www.wider.unu.edu/publications/working-papers/research-papers/2006/en_GB/rp2006-142/*.

Sperling, E. 2006. What's Worrying Engineers? Electronic News, April 6, 2006. Available online at *http://www.edn.com/article/CA6322921.html?ref=nbra&text=worrying+engineers*.

Sturgeon, T.J. 2006. Services Offshoring Working Group Final Report. Cambridge, Mass.: MIT Industrial Performance Center. Available online at *http://web.mit.edu/ipc/publications/pdf/06-006.pdf*.

Thursby, J., and M. Thursby. 2006. Here or There? A Survey on the Factors in Multinational R&D Location. Washington, D.C.: The National Academies Press. Available online at *http://www.nap.edu/catalog/11675.html*.

UNCTAD (United Nations Conference on Trade and Development). 2005a. Globalization of R&D and Developing Countries. Proceedings of the Expert Meeting, Geneva, January 24–26, 2005. Available online at *http://www.unctad.org/en/docs/iteiia20056_en.pdf.*

UNCTAD. 2005b. UNCTAD Survey on the Internationalization of R&D: Current Patterns and Prospects on the Internationalization of R&D. December 12. Available online at *http://www.unctad.org/en/docs/webiteiia200512_en.pdf.*

Wadhwa, V., A. Saxenian, B. Rissing, and G. Gereffi. 2007. America's New Immigrant Entrepreneurs. Duke University and University of California at Berkeley. Available online at *http://memp.pratt.duke.edu/downloads/americas_new_immigrant_entrepreneurs.pdf.*

Yorgason, D.R. 2007. Research and development activities of U.S. multinational companies. Survey of Current Business 87(3): 22–39.

3

Effects of Offshoring on Specific Industries

The committee commissioned papers on offshoring in six economic sectors—automobiles, semiconductors, software, personal computer (PC) manufacturing, pharmaceuticals, and construction engineering and services—to gather information for the workshop (see Part 2). The six industrial sectors were selected based on (1) the size and importance of the industry to engineering and the overall economy and (2) the availability of expert authors. Table 3-1 provides a summary in graphic form of the six industries.

The commissioned authors were given a list of questions to use as guidelines (Box 3-1) and were asked to submit abstracts, and then draft papers, in the run-up to the workshop. Following the workshop, the committee developed questions and suggestions for revisions, which the authors incorporated into the final papers. Table 3-2 provides a summary of answers based on the commissioned papers mapped onto the questions in Box 3-1. Summaries of the papers follow.

SOFTWARE-DEVELOPMENT INDUSTRY

The software-development industry was the first to engage in the offshoring of engineering for the purpose of reducing costs. In Chapter 2, we described the beginnings of software-development offshoring, particularly to India, as part of an overall picture of offshoring. In this summary, we describe the current status of software-development offshoring, trends, information gaps, and unanswered questions. Although offshoring of software development was the leading edge of the practice of engineering offshoring, it is still not clear whether offshoring in other industries will follow a similar pattern.

The paper by Dossani and Kenney on offshoring in the software-development sector is focused on India. Although a few other countries, such as Ireland, which adapts software products developed by multinational companies for the European market, are also destinations for offshoring, the scale of activity in India is much greater than elsewhere. The number of workers employed in software development in India is increasing by 30 to 40 percent a year, from about 2 percent of U.S. employment in 1995 to almost 20 percent in 2005 (Table 3-3).

India today specializes in software services, such as the development and maintenance of custom-application software for large clients in several industries, such as insurance and finance. From this base, multinational companies operating in India and Indian domestic firms have moved into other areas, such as product software and embedded software. The increasing technical sophistication of Indian workers and higher value added to products are being driven by investments by U.S.-based companies (Table 3-4). Dramatic increases in exports from India to countries all over the world indicate that software development there is now targeted at the global market.

The first offshoring of software development by U.S. firms to India had one common characteristic—the work being offshored was modular and did not require regular contact with customers (Dossani and Kenney, this volume). Several U.S. companies began by contracting with a vendor to perform this non-integral work. After a period of time, they decided to set up Indian subsidiaries to perform more integral work. At first, although costs were lower in India, it was more difficult to hire an equivalent team there than in Silicon Valley. Nevertheless, as these companies gained experience in managing offshoring relationships, the barriers

TABLE 3-1 Data on Six Industries, 2002 (except where indicated)

	Computer Systems Design and Related Services NAICS: 5415	Software NAICS: 5112	Semiconductors NAICS: 3344	Automobiles NAICS: 3361-3363	Construction Engineering/ Services NAICS: 23, 3413	Pharmaceuticals NAICS: 3254	PC Manufacturing NAICS: 3341	Total for 6 Industries[a]	U.S. Total
Value-added ($ billions)	173.5	103.5	110.4	469.8	1,358.4	140.6	73.7	2,429.9	10,469.6
Employment	1,107,613	356,708	437,906	1,078,271	8,459,885	248,947	150,751	11,840,081	114,135,000[b]
R&D performed ($ billions)	11.9	12.9	11.9	16 (est.)[c]	10.7	10.1[d]	3	76.5	193.9
R&D scientists and engineers[e]	90,800	80,800	73,000	83,200	n/a	51,800	15,100	394,700	1,066,100

[a]The software industry is represented by two NAICS codes, 5414 and 51112, which clearly do not map exactly onto the industry sectors covered in the commissioned papers, particularly for software (figures here understate the revenue, employment, and R&D of interest) and PC manufacturing (figures here overstate the revenue, employment, and R&D of interest).

[b]Total private sector employment.

[c]In recent years, the auto industry R&D total has not been reported by NSF because it would disclose the total for an individual firm. $16 billion is a rough estimate obtained by subtracting the R&D performed by the aerospace industry from the total R&D for the transportation equipment sector.

[d]2001.

[e]R&D scientists and engineers is not an ideal proxy for the population we are interested in, but this data is collected by NAICS code and allows an apples to apples comparison. Note that Moavenzadeh (this volume) gives an estimate of 189,000 engineers in the auto industry for the relevant NAICS codes.

Sources: Bureau of the Census, 2004 (for value-added and employment); NSB, 2006 (for R&D performed); and Hecker, 2005 (for R&D scientists and engineers).

began to come down. For example, Broadcom, a software-intensive semiconductor company, reports that its team in Bangalore is now as productive as its teams in San Jose and Irvine, with costs in India running about one-third of those in the United States.

As the institutional infrastructure in India has improved, offshoring has become part of the normal way of doing business in the software industry. The diaspora of U.S.-educated Indian entrepreneurs has helped fuel the growth of the Indian tech sector, which is developing in a way that complements Silicon Valley (Saxenian, 2006). One example cited by Dossani and Kenney is Netscaler, a company that turned to offshoring when it was facing a funding crunch. The tactic enabled the firm not only to survive, but also to grow (both in India and the United States).

Aspray et al. (2006) observe that offshoring has become essential to the globalization of the software industry and will undoubtedly continue and increase. In Dossani's workshop presentation, he reported that today, in Indore, which is not a large IT center like Bangalore or Mumbai, wages for engineers who work 12 hours a day, six days a week are about $200 a month. However, in larger centers like Bangalore, salaries for experienced engineers are rising rapidly. For example, in a 2006 survey, "State of the Engineer," published in *EE Times*, the mean salary for Indian respondents was $38,500. However, as the history of Silicon Valley shows, higher costs are not necessarily a barrier to innovation-fueled growth (Saxenian, 2006). Despite very high costs for skilled labor, Silicon Valley has remained a prime location for innovative start-ups.

In a workshop presentation, Alfred Spector, a consultant and NAE member, outlined three possible scenarios for the future of software-development offshoring (Spector, this volume). In the first scenario, offshoring frees up U.S. talent and money, which can then be focused on higher value-added activities, such as testing, which then becomes much more efficient. In the second scenario, the rise of India and other offshoring destinations in certain sub-disciplines leads to a loss of U.S. jobs in those sub-disciplines, but, again, frees up talent and other resources for the creation of new sub-disciplines or super-disciplines that keep U.S. software innovation strong overall. In the third scenario, when U.S. students learn that certain activities are being moved offshore, they conclude that opportunities for software innovation in the United States are drying up and decide not to pursue careers in those areas; this leads to atrophy in the U.S. talent and skills base.[1]

The three scenarios are not mutually exclusive—the United States might maintain its leadership position in some aspects of software but lose it in others. Spector says that the

[1]One reviewer of this report suggested tracking metrics related to software innovation over time to determine which of these scenarios is being realized.

TABLE 3-2 Comparison of the Industry Sectors Covered by the Commissioned Papers

	Software	Semiconductors	Automobiles	Construction Engineering/ Services	Pharmaceuticals	PC Manufacturing
1. Nature of engineering work	Scope of work that can be spatially disaggregated is growing.	Disaggregated business models, functional integration in products.	Increasing pressure to increase efficiency, more open innovation process.	Supply of workers in the industry is a problem.	Increasingly difficult environment for business models based on blockbuster drugs.	Disaggregated business mode grew up in the 1990s.
2. Current status regarding globalization	Strong capabilities in several countries, distributed development increasingly common.	Globalization has complemented U.S. innovation/market leadership.	Successive waves of globalization, "build where you sell," emergence of global suppliers.	Large project sector more globalized than building/ residential sector.	Increasing consolidation, globalization of companies and markets.	Engineering and manufacturing increasingly concentrated in China.
3. U.S. engineering workforce	Increasing, expected to grow over the next decade.	Sustained growth over time, less opportunity for older and less-skilled, increase in foreign-born.	Total employment down over the long-term, same is true for engineers.	Aging—low starting salaries discourage U.S. civil engineering grads.	Appears to be growing, though life sciences may be growing faster than engineering.	Fairly small
4. Countries where work is expanding	India in particular, evidence of growth in other countries.	India	China, India, wherever the automotive market is expanding.	Large range of offshoring destinations, in addition to India and China, Eastern Europe is attracting work.	China, India, United States still attracts innovation investment.	China, Taiwan
5. Offshoring occurring	Yes, driven by cost reduction, extent of high-value job losses uncertain.	Yes, cost reduction a primary motivator.	Yes, both through global optimization of platform development and through offshoring of routine tasks; also onshoring.	Yes, growth of global teams in the large project sector.	Yes, began with clinical trials and is moving up the value chain, but limits on end-to-end; also significant onshoring.	Yes, only limited engineering work remains in the United States.
6. Work that is more or less vulnerable	More vulnerable: standardized service and maintenance; Less vulnerable: Interface with final customer.	Product definition is less vulnerable.	Less vulnerable: Work on vehicle types where the United States is the leading market (e.g. large pick-ups); work where high degree of domain knowledge is needed.	Less vulnerable: Work where high degree of interaction with the customer is necessary.	More vulnerable: clinical trials; Less vulnerable: the most sophisticated R&D.	Less vulnerable; high level definition of product characteristics; most other engineering work is gone already.
7. Future outlook	Diversification of destination countries, increase in value-added of offshored work.	Continued globalization of engineering work.	Fortunes of leading global OEMs diverging, U.S. engineering fortunes have more to do with competitive success of companies than offshoring per se.	Will increase, although there are limitations on offshoring due to licensing, government procurement regulations, national/homeland security concerns.	U.S. engineering employment not likely to be impacted by offshoring.	Companies that can innovate will need at least some U.S. engineers; Taiwanese engineering will be offshored to China.

BOX 3-1
NAE Offshoring Project: Issues and Questions to be Addressed in the Commissioned Papers

1. What is the nature of engineering work[a] in the industry, and how is it changing? Why is it changing? What are the typical entry level skills and credentials required of engineers? How do various countries compare in the production of qualified engineers, and in the institutions that provide skills and credentials?
2. What is the current situation with regards to globalization of the industry? How globalized is the industry in terms of manufacturing, competition (e.g., do firms based in one or a few countries dominate certain market segments?), and capability (e.g., are certain engineering capabilities available in only one or a few countries)?
3. What do we know about the U.S. engineering workforce in this industry from statistics and other data? Is the engineering workforce growing, shrinking, stable, aging, or we don't know? Are wages rising at the same pace as the overall engineering workforce? Are there differences between those with graduate, 4-year, and 2-year degrees?
4. In what countries and regions is engineering work expanding in this industry, and why? Is offshoring occurring? If so what are the primary sources and destinations? What roles have multinational corporations and start-ups played? Has government policy played a role? Has engineering work followed production? Are engineering workforces growing, and if so how fast? What are trends in wages? What are the current and projected capacities for educating and training engineers?
5. Is it fair to say that engineering work previously performed in the United States is being offshored, or is there a positive net effect? Are there qualitative differences in the types of engineering jobs that are performed in the U.S. and those performed elsewhere? Are there types of engineering work in which the United States or other countries enjoy distinct advantages?
6. Are there areas of engineering work that are more or less vulnerable to offshoring? What can individual engineers and U.S. institutions do to retain their competitiveness?
7. Can you make projections regarding future offshoring trends? How concerned should U.S. engineers be about offshoring in this industry? Will wages in countries in offshoring destination countries rise to an equilibrium level? Are new destination countries likely to emerge? What factors will determine future outcomes?

[a]"Engineering work" is defined as the full spectrum of research, product and process development, engineering management, manufacturing engineering, etc.

growth of the open-source movement and other advances in underlying technologies will also affect how offshoring and regional capabilities evolve.

As in other industries, the growth of offshoring in software development thus far has been led by U.S. companies. Japanese companies are much less inclined to offshore software work (Aspray et al., 2006). Western Europe-based firms fall somewhere in between; of these, U.K.-based companies account for the largest share of offshoring.

One technological trend that will challenge software developers in the future, with uncertain implications for offshoring, is the growing popularity of multicore processors and multiple-processor systems. These technologies offer significant advantages in hardware design and more rapid processing, without the heat limitations of single processors. However, multicore designs require software designers who can deal with concurrency and develop new programs in which tasks can be broken into multiple parts that can be processed separately and reassembled later (Krazit, 2005). Because these skills may not be available in the usual offshoring destination countries, relatively more engineering work may become available in the United States.

Some concerns have been raised about whether the globalization of software might be a serious threat to national security (Hamm and Kopecki, 2006). For example, accidental defects or maliciously placed code might compromise the security of U.S. Department of Defense networks. The Defense Science Board is currently completing a study on how the department should address these concerns.

TABLE 3-3 Increases in Offshoring of Software Production in India

Employment	1995	2005
United States	1.5 m	2.6 m
India	27,500	513,000

Source: Dossani and Kenney, this volume.

TABLE 3-4 Rising Sophistication of Technical Work in India

	2001	2002	2003	2004	2005	2006 (E)
Computer-aided design (CAD) and computer-aided manufacturing (CAM) (CAD/CAM) ($B)	3.65	4.40	4.87	5.98	7.67	10.16
Total software exports ($billions)	5.30	6.16	7.10	9.80	13.10	17.10
Share of CAD/CAM (%)	68.90	71.40	68.60	61.00	58.50	59.60
Share of foreign firms' revenue (%)	14.50	22.00	26.00	31.00	31.00	n/a

Source: Dossani and Kenney, this volume.

AUTOMOTIVE INDUSTRY

The paper by John Moavenzadeh, executive director of the International Motor Vehicle Program, on engineering work in the automotive industry begins with a description of the two main categories of engineers—manufacturing engineers and product engineers (the majority). Manufacturing engineers typically work at production facilities, while product engineers typically work at corporate engineering and design facilities. Product engineering can be divided into several categories: product design, development, testing, and advanced engineering. A significant percentage of product engineers work for automotive suppliers rather than for original equipment manufacturers (OEMs), such as Ford, Toyota, and Volkswagen.

Moavenzadeh describes the difficulty of estimating the size of the automotive engineering workforce in the United States based on official statistics, which are not specific to the engineering categories in the industry. By inference and extrapolation, he estimates that at least 160,000 engineers and technicians support OEMs and suppliers in the U.S. automotive industry (Tables 3-5a,b).

The automotive industry ranks second among U.S. industries in terms of overall spending on R&D. Six of the top 20 companies that spend the most on global R&D are automotive OEMs. Engineering and product-development productivity levels differ for OEMs based in different parts of the world; Japanese OEMs are more productive, for example, than OEMs based in the United States and Europe.

From its beginnings in the late nineteenth and early twentieth centuries, the automotive industry has been international. In the first half of the twentieth century, for example, Ford and General Motors both had a large number of overseas assembly plants. Over time, in some of the larger markets, subsidiaries, which operated almost as separate companies, were established to design and build cars specifically for those markets.

Since the 1960s, the auto industry has "undergone a second wave of globalization," fueled by changes in the U.S. market, which is still the largest and most open market in the world (Moavenzadeh, this volume). One of those changes was the growth of the Japanese auto industry. At first Japanese companies in the United States relied exclusively on exports from Japan. Gradually, however, they built manufacturing and then engineering capabilities in the United States and Europe. These so-called "transplants" now account for more than 30 percent of U.S. auto production.

Today more than half of General Motors employees are outside the United States, and companies such as Volkswagen, Hyundai-Kia, and Honda assemble more than half of their vehicles outside their home countries. The supplier base is similarly distributed, especially tier-one suppliers, which provide interiors and other components that require R&D and production closely coordinated with OEMs.

Automotive manufacturers manage their production and engineering "footprint" based on a number of factors, including customers (i.e., the location of the market); capability (i.e., the best way to leverage available talent); cost (i.e., labor costs and integration costs at various locations); and government (i.e., trade and investment policies).

The most important factor, though, is market growth (Moavenzadeh, this volume). The United States, Japan, and Europe have large, but already mature markets that are not growing very rapidly, whereas large developing economies

TABLE 3-5a BLS Data Showing Automotive Engineers in the United States[a]

Occupational Code	NAICS 3361: Motor Vehicle Manufacturing	NAICS: Motor Vehicle Body and Trailer Manufacturing	NAICS 3363: Motor Vehicle Parts Manufacturing	Total of All Three NAICS Codes
Engineering managers	610	570	3,960	5,140
Industrial engineers	3,390	1,240	14,460	19,090
Mechanical engineers	1,920	1,360	9,300	12,580
Electrical engineers	150	110	910	1,170
Engineers, all other	n/a	180	7,200	7,380
Total	6,070	3,460	35,830	45,360
All Occupations	256,700	168,840	693,120	1,118,600

[a]Does not include most product engineers.

TABLE 3-5b Bottom-Up Estimate of Engineers and Technicians Employed by OEMs

Company	Current Number of Engineers and Technicians	Projection
General Motors	11,500	Decreasing
Ford Motor Company	12,000	Decreasing
DaimlerChrysler	6,500	Steady
Japanese companies	3,593	Increasing rapidly
Korean companies (Hyundai-Kia)	200	Increasing rapidly
German companies (BMW)	150	Increasing
TOTAL	About 34,000	

such as China and India have markets that have grown rapidly in recent years and are expected to continue growing. Thus automakers from all over the world are trying to make inroads into those markets. Ford and General Motors have been particularly active, building manufacturing capacity as well as engineering capability in China, Latin America, and elsewhere.

In addition to market factors, cost factors tend to encourage OEMs and automotive suppliers to locate engineering activities in the developing world, especially in China (Figure 3-1). Moavenzadeh estimates that, whereas a fully loaded, experienced engineer in the United States might cost \$100,000 a year, an equivalent engineer in China might cost \$15,000 a year. However, Chinese engineers are reportedly less productive, which may reflect their lack of domain knowledge (many Chinese have never driven a car). As the Chinese economy grows and automobiles become more common, however, we can expect Chinese engineers to become more competitive and more productive.

Even with the current productivity gap, a number of engineering tasks can be offshored fairly easy. As was the case with software development, these are modular tasks that do not require customer contact, such as developing engineering bills of materials, performing failure modes effects analyses, performing routine stress analyses, developing heat-transfer calculations, and generating tool designs from part specifications.

The cost incentives for developing automotive-engineering capabilities in developing economies such as China and India can work in two ways. The first, engineering connected with the manufacturing of parts exported to the United States and elsewhere, involves different motivations and impacts from offshore engineering not connected with manufacturing. The former is an important aspect of globalization in the automotive industry, particularly in the rise of China's auto industry. The U.S. trade deficit with China in auto parts was \$4.8 billion in 2005 and has increased rapidly since then (Moavenzadeh, this volume). Imports from China, whether manufactured by subsidiaries of suppliers based in Europe or the United States, China-based manufacturers, or joint ventures, tend to be less sophisticated products, such as radios, brake components, and after-market aluminum wheels.

The second way cost differentials can provide incentives for offshoring is related to the offshoring of engineering services; companies either contract foreign firms or build their own subsidiaries to perform engineering tasks offshore. In addition to China, India is well positioned as a destination location for this sort of work. ValueNotes (2006), a research consultant company, predicts that offshoring of automotive engineering and design services will increase from the 2005 level of \$270 to \$300 million globally to more than \$1 billion in 2010. Automotive engineering services in India at subsidiaries of global suppliers, such as Delphi, and subsidiaries of Indian OEMs, are expected to increase at an annual rate of 30 percent during this period.

Because China has attracted so much attention from the global auto industry as a growth market and source of components, we now look more closely at the current state of China's engineering capability and the potential effect of offshoring on China's global competitiveness. First of all, the Chinese government has used a number of stratagems over the years to force or encourage the formation of joint

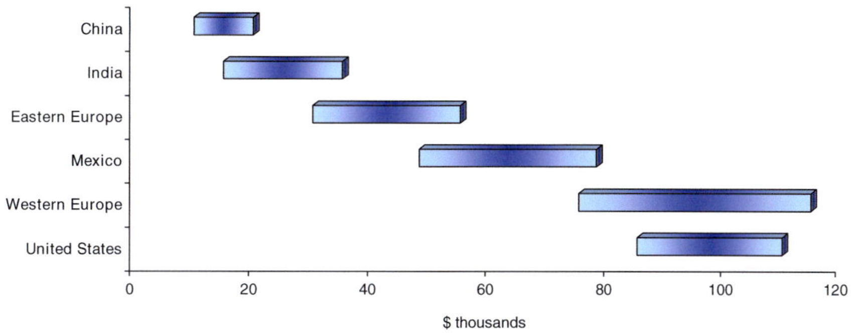

FIGURE 3-1 Engineering labor rates vary widely, as shown by the annual cost of an automotive engineer with 5 to 10 years experience. Source: Moavenzadeh, this volume.

ventures, and subsequent technology transfer from global auto companies to domestic manufacturers (Zhao et al., 2005) However, so far these joint ventures have not led to the transfer of skills the Chinese government had anticipated, mostly because Chinese engineers, who were expected to rotate back to domestic parent companies, have not done so because of the large salary differentials.

China's automotive R&D capability is currently far behind that of countries that build cars for the most sophisticated global markets (Zhao et al., 2005), and it will probably take years for China to assimilate the management of automotive-development processes. R&D management is also in an early stage, and several Chinese auto companies have hired foreign executives at very senior levels to run them. R&D currently being done by joint venture firms mainly involves adapting, or "localizing," foreign technology and designs for the Chinese market.

Notwithstanding these barriers, China will continue to move toward the top tier of auto-manufacturing nations. China's growing exports of automotive parts and the slow, but not insignificant, skills transfer occurring through joint ventures with foreign OEMs are providing an excellent foundation for the development of a world-class automotive technology base. In addition, the Chinese government funds three university centers that conduct applied automotive research for Chinese OEMs.

China's greatest asset is the continuing growth of its domestic auto market. Firms based in the United States, Japan, and Europe have adopted different approaches (some companies in different parts of the same region also differ) to entering the Chinese market. General Motors and Ford, as well as several major Europe-based OEMs, have been aggressively building engineering and manufacturing capability in emerging markets as a way to establish and build market share there. However, to date, most Japan-based OEMs have made efforts to enter the Chinese market through exports rather than through joint-venture manufacturing, although there are indications that this may be changing (Business Week, 2006). The effects of these differences on global automotive competition are topics for future studies.

For U.S.-based OEMs, Ford and General Motors in particular, the most difficult problem today is not offshoring itself but coordinating and optimizing global R&D and engineering operations (Moavenzadeh, this volume). The current goal is to coordinate global programs to produce vehicles with similar fundamental architectures that can be easily modified to meet local customer demands and regulatory requirements. Reducing the number of vehicle architectures will reduce cost, improve speed-to-market, and hopefully enable OEMs to meet the demands of particular markets. At General Motors, for example, Korea is the center of expertise for small-car development, and the United States is the center for full-sized truck development (Cohoon, 2006).

The overall picture of offshoring in the auto industry would not be complete without taking into account the phenomenon of "onshoring," that is, foreign-based OEMs and suppliers building engineering capability in the United States (Table 3-6). Japanese OEMs employ more U.S. engineers than Europe-based automakers because of their much larger manufacturing presence in North America. Although long-term career prospects for U.S. auto engineers are highly correlated with the fortunes of U.S.-based OEMs, especially in southeastern Michigan and a few other areas, onshoring raises the possibility that, as long as the United States remains a leading auto market, OEMs, regardless of nationality, will maintain engineering capability here.

THE PHARMACEUTICAL INDUSTRY

The pharmaceutical industry, including the biotechnology sector, has several unique features, as does the nature of engineering work in pharmaceuticals. The value chain in this industry runs from discovery (including target identification, lead discovery, and optimization) through clinical development to manufacturing to marketing and distribution (Pore, Pu, Pernenkil, and Cooney, this volume).

Pharmaceutical companies have very strong incentives for ensuring that the science-based discovery process is as efficient as possible. Bringing a drug from the concept stage to the marketing stage currently costs about $800 million, takes 8 to 12 years, and requires the testing of 5,000 compounds for every drug that is actually approved (McKinnon et al., 2004). Today's "big pharma" companies are squeezed between a business model that emphasizes blockbuster

TABLE 3-6 Employment in Foreign-Brand R&D and Design Facilities in the United States, 2006

Company	Location(s)	Established	Employees
BMW	Spartanburg, NC; Woodcliff Lake, NJ; Oxnard, CA; Palo Alto, CA	1982	70
Honda	Torrance, CA; Marysville, OH	1975	1300
Hyundai	Ann Arbor, MI	1986	150
Isuzu	Cerritos, CA; Plymouth, MI	1985	100
Mazda	Irvine, CA; Ann Arbor, MI; Flat Rock, MI	1972	100
Mercedes-Benz	Palo Alto, CA; Sacramento, CA; Portland, OR	1995	50
Mitsubishi	Ann Arbor, MI	1983	130
Nissan	Farmington Hills, MI	1983	980
Subaru	Ann Arbor, MI; Lafayette, IN; Cypress, CA	1986	30
Toyota	Gardena, CA; Berkeley, CA; Ann Arbor, MI; Plymouth, MI; Lexington, KY; Cambridge, MA; Wittmann, AZ;	1977	950

Source: Moavenzadeh, this volume.

products, which entail high risks and high fixed costs (i.e., R&D and marketing), and pricing pressure and competition in key product classes (Campbell et al., 2005). Some analysts predict that the industry will experience slower revenue and profit growth in the future because of a slowdown in the new-product pipeline. The industry is also very concentrated; in 2004, the top 10 global companies accounted for almost half of global sales (Gray, 2005).

The fill, finish, formulation, and packaging processes have been globalized for some time and serve global markets. The expiration of patents and consequent competition from generics have increased incentives to control costs by moving manufacturing overseas, even for products manufactured for the U.S. market.

However, the discovery of active pharmaceutical ingredients, the core innovative activity of pharmaceutical companies, has traditionally been centralized at a few research facilities in the home country and, perhaps, a very few other global centers of pharmaceutical innovation. Drug and process development, which involve more engineering than drug discovery, have been similarly concentrated.

The commissioned paper on this industry focuses on China and India as offshoring destinations. Both countries have the advantages of market potential, low costs, multiple R&D shifts in a day, a large number of graduates in chemistry and biology, government research support, and tax incentives. In addition, they have large numbers of treatment-naïve patients, which is an advantage for conducting clinical trials.

China and India also have significant disadvantages, including regulatory barriers (especially in India). In addition, as discussed in Chapter 2, only a fraction of the trained workers are qualified to work in the environment of a multinational company (10 percent in China and 25 percent in India). Another barrier is the uncertainty of protections of intellectual property. Evidence shows that no offshored R&D in emerging economies, in any industry, involves cutting-edge research (Thursby and Thursby, 2006).

McKinsey Global Institute (2005) estimated that global pharmaceutical companies had offshored about 10,000 full-time employees by 2003, almost three-quarters of them working in R&D. As noted in Table 3-1, 51,800 R&D scientists and engineers were working in the U.S. pharmaceutical industry in 2002.

Perhaps the most significant activity in emerging economies is clinical trials, with India the preferred destination. In a recent estimate, the value of the current outsourcing market for clinical trials was $158 million and was predicted to increase to more than $500 million in the next few years (O'Conner, 2006). In addition to cost savings for clinical trials, global pharmaceutical companies also save time because patients there can be recruited more quickly than in developed markets.

China is the preferred location for R&D in advanced proteomics (the systematic, automated study of protein structure and function) and molecular biology, while India is the preferred location for lead optimization (the assessment of a family of candidates and the evolution of the ones with the greatest chance of success). Because these capabilities are fragmented across the value chain, however, no single location can provide end-to-end solutions.

China and India have also become leading global locations for manufacturing of pharmaceuticals, which ultimately contributes to the innovative capabilities and engineering talent base in those countries. For example, China is the world's largest producer of active pharmaceutical ingredients, with sales of $4.4 billion in 2005; India is the third largest producer, with sales of $2 billion. India has the second highest number of FDA-approved manufacturing facilities, after the United States. Overall, India's production costs in pharmaceuticals are about half those of the United States, including labor, raw materials, capital costs, and regulatory costs.

In general, the impacts of offshoring on U.S. engineering (and science) capabilities in the pharmaceutical industry have not been significant so far. Even with rapid growth, McKinsey's projections of the number of jobs that will be offshored in the next few years is small compared to the number of U.S. engineers working in those areas.

In addition, the trend toward R&D or engineering offshoring in the U.S. pharmaceutical industry may be more than offset by significant onshoring. For example, major European pharmaceutical companies such as Novartis and GlaxoSmithKline have shifted much of their R&D and manufacturing activities to the United States in recent years. The large U.S. talent base and the absence of price regulation are the major attractors. In addition, companies based in India that have emerged as global leaders in the generic drug market are beginning to form joint ventures with, and even acquire, U.S. companies, with the goal of building capabilities in marketing and innovation. These trends bear close watching.

PERSONAL COMPUTER MANUFACTURING

PC manufacturing is a $230 billion industry that includes desktops, notebooks, PC-based servers, and hand-held computing devices, such as personal digital assistants (PDAs), personal music players, and smart phones.[2] PCs also drive the sale of PC software (a $225 billion market), IT services, and other hardware, such as peripherals, storage, and networking equipment (Dedrick and Kraemer, this volume).

The United States, which is the leading market for PCs, is home to some of the top PC vendors, such as Dell and Hewlett-Packard. Companies from China (Lenovo, which acquired IBM's PC business a few years ago), Taiwan (Acer), and Japan (Fujitsu, Toshiba, Sony) are also among top PC manufacturers.

[2]Hand-held devices are grouped with PCs here because the design and development processes are very similar. However, the issues related to software development are very different from those related to PCs.

The value chain in the PC manufacturing industry is highly disaggregated and globalized. Two suppliers, Intel and Microsoft, set the most widely adopted standards for the microprocessor and operating system. Other components, including storage, displays, semiconductors, and wireless networking components, are core technologies of the PC manufacturing industry. Most of the manufacturing and physical engineering for notebooks and desktop machines is done by contract manufacturers and original design manufacturers (ODMs). U.S.-based firms originally turned to Taiwan-based companies to take advantage of lower costs and to avoid becoming dependent on Japanese companies that could become competitors. Since then, Taiwan-based ODMs have shifted almost all manufacturing to China; they are now shifting much of the engineering to China as well.

The main tasks for PC firms are to define and anticipate the changing needs of customers, integrate innovations of suppliers into well designed product packages that meet those needs, and bring the packages to market quickly at an attractive price. Therefore, the focus of PC manufacturing companies is on product development rather than R&D.

The engineering tasks for PC manufacturing vary with the product category. Desktops, for example, require the integration of components into the chassis. Although it may take as long as nine months to design a new chassis, it takes as little as two weeks to design specific models based on that chassis. Notebook PCs involve more complex engineering tasks, such as the optimization of design elements that require trade-offs in weight, sturdiness, heat generation, energy efficiency, and so forth to achieve an ideal mix. Newer product categories, such as smart hand-held devices and blade servers, present many engineering challenges, first because no dominant technology standards have been established for devices such as iPods and PDAs, and second, because some products are unique to particular companies.

The disaggregated business model of the PC industry, in which significant aspects of engineering are offshored, has enabled Dell, Hewlett-Packard, and other leading companies to hold down costs and remain competitive in an industry that has rapid product cycles. Apple has even used a disaggregated business model to create the iPod, a new kind of product that straddles the line between IT and consumer electronics.

The key to long-term success in the PC industry appears to be protecting the interface with customers and the resulting information flow. Knowledge gained from customers feeds into product definition, high-level design, and the most sophisticated engineering tasks. Thus the effective use of offshoring can enable firms to sustain their U.S. operations and employment levels with U.S. employees who work mainly in non-engineering jobs. As a result, not many engineering jobs in the United States remain in this industry, and the ones that do require high levels of skill and experience, as well as an ability to innovate.

Overall employment in the U.S. computer manufacturing industry, of which the PC industry is a part, appears to have remained relatively stable in recent years, although the composition of the workforce has changed. Employment in electronics and electrical engineering has gone down; employment in applications-software engineering has gone up; and employment in other categories has remained more or less stable (Dedrick and Kraemer, this volume). Table 3-7 shows the distribution of engineering jobs in the computer industry as of 2005.

Because of changes in classifications, it is difficult to track the events of the 1990s, when the trend toward offshoring in PC manufacturing and engineering developed. Table 3-8 shows engineering salaries in the computer industry. Table 3-9 shows the supply and demand for engineering skills at PC and related industry firms.

As the employment numbers suggest, there is a growing need for engineers with software skills and knowledge of both hardware and software in the manufacturing processes still based in the United States. Also, U.S. firms are looking for experienced engineers who can be productive immediately. Thus only a few of the firms interviewed by Dedrick and Kraemer report making entry level hires. Management skills are also in high demand. Taken together, these trends indicate that employment in PC manufacturing could be difficult for young U.S. engineers with little or no job experience.

Both U.S. and Taiwanese PC companies are now looking to China for engineering talent. According to Dedrick and Kraemer's interviews with managers, Chinese engineers today do not have analytical skills and market knowledge comparable to those of experienced U.S. engineers. However, investments in training Chinese engineers have paid significant dividends, even though turnover remains high and salaries are rising rapidly. The largest Taiwanese ODMs have been focusing their efforts on training engineers in the Shanghai/Suzhou region, which is now the hub of notebook PC manufacturing.

As in the auto industry, PC firms based in different countries have adopted different approaches to offshoring. For example, Japanese PC makers, whose efforts are focused mainly on satisfying the demand in their domestic market, do not offshore manufacturing or engineering jobs.

CONSTRUCTION ENGINEERING AND SERVICES

Construction is a $4 trillion industry, with about one-fourth of that in the United States. Conceptually, the industry can be divided into two sectors. Engineering, procurement, and construction (EPC), which involves the construction of industrial and infrastructure facilities, is made up of large firms that employ many engineers. Architecture, engineering, and construction (AEC), which involves construction of buildings and residential facilities, is made up mostly of smaller firms (Messner, this volume). The description that follows addresses offshoring in both sectors, although the companies (mostly very small) that make up the residential

TABLE 3-7 Engineering Jobs in the U.S. Computer Industry, 2005

	2005
Computer software engineers—applications	12,800
Computer software engineers—systems software	18,240
Computer hardware engineers	12,940
Electrical engineers	2,900
Electronics engineers, except computer	3,710
Industrial engineers	3,430
Mechanical engineers	2,280
Engineering managers	5,630
Industrial designers	180
Total	62,110

Source: BLS, 2007.

TABLE 3-8 Engineering Salaries in the U.S. Computer Industry, 2005

Computer software engineers—applications	$94,760
Computer software engineers—systems software	$92,030
Computer hardware engineers	$94,690
Electrical engineers	$84,820
Electronics engineers, except computer	$86,330
Industrial engineers	$77,710
Mechanical engineers	$78,740
Engineering managers	$130,020
Industrial designers	$94,800

Source: BLS, 2007.

TABLE 3-9 Survey Results for Jobs in PC and Related Industries

Engineering Job Category	Major Activity Where This Skill Is Used	Demand for Engineers	Availability in U.S.	Availability in Other Locations Where Activity Takes Place[a]	Cost and Quality Relative to U.S.[a]
Engineering managers	R&D, design, development	Stable or growing	Tight	Tight or enough	Lower cost, lower quality
Engineering product managers	Design, development	Stable	Tight or enough	Tight or enough	Lower cost, same quality
Hardware engineers	Design, development	Stable	Tight or enough	Enough	Lower cost, same or lower quality
Electrical engineers	R&D, design, development	Falling or growing	Tight or enough	Enough	Lower cost, same or lower quality
Electronic engineers	Development	Falling	Tight or enough	Enough	Lower cost, same or lower quality
Mechanical engineers	R&D, design, development	Stable or growing	Tight or enough	Enough	Lower cost, same or lower quality
Software engineers	R&D, design, development	Growing	Tight	Tight or enough	Lower cost, same or lower quality
Industrial engineers	Manufacturing	n/a[b]	n/a[b]	Enough	Lower cost, same quality
Industrial designers	Design	Stable	Enough	Enough	Lower cost, lower quality

Note: Names of firms are confidential. Four were personal computing companies. One was a component supplier.

[a]Responses regarding availability, cost, and quality of some skills in other locations vary by firm, depending on where they perform these activities. We report one response when there was general consensus, more than one if there were different responses. Other locations include Singapore, Taiwan, Malaysia, Ireland.

[b]Firms interviewed had no manufacturing in the United States, so demand and availability of industrial engineers was not relevant.

Source: Dedrick and Kraemer, this volume.

portion of the AEC sector (about half of the U.S. construction market) do not engage in offshoring. In contrast to industries in which the top 20 global companies account for a large percentage of the market (e.g., pharmaceuticals and PC manufacturing), construction is highly decentralized. The 400 top U.S. contractors account for less than 20 percent of the market.

Engineers in the construction industry are involved in all phases of the delivery and operation of facilities. Factors that influence offshoring include the uniqueness of projects, the extensive local knowledge necessary to meet local codes and conditions, the active involvement of owners in most projects, and the desire of owners to keep information about a project from being widely disseminated, particularly overseas.

Like data for other industries, the data for the construction industry are not sufficient to provide a clear picture of the current status of offshoring. We can say that offshoring in construction engineering and services is occurring, but firms are also aggressively hiring civil engineers (the largest engineering discipline) and other design professionals (e.g., architects) in the United States. There is no evidence that offshoring has had a significant impact on the employment of engineers in the U.S. construction industry.

The following discussion is based on available statistics supplemented by information from surveys and interviews.

The focus of this analysis is on civil engineers (although small numbers of other types of engineers also work in the construction field) and architects (design professionals whose work is subject to offshoring). The current demand for engineers and architects in the United States is high because the civil engineering workforce is aging, many engineers are retiring, and, until recently, the construction market was growing rapidly.

Firms that were interviewed for this study in the EPC sector, the most active sector in offshoring, said that offshoring has had little impact on the size of their U.S. workforce (Messner, this volume). However, India and China are again the primary offshoring destinations, and the large wage differentials are shrinking rapidly as salaries in some places in developing countries rise, particularly in Mumbai, India, and other specific locations.

EPC firms that are active in international markets have been offshoring engineering work for 15 years or more. The vast majority, however, indicate that they coordinate work among locations to meet the needs of specific projects (Figure 3-2). Several steps in the construction-engineering process have been subject to large-scale offshoring. These include the development of 3D models during the design process, the conversion of 2D sketches to CAD models, and the development of engineering shop drawings for mechanical and steel subcontractors; there is also some offshoring in the IT sector. Cost reduction was the reason cited most often (followed by better quality) for offshoring among the EPC firms surveyed (Table 3-10). There was some difference of opinion about whether offshoring also reduced engineering time.

In 2004, the United States had an official trade surplus for AEC services of almost $3 billion, including a bilateral surplus with India. This number does not include interactions between a U.S. company and its Indian subsidiary, but does include outsourced work. Work that is offshored in the AEC sector includes the transformation of hand-drafted documents into 2D CAD or 3D CAD models and some engineering tasks for building projects, such as engineering of the foundation, structure, and technical systems.

TABLE 3-10 Impact of Offshoring on Projects in the EPC Sector

Impact on:	Percentage of Responses	Opinion
Engineering cost	48	More than 10% reduction
Construction cost	75	No impact
Engineering time	48	No impact
Overall project delivery time	59	No impact
Engineering quality	65	No impact
Construction quality	72	No impact

Overall, however, offshoring in the AEC sector has been limited for several reasons, such as the small size of most AEC firms, the need to protect sensitive or secure information for some projects, the need for local knowledge or interaction with the owner for some projects, and a poorly developed institutional infrastructure for construction engineering and services in potential destination countries. Some tasks are being automated through new software tools rather than being offshored.

More than 90 percent of survey respondents in the EPC sector said they thought offshoring would increase in the future (Figure 3-3). Some said they thought increasing offshoring would lead to lower quality designs, but others, in companies that have established operational low-cost engineering centers abroad, said they believe that with effective organization and management, they will be able to maintain quality and lower their costs. With lower costs, they said, they can produce more detailed designs than would be possible if the work were performed solely in the United States.

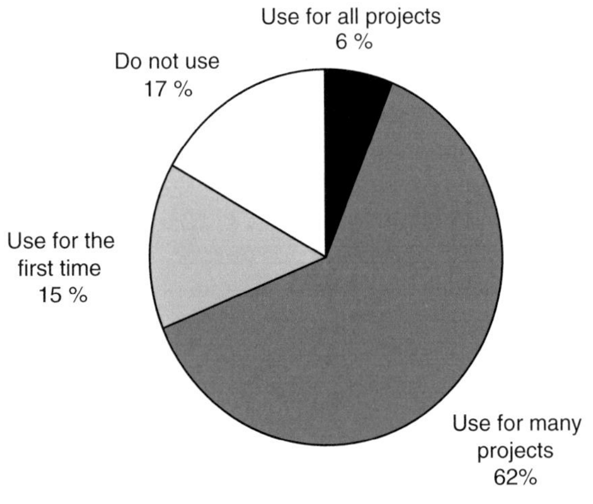

FIGURE 3-2 Use of global teams by firms in the EPC sector. Source: Adapted from CII Project Team 211 Survey, 2004.

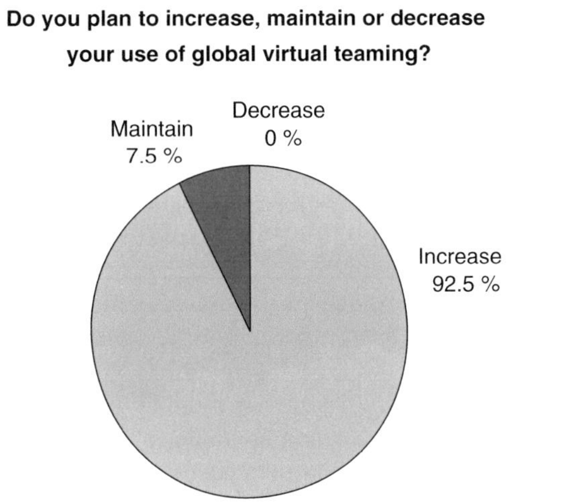

FIGURE 3-3 EPC contractors' perceptions. Source: Adapted from CII Project Team 211 Survey, 2004.

Concerns have been raised about potential security risks for U.S. buildings, particularly for critical infrastructure facilities, when detailed plans are developed and disseminated outside the United States (ASCE, 2005). Industry groups such as COFPAES and AIA continue to monitor this situation.

SEMICONDUCTORS

According to the authors of the paper on offshoring in the semiconductor industry, Clair Brown, director of the Center for Work, Technology, and Society at University of California Berkeley, and Greg Linden, a research associate at the center, the long-term trend of globalization of technology has had a significant impact on the nature of engineering work in the United States in this industry. Offshoring of engineering is increasing in all three major stages of semiconductor production—design, fabrication, and assembly and packaging.

For more than three decades, the number of transistors per unit of area has increased exponentially. Along with these advances, the cost of fabrication facilities (fabs) has also increased steadily; today a 300-mm wafer fab of minimally efficient scale costs about $3 billion. The demands on designers have also increased as they must find ways of using available "real estate," or space, on a device.

In short, projects have become increasingly complex with significant implications for the engineering labor market and for offshoring. The need for the efficient integration of system-level components has led to a greater emphasis on system software, preferably generated in parallel with chip design. According to Brown and Linden, "Software can now account for as much as half the engineering hours involved in a large chip development project" (this volume).

U.S.-based companies account for more than half of global industry revenues; Intel alone, the largest company, accounts for 15 percent. Texas Instruments is the only other U.S.-based firm in the global top 10, but a number of rapidly growing medium-sized companies are in the top 50. A number of these firms are "fabless," meaning they do not manufacture their own devices but contract out fabrication and assembly/packaging to "foundries," such as Taiwan Semiconductor Manufacturing Company (TSMC).

In the 1970s, assembly and packaging were shifted to Southeast Asia. In the 1990s, with the emergence of the Korean and Taiwanese semiconductor industries, the number and dispersion of fabrication facilities was accelerated. U.S.-based companies have shown a willingness to locate new fabrication facilities in various countries, as well as in the United States, in response to the size and potential of the market, tax advantages, and other incentives. One recent, widely discussed example is Intel's plan to build a $2.5 billion fabrication facility in China (which would not include Intel's leading chip designs) (IHT, 2007).

In contrast to software, for which work is often outsourced to other companies, most offshore design work for semiconductors is done by subsidiaries. Larger U.S. chip companies have established design centers around the world, mostly in Asia. Sometimes the goal is to capture specialized skills that are available at the offshore location, such as knowledge of wireless networking technology in Scandinavia. Sometimes the goal is to capitalize on government policies, such as in China, where the government encourages, sometimes requires, direct investment in return for market access.

The motivation for offshoring design work to India, the most popular destination, has been primarily to reduce costs. Of the top 20 U.S. semiconductor companies, 18 have established design centers in India; nine of those have opened since 2004. The size of these design centers varies widely, from 100 or so engineers in the smaller centers to 3,000 engineers at Intel's design center. The flood of investment has led to challenges, such as finding enough trained engineers, coping with the high turnover rate, and meeting demands for rapidly rising salaries.

The semiconductor industry has shown that design offshoring arrangements can be managed effectively, even though the costs of coordination and communication tend to offset some of the cost reductions in other areas. In reality, savings may come to 25 to 50 percent, rather than the 80 to 90 percent suggested by salary comparisons alone (Brown and Linden, this volume). Nevertheless, offshoring of the design phase is now a fundamental, expected feature of the business model for new U.S. semiconductor companies, which are also likely to be fabless. U.S.-based semiconductor start-ups, especially those with a founder or co-founder born in India, increasingly include design offshoring as part of their business plans (Saxenian, 2006).

The impact of offshoring on semiconductor engineers and engineering organizations in the United States is difficult to determine exactly because, once again, the data are not definitive. However, based on government statistics, the Semiconductor Industry Association annual survey, and other sources, the U.S. workforce in semiconductor engineering has recovered from a drop during the tech bust several years ago and is now growing. Overseas employment, however, is growing faster (see Table 3-11). Overall, we can say that the availability of offshore design and manufacturing capability has made it possible for the creation and growth of new, innovative, U.S. semiconductor firms.

Whether offshoring has had a negative impact on wages or on certain segments of the engineering workforce are questions that remain to be answered. Brown and Linden document several disturbing trends to which offshoring may contribute but which have multiple causes, including rapid changes in technology in the semiconductor industry, the high reliance in recent years on H1-B visa holders, the lingering effects of the tech bust, and so forth. One of these trends is that a substantial number, perhaps 10 percent, of older engineers has experienced long periods of unemployment.

A second trend is that many U.S. citizens and permanent residents appear to have decided that graduate engineering

TABLE 3-11 Engineers at U.S. Chip Firms by Location (in thousands)

	1999	2000	2001	2002	2003	2004	2005
United States	61.9	76.1	72.6	72.9	72.0	66.6	83.2
Offshore	17.5	20.0	27.2	29.8	30.9	34.6	42.2
Percentage in the United States	77.9	79.2	72.7	70.9	69.9	65.8	66.3

Source: Brown and Linden, Table 11, this volume.

degrees do not provide enough return on investment to justify the time and expense of pursuing them.[3] Companies are still hiring foreign students who earn degrees from U.S. institutions, however, who can work at U.S. companies on H1-B visas. Interestingly, when the number of H-1B visas was cut recently, U.S. semiconductor companies reacted by sending their visa-holding Indian and Chinese employees back to their home countries to help manage and develop subsidiaries there.

Although offshoring has had positive impacts on destination countries, India and China face challenges in upgrading their educational systems to produce more engineering graduates and making their infrastructure, such as electrical power, more reliable. Overall, however, experience in the semiconductor industry shows that offshoring is likely to become a trend for engineering work everywhere. For example, both Taiwan and India are now offshoring work to China.

Because semiconductors are critical components of many modern weapons systems, the U.S. Department of Defense (DOD) is justifiably concerned about maintaining access to semiconductors and related capabilities in design and manufacturing. In 2005, a task force of the Defense Science Board released a report recommending that DOD not only take steps to track the military's needs and ensure that "trusted microelectronics components" are available, but also spearhead a broad national effort to ensure that leading-edge microelectronics skills and capabilities remain in the United States (DSB, 2005).

It should be noted that there has also been significant "onshoring" in semiconductor design. Foreign-based firms like Philips, Hitachi, and Toshiba, for example, maintain extensive design operations in the United States.

REFERENCES

ASCE (American Society of Civil Engineers). 2005. Offshoring of Engineering Services. Policy Statement. Available online at *http://www.asce.org/pressroom/news/policy_details.cfm?hdlid=507*.

Aspray, W., F. Mayadas, and M.Y. Vardi, eds. 2006. Globalization and Offshoring of Software: A Report of the ACM Job Migration Task Force. New York: Association for Computing Machinery. Available online at *http://www.acm.org/globalizationreport/*.

BLS (Bureau of Labor Statistics). 2007. Occupational Employment Statistics, 2007. Available online at *http://www.bls.gov/oes/home.htm*.

Bureau of the Census. 2004. Economic Census 2002. Available online at *http://www.census.gov/econ/census02/*.

BusinessWeek. 2006. Toyota in China: Full Speed Ahead. BusinessWeek, March 9.

Campbell, A., G. Rotz, and K. Worzel. 2005. Getting Fit in Pharma: From Periodic Cost-Cutting to Continuous Productivity Improvement. New York: Marakon Associates. Available online at *http://www.marakon.com/ideas_pdf/id_051205_campbell.pdf*.

CII (Construction Industry Institute). 2004. Planning a Global Virtual Engineering Team: A Tool for Success. Research Summary 211-1. Austin, Tex.: CII.

Cohoon, J.D. 2006. Remarks at the Workshop on the Offshoring of Engineering. National Academy of Engineering, Washington, D.C., October 24, 2006.

DSB (Defense Science Board). 2005. High Performance Microchip Supply. February. Available online at *http://www.acq.osd.mil/dsb/reports/2005-02-HPMS_Report_Final.pdf*.

Gray, N. 2005. PharmExec 50: Untying the Gordian Knot. Special Report. Pharmaceutical Executive (May): 84–100.

Hamm, S., and D. Kopecki. 2006. Tech's Threat to National Security. BusinessWeek.com. November 2. Available online at *http://www.businessweek.com/technology/content/nov2006/tc20061102_797312.htm?campaign_id=rss_tech*.

Hecker, D.E. 2005. High-technology employment: a NAICS-based update. Monthly Labor Review 128(7): 57–72.

IHT (*International Herald Tribune*). 2007. China approves $2.5 billion Intel Corp. chip plant amid booming demand in country. IHT, March 13.

Krazit, T. 2005. Microsoft: Multicore Chips Changing PC Software. InfoWorld, October 26. Available online at *http://www.infoworld.com/article/05/10/26/HNmicrosoftchips_1.html*.

McKinnon, R., K. Worzel, G. Rotz, and H. Williams. 2004. Crisis? What Crisis? A Fresh Diagnosis of Big Pharma's R&D Productivity Crunch. New York: Marakon Associates. Available online at *http://www.marakon.com//ideas_ pdf/id_041104_mckinnon.pdf*.

McKinsey Global Institute. 2005. The Emerging Global Labor Market. New York: McKinsey & Company. Available online at *www.mckinsey.com/mgi/publications/emerginggloballabormarket/index.asp*.

NSB (National Science Board). 2006. Science and Engineering Indicators. Arlington, Va.: National Science Foundation.

O'Conner, A. 2006. India a growing option for low-cost drug trials. The Australian, October 11. Available online at *http://www.theaustralian.news.com.au/story/0,20867,20557947-12332,00.html*.

Saxenian, A. 2006. International Mobility of Engineers and the Rise of Entrepreneurship in the Periphery. Research Paper 2006/142. United Nations University World Institute for Development Economics Research. Available online at *http://www.wider.unu.edu/publications/publications.htm*.

Thursby, J., and M. Thursby. 2006. Here or There? A Survey on the Factors in Multinational R&D Location. Washington, D.C.: The National Academies Press. Available online at *http://www.nap.edu/catalog/11675.html*.

ValueNotes. 2006. Offshoring of Automotive Design & Engineering Services to India. July.

Zhao, Z., A. Jaideep, and W. Mitchell. 2005. A Dual Networks Perspective on Inter-Organizational Transfer of R&D Capabilities: International Joint Ventures in the Chinese Automotive Industry. Journal of Management Studies 42(:1): 127–160.

[3]This perception that pursuing a graduate degree in engineering is not a great investment might well exist in other fields, but Brown and Linden were the only authors to address it in detail.

4

Workshop Findings and Discussion

The NAE workshop on the offshoring of engineering addressed the effects on several key industries but was not a comprehensive examination of offshoring in all industries or all aspects of engineering. The study committee took into account engineering employment, salaries, and education and recent debates about the policy implications of offshoring. The committee also identified areas for future study.

The committee understands "globalization" to be a widespread, long-standing process whereby national economies and business activities are becoming increasingly integrated and interdependent, mainly through expanded trade, capital flows, and foreign direct investment. "Offshoring" refers to a more recent phenomenon of work relocated and diffused across national borders, enabled by advances in communications technology and changes in management practices.

Although a wide range of services work is being offshored, this workshop and report focus only on engineering and the impact of business practices, such as the international diffusion of corporate R&D, and the movement of engineering work as a result of the relocation of manufacturing activity. The report also includes examples of "onshoring," engineering work moved to the United States from abroad. A more detailed explanation of the committee's working definition of offshoring is provided in Chapter 2.

The workshop discussions and commissioned papers on six specific industries show how the offshoring of engineering work affects the U.S. engineering enterprise. Clearly, less complex work that does not require interaction with customers is offshored first. However, evidence shows that the level of sophistication increases over time, except in industries where R&D is being diffused, such as pharmaceuticals, or when the relocation of product and manufacturing engineering is closely tied to the relocation of manufacturing facilities, as in the automotive industry. In the semiconductor and software industries, the increase in offshoring in the last five years has led to a complete transformation of the business models for those industries. In other industries, such as construction engineering and services, the impact of offshoring has been less obvious.

The findings and discussion that follow are not arranged in order of priority.

TRENDS AND IMPACTS

Effects by Industry Sector

FINDING 1. *The offshoring of engineering, an inevitable aspect of globalization, has significantly impacted the U.S. engineering enterprise. However, the effects of globalization and offshoring have been uneven, and disparities among industry sectors and engineering sectors are likely to continue.*

Offshoring has increased most rapidly in information technology (IT)-related industries such as software, semiconductors, and PC manufacturing. As Ralph Wyndrum, then president of IEEE-USA, points out in his paper on the implications of offshoring for the engineering workforce and profession, "virtually all bids for commercial work now include an offshore component . . ." (this volume). For both established U.S.-based IT firms and start-ups, the location of at least some engineering work in India or China is now taken for granted. In fact, in most IT-related sectors (e.g., semiconductors, software, and PC manufacturing), the offshoring of engineering work is an established part of the business model for U.S.-based companies.

The Indian software industry, which employed more than 500,000 people and exported $17 billion in 2005, has grown at an annual rate of 30 to 40 percent over the past decade (Dossani, this volume). At the same time, the proportion of software exports accounted for by foreign (predominantly U.S.-based) companies has increased, as has the sophistication of the product mix. Thus the Indian software industry is tightly integrated with U.S.-based software development.

In the semiconductor industry, some steps (e.g., assembly, packaging, and testing) in the manufacturing process have long been globalized. In recent years, more sophisticated steps, such as wafer fabrication, have followed suit. Offshoring of semiconductor design is also increasing rapidly. In fact, 18 of the top 20 U.S.-based companies have opened design centers in India, nine of them since 2004 (Brown and Linden, this volume).

The manufacture of PCs and many PC components was moved from the United States to Taiwan more than a decade ago. Since then, it has been moved again, almost exclusively to China (Dedrick and Kraemer, this volume). In addition, much of the product design and engineering for PCs is now done by original design manufacturers based mainly in Taiwan.

Some engineering activities in the automotive industry and construction engineering and services industry have long been internationalized. U.S.-based auto companies have traditionally followed an imperative of manufacturing where they sell, and they often design and develop vehicles for specific markets (Moavenzadeh, this volume). Thus the employment of significant numbers of engineers abroad by U.S.-based companies in the auto industry is nothing new. Similarly, construction engineering and services firms that operate globally have always required engineering help in the countries where projects are located (Messner, this volume).

Nevertheless, the offshoring of less complex engineering work is increasing in both of these industry sectors. In the auto industry, some companies are trying to boost the productivity of their global engineering workforces by organizing distributed teams around global tasks. For example, global engineering leadership for a certain category of vehicle may be located in a specific country (e.g., full-size trucks in the United States, compact cars in Korea). Engineering teams in several countries contribute to the design of specific models.

Finally, the trend toward globalization of R&D in a range of other industries, including pharmaceuticals, is almost certain to gain momentum in coming years. For example, well over half of the more than 200 U.S.- and Europe-based companies that responded to a recent survey anticipate increasing technical employment in China, India, and other locations in Asia in the next three years (Thursby and Thursby, 2006).

The Need for Data

FINDING 2. *More and better data on offshoring and other issues discussed in this report, such as the effects on the engineering workforce and engineering education, are necessary for discerning overall trends. As has been pointed out in other recent reports, better U.S. and international statistics on trade in services and employment would give us a much better grasp of basic trends.*

Although various surveys, projections, and analyses by consulting companies, academics, and others can shed some light on the situation, significant data gaps have kept policy makers and the public from getting an accurate read on what is actually occurring in the international trade in services and offshoring. Several recent reports (GAO, 2005a,b; NAPA, 2006; Sturgeon, 2006, etc.) have pointed out deficiencies in U.S. government statistics. For example, trade statistics track many fewer categories of service products than manufactured goods, even though services now constitute a much larger share of the U.S. economy than manufacturing. In addition, current employment statistics make it impossible to track employment by occupation over time.

Statistics on the science, technology, engineering, and mathematics workforce could also be improved (Ellis et al., 2007). One improvement would be for agencies that collect and publish these data to adjust the classifications and coding so that occupations are easier to identify and track. An example of the problem, cited by Ellis et al. (2007), is the difficulty of tracking postsecondary teachers, who are usually subsumed in the general category of educators. Thus tracking jobs in engineering is difficult because, in some fields, academics make up a large percentage of the total workforce. In addition, more information on citizenship and the migration of engineers would make it easier to understand offshoring and discern other trends in the engineering workforce.

A study of offshoring in specific industries is no doubt valuable, but we must remain cognizant of the lack of timely, comprehensive data. We must also keep in mind that, even if we had all relevant information, it would represent only a snapshot in time. Thus all estimates or projections include considerable uncertainties, as offshoring continues to change!

Although many basic questions about offshoring, particularly questions specific to engineering, cannot be answered definitively, a review of the literature on offshoring, and trade in services generally, reveals several points of rough consensus. The combination of technological advances, innovations in management techniques, and the accessibility of overseas talent has made a growing number of services jobs vulnerable to offshoring. Estimates of the number of vulnerable U.S. jobs vary considerably, from the most common estimates of around 10 percent of the current workforce (NAPA, 2006)[1] up to 40 million (Blinder, 2006).

[1] In April 2007, for example, U.S. employment stood at 145.8 million, 10 percent of which is 14.6 million.

Estimates of the number of jobs that will actually be offshored vary as well. These estimates are generally expressed as the number of jobs offshored over a period of time (NAPA, 2006). For example, Forrester estimates that 3.4 million jobs will be offshored from 2005 to 2015 (340,000 per year); Goldman Sachs estimates that 6 million will be offshored from 2003 to 2013 (600,000 per year) (GAO, 2004). Thus, despite a consensus that offshoring is significant and increasing, it is impossible to say what the net impacts on U.S. employment have been or will be.

Even if the number of jobs offshored is at the high end of estimates, only a small percentage of overall jobs in the services sector would be lost (or gained) when trends in the domestic U.S. economy are factored in. After the collapse of the tech bubble and during the slow recovery that followed, some U.S.-based companies announced large-scale layoffs in the United States, at the same time launching new operations overseas, particularly in India. However, since 2005, the U.S. tech economy has stabilized and recovered, and there have been fewer cases like these. Thus, overall, there may still be net job creation in the United States in many occupations that will be subject to widespread offshoring over the long-term.

Even though engineering is on almost every list of occupations vulnerable to offshoring, the uncertainties in estimates of offshoring of engineering are even greater than for offshoring in general. For example, McKinsey Global Institute (2005), based on its global analysis, estimates that more than half of engineering positions in the industries it examined could be performed anywhere in the world. NASSCOM (2006) projects that the Indian engineering services offshoring industry will grow from about $1.5 billion today to $30 to $60 billion by 2020.

Yet it would be unwarranted to conclude that half of the 1.5 to 2 million current U.S. engineers are in danger of losing their jobs in the next few years. For one thing, the Bureau of Labor Statistics estimates that the U.S. engineering workforce will grow by 13 percent between 2004 and 2014, roughly in keeping with the projected growth of the total U.S. workforce (CPST, 2006). For another, offshoring will be limited by the supply of talent available in destination countries. Although emerging economies such as India and China are turning out large numbers of young engineers and are taking steps to increase their numbers and improve their quality, the speed at which these improvements can be made is limited. McKinsey (2005) estimates that only 15 to 20 percent of young engineers in developing countries are currently qualified to work in international companies. Finally, developments in the United States will play an important role. For example, U.S. engineering education may or may not evolve in ways that support engineering as a profession that can attract more of the best and brightest U.S. students.

The emergence of offshoring signaled the beginning of an era in which a broad swath of the U.S. workforce, including engineers and workers in many other services professions, became subject to international competition. Based on a comprehensive, up-to-date understanding of trends in offshoring, the United States can remain a premier location for engineering activity, and the engineering enterprise can adapt and renew itself. However, for the United States to develop policies to preserve its economic vitality and avoid adopting policies that are counterproductive, policy makers must have a clear understanding of what is happening and why.

At the organizational level, the institutions and associations that educate and rely on engineers also need to understand trends in offshoring as a basis for developing new approaches to defining necessary skills and training engineers for careers in a globalized world. On the personal level, individual engineers must have the information they need to determine the most promising career paths and prepare themselves accordingly.

Realistically, it may be some time before even glaring data deficiencies are addressed. In addition, much of the offshoring activity by companies is inherently difficult to track through trade statistics. As a result, although industry-specific analyses of the type commissioned for this workshop will continue to be important sources of information about offshoring and globalization, they can provide only a snapshot. Further studies will be necessary as engineering offshoring evolves.

Winners and Losers

***FINDING 3.** Offshoring appears to have contributed to the competitive advantage of U.S.-based firms in a variety of industries, and the negative impacts of offshoring on U.S. engineering appear to have been relatively modest to date. However, the negative effects have been much more severe in some industry sectors and for some jobs than others.*

Cost reduction is often an important factor in the initial offshoring decision, particularly for IT-related companies. Another consideration is the need to compete in new or rapidly growing markets. For individual businesses, decisions about where to locate engineering activities are made on the basis of both value and the potential for market growth—similar to the way decisions concerning access to capital and other resources are made. The second factor has been very important to foreign-based firms locating engineering activities in the United States. This so-called "onshoring" is an important part of the overall picture of globalization.

Although some kinds of offshoring have appeared only recently, disaggregated business models have a long history in several U.S. industries. For example, "fabless" semiconductor companies that contract out manufacturing first appeared in the 1990s. U.S. firms developed this model, and fabless companies (e.g., Broadcom) are among the most successful and fastest growing semiconductor companies in the past decade. The "foundry" industry, which fabricates semiconductors on a contract basis, emerged in Taiwan and is now expanding in China.

Similarly, branded firms in the U.S. PC industry began offshoring manufacturing long ago; over time, they have also offshored significant engineering tasks. Careful coordination and management of offshoring has enabled companies such as Dell and Hewlett-Packard to hold down costs and remain competitive in an industry with rapid product cycles. Apple used a disaggregated business model to create a new category of products that straddle the line between IT and consumer electronics (e.g., the iPod), and, in the process, accelerate its growth.

The key to long-term success for companies shifting toward globalization of engineering activities appears to be protecting the interface with customers and the resulting information flow. Knowledge from customers feeds into product definition, high-level design, and the most sophisticated engineering tasks. The effective use of offshoring has helped many PC firms sustain their U.S. operations.

Like the initial movement of manufacturing activity to overseas locations, much of the upsurge in offshoring of design and engineering work has been motivated by attempts to keep costs under control, or even to reduce them. This is clearly the case in IT-related sectors that have offshored engineering and other functions. However, decisions about locating R&D facilities overseas, even in emerging economies such as China, are often also influenced by other factors, including the desire for companies to establish a full-spectrum presence in rapidly growing markets (Thursby and Thursby, 2006). In addition, some companies are trying to access specific expertise with their R&D investments.

The "onshoring" of R&D and other engineering work—multinational companies based in Europe or Asia establishing or acquiring operations in the United States—is a significant trend in the pharmaceutical and automotive industries. Companies such as Toyota and Honda are expanding their engineering employment in the United States. Even companies based in India and China are beginning to make R&D investments in this country.

In general, the lack of comprehensive, accurate data makes it difficult to measure the net impact of offshoring on engineering jobs and salaries in recent years. Remember that the upsurge in offshoring coincided with a downturn in the U.S. economy that hit the tech sector particularly hard (the dotcom bust). During the first half of the 2000s, unemployment in subsectors of the engineering workforce rose to record levels, while salaries remained flat or even declined. By 2005, employment and salaries had begun to recover, and by early 2007, the unemployment rate for engineering and architectural occupations had fallen considerably. At the same time, offshoring apparently continued to expand. Clearly, it is difficult to separate the impacts of broad economic changes, offshoring, and related trends in globalization, such as increased immigration.

Even if the net impact of offshoring on employment in the engineering workforce is relatively small, there are still some winners and some losers. When certain routine engineering tasks are moved to India or China, U.S. engineers who previously performed those tasks might lose their jobs. At the same time, more jobs may be created for U.S. engineers performing higher level tasks. Hira and Hira (2005) describe the difficulties faced by tech workers displaced by offshoring or immigration. The negative individual and social impacts of mass layoffs in general, not necessarily in engineering, are described by Uchitelle (2006). The issues for engineering education and public policy raised by this displacement are discussed in more detail below.

IMPLICATIONS FOR ENGINEERING EDUCATION

FINDING 4. *Engineering education at the undergraduate and graduate levels has been a major source of strength for the U.S. engineering enterprise. Even today, engineers educated in the United States remain among the best trained and most flexible in the world. At a time when other nations are making significant efforts to upgrade their engineering education capabilities, the United States will be challenged to sustain engineering education as a national asset.*

Based on workshop discussions, both industry and academic participants believe that U.S. engineering education will continue to be a valuable asset as the U.S. engineering enterprise adapts to new global realities. Charles Vest, NAE president and President Emeritus of MIT, presents the overall case for U.S. engineering education (this volume), while Ted Rappaport, director of the Wireless Networking and Communications Group at the University of Texas at Austin, details the importance of academic engineering research and education to the key field of network systems (this volume).

Nevertheless, other countries and regions are working hard to upgrade their engineering education capabilities and adapt them to new global realities. For example, European countries are working toward standardizing degree programs so that engineers at a certain degree level in, say, Spain will have skills and attributes similar to those of engineers at that level, or its equivalent, in, say, Sweden.

Emerging countries, most prominently China and India, which are the prime destinations for offshoring, are taking substantial steps to increase their capacities for delivering high-quality engineers. China has adopted a top-down, directive approach, while India has adopted a more market-oriented, bottom-up approach (Wadhwa et al., 2007a,b). China appears to be focusing on increasing the number of Ph.D. and master's-level engineers and scientists to meet future R&D needs. India is producing more graduates with the skills and aptitudes appropriate for the jobs being created there today. The number of graduate engineering degrees in China has increased from about one-fourth as many as in the United States in 1995 to a higher number than in the United

States today. At the same time, large numbers of students from China and India continue to come to the United States for graduate engineering education.

As offshoring and the general globalization of engineering continue, more and more engineering work will be performed by multi-country teams and in other international contexts. Speakers at the workshop and the authors of the commissioned papers identified several areas in which U.S. engineering educators might consider new approaches, particularly in communication and management skills, to meet the demands of a globalized engineering environment. These areas of change were compatible with the findings and recommendations in several recent reports by NAE and others and statements by professional societies.

For example, a 2004 report by the American Society of Civil Engineers on the "body of knowledge" necessary for civil engineers concludes that management and communication skills should be part of the engineering curriculum. In two reports issued by the NAE Engineer of 2020 Project, in 2004 and 2005, just as attention was turning to offshoring, a key observation was that "good engineering will require good communication." In fact, all recent studies on this subject recognize that the nature of engineering is changing in ways that will require engineers to work across sectoral and disciplinary boundaries as well as national borders. Interacting effectively with and being accountable to diverse, global customers means that engineers will have to "listen effectively as well as communicate through oral, visual, and written mechanisms" (NAE, 2004, 2005).

Opportunities for leadership will increase but will require new levels of sophistication (NAE, 2004). Past experience has shown that engineers who have mastered business and management principles are often rewarded with leadership roles. The study committee of the present report urges U.S. engineering educators to prepare students to tackle these global challenges.

At the same time, many engineers, and the profession as a whole, have sometimes been ambivalent about engineers moving into management, policy making, and other fields, rather than remaining in technical roles. In addition, the financial compensation for engineers in non-technical roles tends to be higher than for those who remain in technical positions. As a result, professional societies and other engineering leaders have urged employers to ensure that there are attractive career tracks for mainstream engineers.

Maintaining technical career tracks and technical currency through continuing education (discussed below) will be challenging in an environment where offshoring and globalization are changing the nature of engineering work. Engineering educators and the engineering profession will have to monitor the global marketplace for engineers and consider how well current educational approaches are preparing students to meet the demands of that marketplace. Understanding deficits in skill sets and addressing them, increasing the participation of women and minorities, and providing more varied and realistic career paths for students with engineering degrees will be crucial for the future.

FINDING 5. *Although individual engineers must ultimately take responsibility for their own careers, industry, government, universities, professional societies, and other groups with a stake in the U.S. engineering enterprise should consider supporting programs and other approaches to helping engineers manage their careers, renew and update their skills, and sustain their capacity to innovate, create, and compete.*

A continuing subject of discussion both during the workshop and the steering committee meetings was the effects of offshoring on individual engineers. As companies and other organizations grow and shrink and as jobs are gained and lost, the environment for engineering work is changing significantly. Engineers who are proactive in keeping their skills up to date and are able to take advantage of the trend toward more frequent job and career shifts have adapted well to these changes and are much less vulnerable to the negative effects of offshoring.

Those who are not as skilled or proactive are faced with job insecurity and slow wage growth. During the workshop discussions and in his paper, Ralph Wyndrum, then president of IEEE-USA, the largest U.S. professional engineering society, called for renewed efforts on the part of all stakeholders in U.S. engineering—educators, professional societies, employers, government, and engineers themselves—to address the needs of mid-career engineers faced with the prospect of developing new skills and abilities for a constantly changing job market (in this volume). This support could be an important factor in determining whether U.S. engineering retains its global leadership position.

Both of the NAE 2020 reports highlighted the need for lifelong learning and that the engineering education system must do more to help students become self-learners. In addition to the challenge of global competition, the body of knowledge in engineering is expanding exponentially and pressuring engineers to keep up by becoming ongoing learners. One challenge that may arise for employers will be balancing the benefits of "up-skilling" their workers with the risks of making their employees highly desirable to other companies. As one workshop participant noted, "poaching" by competitors can be a problem for companies that maintain a highly trained staff. Everyone agreed, however, that choosing not to up-skill workers is not the solution to this problem.

Most professional societies and many engineering schools already offer continuing education programs for mid-career professionals. To determine what else might be done, it would be helpful to have an in-depth assessment of current efforts to determine if mid-career engineers are taking advantage of these programs and, perhaps, to suggest incentives that might encourage further participation. An inventory of lifelong-learning programs and participation would also be a valuable tool for policy makers and private organizations.

During the workshop discussions, one participant suggested that an engineering degree should come with a warranty, or a coupon, for "free upgrades." Another model would be "executive" technical degrees similar to executive MBAs. Others have suggested the creation of an 8- to 12-year learning model that would be the shared responsibility of universities and employers. A useful model might be military academies and services, which have adopted a systematic approach to continuing education. In the final analysis, government may have to help provide incentives and wherewithal for individual mid-career engineers to take advantage of opportunities for learning.

This study and workshop highlighted the need for more discussion of these issues, not only in connection with engineering education, but also for the future of U.S. leadership in a global economy. Many have called for assistance to engineers and other service workers whose jobs are displaced by offshoring, including, perhaps, public subsidies for continuing education. Approaches that have been discussed include (1) expanding eligibility for Trade Adjustment Assistance to engineers and other service-industry workers who lose their jobs and (2) providing some form of wage insurance to help displaced workers who are forced to take lower paying jobs.

FINDING 6. *Over the past several decades, engineering has become less attractive to U.S. students as a field of study and as a career compared to some other professions. Although it is widely assumed that globalization and offshoring are contributing to this relative decline in popularity, it is impossible to know how important globalization is compared to other factors. A great deal more needs to be understood about the relationship between offshoring and the attractiveness of engineering as a career.*

Concerns about whether offshoring has a negative effect on the public perception of engineering and whether this perception causes fewer of the "best and brightest" to pursue engineering careers were raised numerous times at the workshop and in several of the commissioned papers (e.g., Stevens and Rappaport in this volume). Certainly, engineering is a less popular undergraduate major than it was. Between 1983 and 2002, the number of bachelor's degrees in engineering declined by about 16 percent (NSB, 2006). During that same period, the overall number of bachelor's degrees increased by about 33 percent. Thus engineering degrees as a percentage of the total declined from 7.4 percent to 4.6 percent.[2] We have anecdotal evidence, but very limited data at this point, to determine if offshoring and globalization are contributing to the decline and, if so, to what extent. As noted in Chapter 2, the growth of offshoring coincided with the dot-com bust and a downturn in the overall economy.

Questions about the negative perceptions of engineering persist, as do questions about whether, and what, the engineering profession should do about them. Other professions, such as medicine, business (at the graduate level), and law continue to attract ambitious, bright students at least partly because of real or perceived high payback, and the lack of early high payback may discourage students from going into some engineering fields. However, payback is not always predictable. For example, graduates in civil engineering who chose to go into the field of information technology in the late 1990s because of higher starting pay might have regretted that decision a few years later.

Engineering managers trying to "connect" with the "millennial generation" and communicate the excitement of engineering careers must address not only salaries and job security, but must also convey the excitement of opportunities for engineers to make a difference and improve people's quality of lives (Stevens, this volume). Changing public perception of engineering may require a systematic, well thought out campaign of public education. This was the conclusion of a study committee of a 2002 NAE report (Davis and Gibbin, 2002).

IMPLICATIONS FOR PUBLIC POLICY

FINDING 7. *For the United States, attracting and retaining world-class engineering activities in an increasingly competitive global environment will require that core U.S. strengths be sustained. Perhaps the most critical task in doing so will be to avoid complacency.*

Several speakers at the workshop addressed the issues raised by globalization for U.S. engineering. In the opinion of Charles Vest, even in a "flatter" world, the United States and U.S. engineers enjoy significant advantages. The biggest threat to our future success, he believes, is complacency (Vest, this volume). Robert Galvin, Chairman Emeritus of Motorola Inc., described how addressing global challenges in energy and transportation would create engineering jobs in the United States (this volume). Ted Rappaport, stressed the importance of public investment in network systems and other critical areas (this volume). Overall, the speakers agreed that public and private efforts to tackle large, even global, problems could help create entire new industries and would go a long way toward creating new opportunities for U.S. engineers. A number of recent reports have also explored what the United States can, indeed must, do to maintain its lead in science and engineering (COSEPUP, 2005; Council on Competitiveness, 2005).

FINDING 8. *Plausible scenarios have been developed showing that offshoring either helps, is neutral, or hurts engineering in the United States. Only continued discussions*

[2]When engineering and computer science degrees are combined, the decline in popularity decreases, from about 10 percent to 8.4 percent.

and further studies will lead to a thorough understanding of the potential benefits and costs of offshoring.

For the American public and the U.S. economy as a whole, the offshoring phenomenon, and offshoring of engineering in particular, could have a number of costs and benefits. The workshop discussions and presentations shed some light on the magnitude and likelihood of these costs and benefits. Of course, a clearer picture will emerge over time.

Potential Benefits

U.S.-based companies that manage offshoring effectively appear to be benefiting in terms of competitiveness and profitability. Having access to skilled engineers at lower cost may create a climate for faster, more cost-effective innovation and might even ultimately lead to higher employment levels in the United States.

Offshoring has become an accepted component of the business model in some industries, particularly IT-related industries and electronics (Wyndrum, this volume). Start-up companies that combine U.S. project management and market savvy, Indian engineering implementation, and, perhaps, Chinese manufacturing capabilities, can stretch the dollars of venture capitalists and improve the odds of their survival and ultimate success. At the same time, these companies increase the potential for innovation and the development of new products that might not have appeared as quickly, or even at all, without offshoring.

It has long been assumed that globalization and trade will ultimately deliver net benefits to the U.S. economy. According to one analysis, globalization since World War II has increased the U.S. GDP anywhere from $800 billion to $1.4 trillion per year ($7,000 to $13,000 per household) (Bradford et al., 2006). As a continuation of globalization, offshoring might also deliver net economic benefits. However, many questions are being raised about whether this will happen.

Offshoring is delivering economic benefits to several emerging economies, particularly India and China. The argument has been made that long-term U.S. interests will be served as these countries and other developing economies become integrated into the global economy and living standards rise, even though this will inevitably lead to better engineering capabilities in these countries. If U.S. engineering capability can be sustained, the emerging global networks will be open to participation by Americans and American organizations. In that case, globalization would present a "win-win" situation, because U.S. engineers would also benefit directly through expanded markets for their skills.

Potential Costs

There are many possible downsides to this scenario. Some have argued that offshoring and other forms of trade can undermine U.S. economic strength and national interests (e.g., Gomory and Baumol, 2001). Some prominent economists have raised concerns that the distributional impacts of offshoring on engineers and other service-sector workers in the United States will pose a serious challenge to free trade (Blinder, 2006). Others argue that offshoring might lead to a degradation of U.S. engineering capability and that, even if the U.S. engineering enterprise and economy as a whole are better off as a result of offshoring, those who are most vulnerable to competition might suffer severe hardships.

One scenario in which U.S. engineering capability might be damaged through offshoring is if U.S.-based companies attempting to take advantage of lower costs and perceived better value move a large percentage of engineering, R&D, and other activities from the United States to the developing world over a short period of time, giving engineers and others who would lose jobs little time to anticipate or adjust to significant change. Although this is within the realm of possibility, there are reasons to believe that a wholesale shift of engineering work from the United States to China and India is unlikely. First, as described in Chapter 3, the gaps and deficiencies in the science and engineering enterprises of large emerging economies will take time to address. Second, as the engineering capabilities in emerging economies improve, the productivity and pay of their engineers are also likely to rise, thus reducing the cost advantage of offshoring.

In another scenario, U.S. engineering capabilities might atrophy gradually as the result of a combination of depressed wages and job insecurity, which would discourage significant numbers of young Americans from entering the engineering profession. Many studies and debates have attempted to determine the number of engineers and scientists necessary to the U.S. economy, whether or not there is or will be a shortage, the number of engineering graduates produced by the United States compared with China and India, and whether offshoring of engineering and other forms of trade could erode America's ability to innovate, which could leave the country as a whole worse off (e.g., COSEPUP, 2005; Wadhwa et al., 2007a).

These and other questions raised at the workshop will continue to be discussed, and the answers will surely affect the evolution of offshoring.

Incentives and Disincentives

FINDING 9. *As the debate about offshoring continues, it will be important to determine whether current U.S. policies, including immigration policies, provide artificial advantages or incentives for offshoring.*

There have been many calls for changes in policy to provide assistance to engineers and other service workers whose jobs are displaced by offshoring. One mechanism would be to expand the number of people eligible for benefits through Trade Adjustment Assistance. Currently only people whose jobs are lost due to imports of goods are eligible, but many believe people whose jobs are lost because of imports of services should also be eligible (e.g., Kletzer and Rosen, 2005). Assistance might also be provided through some form of wage insurance to help displaced workers adjust if they are forced to take lower paying jobs (Andrews, 2007). Of course, these kinds of policies would have widespread repercussions and must be explored thoroughly before they are adopted.

Immigration Policy

Although a detailed examination of immigration policies for engineers was beyond the scope of this study, immigration issues are closely related to offshoring. Wyndrum argues that several major participants in offshoring use H-1B and L-1 visa programs to bring in employees, train them in the United States, and then send them back to their countries to expand the company's offshoring operations (this volume). Abuses of these visa programs by recruiting firms have also been reported. Some workshop participants reported that their companies participate in the H-1B program to reduce costs but have found that cost and uncertainties involved in hiring visa holders can offset some of the anticipated savings. Other companies hire visa holders because they are unable to find qualified, highly trained engineers who are U.S. citizens.

There is a good deal of debate and uncertainty about the current and future role of foreign engineering students. Does the United States rely too heavily on foreign engineering talent? Might fewer foreign students study at U.S. engineering schools in the future, and might fewer of those who graduate from U.S. institutions remain in this country? Some analysts have asserted that a growing number of U.S.-educated foreign science and engineering students are returning to their home countries (Newman, 2006). However, an annual survey of foreign engineering students who receive doctorates from U.S. institutions shows that "stay rates" have remained about the same in recent years (NSB, 2006).

Some argue that, as immigration policies become more stringent, the United States will be cutting itself off from a vital source of engineering talent. Clearly, the immigration of scientists and engineers, the training of foreign students, and the overall openness of the United States to foreign talent has been a boon to U.S. engineering and to the larger U.S. economy. Others argue that stricter policies could, in effect, subsidize or provide artificial incentives for offshoring engineering, which would be just as counterproductive and market-distorting as erecting artificial barriers or penalties for offshoring.

All of these questions should be investigated thoroughly as part of the policy making process.

Security Issues

FINDING 10. *Security concerns related to the offshoring of engineering have been raised, specifically for the information technology and construction industries.*

Concerns about national and homeland security related to offshoring have been raised in connection with several of the industries studied at the workshop. For example, offshore construction engineering and services might result in detailed plans and other information about U.S. buildings and critical infrastructure falling into the wrong hands (ASCE, 2005). Similar concerns have been raised about offshoring of engineering work that involves geospatial data (MAPPS, 2006). Legislation was proposed in the last Congress to address the latter issue, and relevant professional societies are working to ensure that sensitive information is protected within the existing legal framework.

Concerns have also been raised about whether the globalization of software development poses a serious threat to national and homeland security, particularly if accidental defects or maliciously placed code could compromise the security of U.S. Department of Defense (DOD) networks (Hamm and Kopecki, 2006). The Defense Science Board (DSB) is currently completing a study on how DOD should address these concerns. DSB previously issued a report raising concerns about the migration of semiconductor technologies offshore and how U.S. military access to critical microelectronics manufacturing capability could be maintained (DSB, 2005).

REFERENCES

Andrews, E. 2007. Why Wage Insurance is Dividing Democrats. New York Times, March 18.

ASCE (American Society of Civil Engineers). 2005. Offshoring of Engineering Services. Policy Statement. Available online at *http://www.asce.org/pressroom/news/policy_details.cfm?hdlid=507.*

Blinder, A.S. 2006. Offshoring: The Next Industrial Revolution? Foreign Affairs (March/April.) Available online at *http://www.foreignaffairs.org/20060301faessay85209/alan-s-blinder/offshoring-the-next-industrial-revolution.html.*

Bradford, S.C., P.L.E. Grieco, and G.C. Hufbauer. 2006. The payoff to America from globalization. The World Economy 29(7): 893–916.

COSEPUP (Committee on Science, Engineering, and Public Policy). 2005. Policy Implications of International Graduate Students and Postdoctoral Students in the United States. Washington, D.C.: The National Academies Press. Available online at *http://newton.nap.edu/catalog/11289.html.*

Council on Competitiveness. 2005. Innovate America: Thriving in a World of Challenge and Change. Washington, D.C.: Council on Competitiveness.

CPST (Commission on Professionals in Science and Technology). 2006. STEM Employment Forecasts and Distributions among Employment Sector: STEM Workforce Data Project: Report No. 7. Available online at *www.cpst.org.*

Davis, L.A., and R.D. Gibbin, eds. 2002. Raising Public Awareness of Engineering. Washington, D.C.: The National Academies Press.

DSB (Defense Science Board). 2005. High Performance Microchip Supply. February. Available online at *http://www.acq.osd.mil/dsb/reports/2005-02-HPMS_Report_Final.pdf.*

Ellis, R.A., E.L. Babco, L.M. Frehill, and N.E. Bell. 2007. Improving Federal Statistics on the Science, Technology, Engineering and Mathematics (STEM) Workforce. White Paper Number 2. Washington, D.C.: Commission on Professionals in Science and Technology.

GAO (Government Accountability Office). 2004. Current Government Data Provide Limited Insight into Offshoring of Services. GAO-04-932. Washington, D.C.: GAO. Available online at *http://www.gao.gov/new.items/d04932.pdf.*

GAO. 2005a. International Trade: U.S. and India Data on Offshoring Show Significant Differences. GAO-06-116. Washington, D.C.: GAO.

GAO. 2005b. Offshoring of Services: An Overview of the Issues. GAO-06-5. Washington, D.C.: GAO. Available online at *www.gao.gov/new.items/d065.pdf.*

Gomory, R., and W.J. Baumol. 2001. Global Trade and Conflicting National Interests. Cambridge, Mass.: MIT Press.

Hamm, S., and D. Kopecki. 2006. Tech's Threat to National Security. BusinessWeek.com. November 2. Available online at *http://www.businessweek.com/technology/content/nov2006/tc20061102_797312.htm?campaign_id=rss_tech.*

Hira, R., and A. Hira. 2005. Outsourcing America: What's Behind Our National Crisis and How We Can Reclaim American Jobs. New York: AMACOM. Available online at *http://www.amazon.com/gp/product/0814408680/qid=1133982951/sr=8-1/ref=pd_bbs_1/002-9456996-1835242?n=507846&s=books&v=glance.*

Kletzer, L.G., and H. Rosen. 2005. Easing the Adjustment Burden on U.S. Workers. Pp. 313–342 in The United States and the World Economy: Foreign Economic Policy for the Next Decade, edited by C.F. Bergsten. Washington, D.C.: Institute for International Economics.

MAPPS (Management Association for Private Photogrammetric Surveyors). 2006. Policy on Globalization of Geospatial Services. Available online at *http://www.mapps.org/InsideMAPPSNews.asp?ARTICLE19659=26569.*

McKinsey Global Institute. 2005. The Emerging Global Labor Market. New York: McKinsey & Company. Available online at *www.mckinsey.com/mgi/publications/emerginggloballabormarket/index.asp.*

NAE (National Academy of Engineering). 2004. The Engineer of 2020: Visions of Engineering in the New Century. Washington, D.C.: The National Academies Press. Available online at *http://newton.nap.edu/catalog/10999.html.*

NAE. 2005. Educating the Engineer of 2020: Adapting Engineering Education to the New Century. Washington, D.C.: The National Academies Press.

NAPA (National Academy of Public Administration). 2006. Off-shoring: An elusive phenomenon. Washington, D.C.: NAPA.

NASSCOM (National Association of Software and Service Companies). 2006. Globalization of Engineering Services: The Next Frontier for India. Washington, D.C.: NASSCOM. Available online at *http://www.nasscom.in/Nasscom/templates/NormalPage.aspx?id=49790.*

Newman, R.J. 2006. Can America Keep Up? U.S. News and World Report, March 19.

NSB (National Science Board). 2006. Science and Engineering Indicators. Arlington, Va.: National Science Foundation.

Sturgeon, T.J. 2006. Services Offshoring Working Group Final Report. Cambridge, Mass.: MIT Industrial Performance Center. Available online at *http://web.mit.edu/ipc/publications/pdf/06-006.pdf.*

Thursby, J., and M. Thursby. 2006. Here or There? A Survey on the Factors in Multinational R&D Location. Washington, D.C.: National Academies Press. Available online at *http://www.nap.edu/catalog/11675.html.*

Uchitelle, Louis. 2006. The Disposable American. New York: Alfred A. Knopf.

Wadhwa, V., G. Gereffi, B. Rissing, and R. Ong. 2007a. Where the Engineers Are. Issues in Science and Technology 23(3): 73–84.

Wadhwa, V., A. Saxenian, B. Rissing, and G. Gereffi. 2007b. America's New Immigrant Entrepreneurs. Duke University and University of California at Berkeley. Available online at *http://memp.pratt.duke.edu/downloads/americas_new_immigrant_entrepreneurs.pdf.*

Additional Reading

Amiti, M., and S.-J. Wei. 2004. Fear of Service Outsourcing: Is it Justified? International Monetary Fund Research Paper WP/04/186. October. Washington, D.C.: IMF. Available online at *www.imf.org/external/pubs/ft/wp/2004/wp04186.pdf*.

This literature review takes a positive view of offshoring.

Antràs, P., L. Garicano, and E. Rossi-Hansberg. 2005. Offshoring in a Knowledge Economy. National Bureau of Economic Research Working Paper 11094. January. Cambridge, Mass.: NBER. Available online at *http://papers.nber.org/papers/w11094.pdf*.

Uses a two-country model of offshoring to explain impacts on wages, consumption, and trade.

Arora, A., and A. Gambardella. 2004. Globalization of the Software Industry: Perspectives and Opportunities for Developed and Developing Countries. National Bureau of Economic Research Working Paper 10538. June. Cambridge, Mass: NBER. Available online at *http://papers.nber.org/papers/w10538.pdf*.

Concludes that offshoring is unlikely to pose a threat to long-term U.S. leadership in S&T.

Bardhan, A.D., and D.M. Jaffee. 2005. R&D, Innovation, and Offshoring. Fisher Center for Real Estate and Urban Economics, University of California at Berkeley Research Report 1005. Available online at *http://* repositories.*cdlib.org/cgi/viewcontent.cgi?article=1035&context=iber/fcreue*.

Uses a firm survey to explore offshoring in the context of the historical development of industrial R&D.

BusinessWeek Online Feature on Commerce Department Offshoring Study. October 2005. Describes a Commerce Department study where the original staff version critical of offshoring was revised by political appointees to portray offshoring as an unalloyed good. Available online at *http://www.*businessweek.*com/magazine/content/05_42/c3955025.htm#ZZZ24WZQGEE*.

Collins, S.M., and L. Brainard, eds. 2006. Offshoring White-Collar Work—The Issues and the Implications: Brookings Trade Forum 2005. Washington, D.C.: Brookings Institution. Available online at *http://www.brookings.edu/press/books/brookingstradeforum2005.htm*.

The 2005 conference and forthcoming conference volume focus on offshoring. Committee member Lori Kletzer has a paper in this volume that is listed separately under coauthor J. Bradford Jensen.

Criscuolo, C., J.E. Haskel, and M.J. Slaughter. 2005. Global Engagement and the Innovation Activities of Firms. NBER Working Paper 11479. June. Cambridge, Mass.: National Bureau of Economic Research. Available online at *http://nber15.nber.org/papers/w11479.pdf*.

Firms that export and MNCs are better innovators.

Dedrick, J., and K.L. Kraemer. 2005. Is Production "Pulling" Knowledge Work to China? A Study of the Global Computer Industry. Personal Computer Industry Center Working Paper. University of California, Irvine. Available online at *http://www.pcic.merage.uci.edu/pubs/2005/isProductionPulling.pdf*.

Dutta, A., and R. Roy. 2005. Offshore Outsourcing: Counteracting Forces and Their Dynamic Effects. Proceedings of the 38th Hawaii International Conference on Systems Sciences. New York: Institute of Electrical and Electronics Engineers. Available online at *http://ieeexplore.ieee.org/iel5/9518/30166/01385770.pdf?tp=&arnumber=1385770&isnumber=30166*.

Models outsourcing in a two-country context.

Economist Intelligence Unit. 2004. Scattering the seeds of invention: the globalisation of research and development. White paper sponsored by Scottish Development International. September. Available online at *http://graphics.eiu.com/files/ad_pdfs/RnD*_GLOBILISATION_*WHITEPAPER.pdf*.

Overview of the globalization of R&D.

Florida, R. 2005. The world is spikey. Atlantic Monthly (October): 48–51. Available online at *http://www.creativeclass.org/acrobat/TheWorldIsSpiky.pdf*.

Frames globalization of innovation in terms of "have" and "have-not" regions.

Florman, S. 2005. My Profession and the Nation: A Worrisome Confrontation. The Bent of Tau Beta Pi. Summer. Available online at *http://www.tbp.org/pages/publications/BENTFeatures/FlormanSu05.pdf*.

Global Insight Inc. 2004. Executive Summary: The Comprehensive Impact of Offshore IT Software and Services Outsourcing on the U.S. Economy and the IT Industry. Arlington, Va.: Information Technology Association of America. Available online at *www.itaa.org/itserv/docs/execsumm.pdf*.

This industry association-commissioned consultant's report says that the benefits outweigh the costs.

Gupta, A.K., and D.E. Westney, eds. 2003. Smart Globalization: Designing Global Strategies, Creating Global Networks. Hoboken, N.J.: Jossey-Bass. Available online at *http://www.amazon.com/gp/product/0787965324/qid=1133984042/sr=1-2/ref=sr_1_2/002-9456996-1835242?s=books&v=glance&n=283155*.

A compendium of recent global business strategy articles.

Lynn, L., and H. Salzman. 2005. The "New" Globalization of Engineering: How the Offshoring of Advanced Engineering Affects Competitiveness and Development. Paper presented at the 21st European Group for Organizational Studies Colloquium. June. Available online at *http://www.urban.org/UploadedPDF/411226_new_globalization.pdf*.

Discussion of engineering offshoring in the context of business innovation and corporate strategy literature.

Markusen, J. 2005. Modeling the Offshoring of White-Collar Services: From Comparative Advantage to the New Theories of Trade and FDI. NBER Working Paper 11827. Cambridge, Mass.: NBER. Available online at *http://nber15.nber.org/papers/w11827.pdf*.

Draws on a number of modeling approaches to analyze offshoring.

Mataloni, R.J. Jr. 2005. U.S. Multinational Companies: Operations in 2003. Survey of Current Business. Available online at *http://www.bea.gov/bea/ARTICLES/2005/07July/0705_MNCs.pdf*.

Mitra, D., and P. Ranjan. 2005. Y2K and Offshoring: The Role of External Economies and Firm Heterogeneity. NBER Working Paper 11718. Cambridge, Mass.: NBER.

Uses economic modeling to explain how Y2K served as a catalyst for IT offshoring. Available online at *http://nber15.nber.org/papers/w11718.pdf*.

OECD (Organization for Economic Cooperation and Development). 2005. OECD Science, Technology and Industry Scorecard 2005: Towards a Knowledge-Based Economy. Paris: OECD. Available online at *http://hermia.sourceoecd.org/vl=2279336/cl=27/nw=1/rpsv/scoreboard/*.

Special focus on globalization of R&D.

Parker, G.G., and E.G. Anderson Jr. 2002. From Buyer to Integrator: The Transformation of the Supply-Chain Manager in the Vertically Disintegrating Firm. Production and Operations Management 11(1): 75–91. Available online at *http://ggparker.net/papers/SCIntegratorPOMpages.pdf*.

Part II

Commissioned Papers and Workshop Presentations

Commissioned Papers

Implications of Globalization for Software Engineering

Rafiq Dossani
Stanford University
and
Martin Kenney
University of California, Davis, and Berkeley Roundtable on the International Economy

ABSTRACT

The offshoring of software engineering, which is more than three decades old, has been at the leading edge of the offshoring of information-technology services. Over the past decade, the pace of offshoring has increased dramatically. This has been due in large part to new communications technologies and the emergence of India as an offshore location. This report describes the evolution of the globalizing software supply chain. We predict that higher value-added work will be an increasing component of offshored software and discuss its implications for employment and innovation in developed countries.

INTRODUCTION

By the end of 2005, 2.9 million people (2.2 percent of the U.S. workforce) were employed in the software industry. The annual growth rate was 7 percent over the previous decade, well ahead of average workforce growth of 1 percent.[1] The Bureau of Labor Statistics (BLS) predicts that the software industry will be among the fastest growing employers in the coming years. Six of the 20 most rapidly growing jobs from 2004 to 2014 are likely to be in high-value software work, including network systems, data-communications analysis and administration, software applications, and systems engineering.

The significant exception to high growth within software is programming, where employment decreased from 570,000 in 1995 to about 450,000 persons in 2005. Programming requires less training than some other software work, and programmers, on average, earn less than software engineers and computer scientists (Table 3). Whereas software engineers and computer scientists should see job growth of over 45 percent and 27 percent respectively between 2004 and 2014, the Bureau of Labor Statistics forecasts less than 5 percent job growth for computer programmers. This rate is below even the economy's average job growth.

This reflects two trends. First, much routine programming is now automated. This has both reduced the programmers' share of work in software creation and increased the average sophistication of the work. Second, even as this has happened, the growth of online collaboration via the Internet and higher capacities at lower costs offshore has increased the offshoreability of programming.

This may be seen from the following information on India, which is now the largest exporter of software after the United States, accounting for 60 percent of non-U.S. software exports. Programming accounts for 60 percent of Indian software exports, down from 90 percent in 1995. Programming is, of course, not a stand-alone function. The work done by the Indian software industry is part of a supply chain, with most of the components still being fulfilled in the developed world.

Indian software employment has grown by 35 percent per annum over the past decade. Software-exporting firms located in India employed 706,000 people in 2006, up from 513,000 in 2005. In 1995, the comparable numbers for the Indian and American software industry were 27,500 and 1.5 million (1.3 percent of the U.S. workforce of 118 million).

[1]Data for this section is from Bureau of Labor Statistics *http://www.bls.gov/oco/oco1002.htm* and *http://www.bls.gov/oes/current/oes_nat.htm#b15-0000*; GAO, 2005; Heeks, 1996; Nasscom, 2006; Ellis and Lowell, 1999.

Two-thirds of India's software exports are to the United States, a share that has remained nearly steady over the past decade.

The impact is perhaps better appreciated by calculating the Indian share of employment within the American supply chain of software. The share of Indian employment has risen from 3 percent of the programmer pool used in American software production in 1995 to over 30 percent in 2005.[2]

Meanwhile, work besides programming has also been offshored. Some of this newer work is even lower-end work than programming, such as installation of software and maintenance of software programs. This has happened largely because of the Internet. However, as will be shown below, new tasks, hitherto considered both difficult to offshore and high value-added relative to the programming function, such as product development and contract R&D for the software industry, have been offshored over the past decade, particularly to India. For example, as of 2006, the world's largest contract R&D firm in software, employing 14,000 persons, is the Indian firm, Wipro. A decade ago, Wipro, like others in the Indian software industry, did not do such work.

This paper fulfills two objectives. First, it explains the genesis of software offshoring. This includes a consideration of why programming was the function that was most commonly offshored right from the earliest stages. Second, it examines the scope for offshoring software work other than programming. This includes a consideration of whether the additional scope is higher or lower value-added, how it is linked to the earlier phase of programming offshoring, and its likely evolutionary trajectory.

The paper proceeds as follows: in the next section, we discuss the current status of the debate on software offshoring. The following section provides a historical overview of developments that led to offshoring in the software industry, with a focus on developments in India. This is followed by a theoretical framework for analyzing how skilled work may be offshored. We conclude with a discussion of the impact of software offshoring on employment and innovation in the United States and other developed countries, and the implications for policy on education.

THE CURRENT STATE OF THE DEBATE

A lively discussion is under way about the impact of globalization on employment and productivity in the American software industry. An assessment published in 2006 by the Association for Computing Machinery (ACM) notes that "attracted by available talent, work quality and, most of all, low-cost companies in high-wage countries, such as the United States and United Kingdom, are increasingly offshoring software and software-service work to . . . low-wage countries." The report concudes that "the globalization of, and offshoring within the software industry will continue and, in fact, increase" (ACM, 2006).

As Bhagwati et al. (2004) and Mankiw and Swagel (2006) have pointed out, the offshoreability of the software industry means, first, that software services are now tradable, whereas in the past they were not. Second, given that international trade is usually beneficial to both trading partners, they conclude, ipso facto, that globalization will have positive implications for the U.S. economy. They argue that workers in the services sector of developed nations will shift to jobs in which they have a comparative advantage, thus ensuring full employment in the long run. As Mankiw and Swagel (2006) note, "Economists see outsourcing[3] as simply a new form of international trade, which as usual creates winners and losers, but involves gains to overall productivity and incomes." By contrast, Samuelson (2004) has cautioned that these gains may largely be captured by developing countries; and Gomory and Baumol (2000) have argued that nationally located high-growth industries are important for national growth because of their spillover effects on overall productivity.

To some, these latter cautions suggest potentially dramatic negative impacts for software-related employment in developed countries. These argue that if software development overseas increases in quantity and, especially in scope, to include the most highly skilled work, the result may be unemployment, even for the most highly skilled software engineers in developed countries (Hira and Hira, 2005). The ACM report and other evidence points to the fact that higher skilled work is already being moved offshore in some fields of software, such as computing research (ACM, 2006; Dossani, 2006; Sridharan, 2004).

There is no comprehensive empirical evidence on software offshoring, primarily because of the poor quality of primary data. See Figure 1 for an example of contradictory data reported by the U.S. Bureau of Economic Analysis and the Indian software industry association, Nasscom. As far as we can tell, there is no systematic evidence yet of significant losses of high-value jobs in the United States to services offshoring. As noted in NAPA (2005), "The number of jobs impacted (by services offshoring in general) appears relatively small, when compared to total annual job losses in the United States."

Other empirical studies offer indirect evidence in support of the NAPA findings. For example, Mann (2006) shows that the elasticity of demand for U.S. exports of services is lower than for U.S. imports of services. If this finding is applicable to software, it would imply that globalization could have positive implications for the U.S. balance of payments.

Landefeld and Mataloni (2004) show that the share of

[2]This has happened even as the number of programmers in the total software pool has stayed relatively steady (rising from just under 600,000 in 1995 to 650,000 in 2005), while declining in share of software employment from 38 percent to 21 percent.

[3]Technically, the correct term is "offshoring."

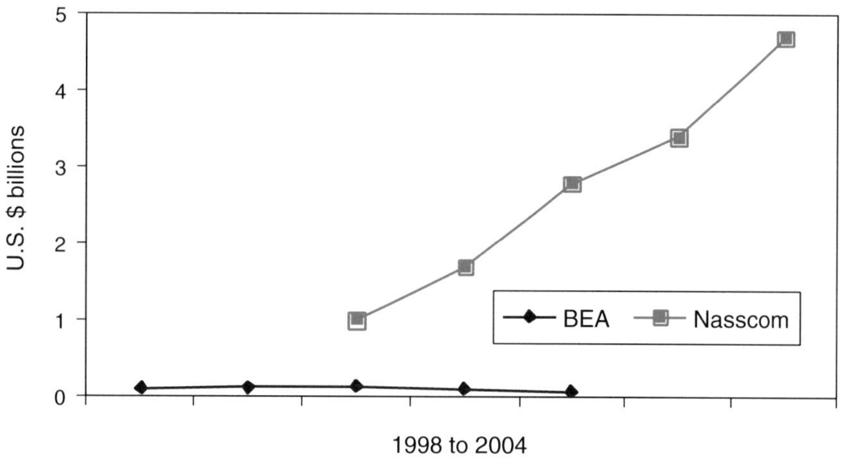

FIGURE 1 BEA and Nasscom figures for software sales from India to the United States ($ billions). Sources: *www.bea.gov* and Nasscom (various years).

imports from subsidiaries of U.S.-based multinationals to the parent country (as a percentage of sales) did not increase from 1997 to 2001. They also find that job creation by the expansion of multinationals overseas is no different from overall job creation. Both findings imply that multinationals that offshore work to their subsidiaries are not responsible for job losses in the United States. Of course, the destination for offshored work might be unaffiliated firms, for which these data have no implications.

According to Hanson et al. (2001), the evidence of offshoring of manufacturing has shown a positive, complementary effect on American jobs from high-value offshoring and a negative, substitution effect from low-value offshoring. In the software industry, the lower value work consists of programming and the higher value work consists of design, consulting, system integration and managed services (Table 3). Hanson's findings—if applicable intra-sectorally to software—imply that the export of low-end work, such as programming, could reduce industry jobs. As Table 2 shows, this is the field with the highest market share in India, suggesting by extrapolation that job losses in the United States may indeed occur as a result.

This kind of indirect evidence has obvious limitations. Quite simply, we do not know if it is applicable to software. From our interviews with firms that have offshored work, we learned that fulfillment of various aspects of software development can be accomplished in spatially distant locations. Many of these firms state that they will increasingly shift their operations to lower cost countries like India and China. This suggests that the logic for software development at any particular location may be being eroded. The data we provided in the introduction on the programming function may be only the first wave of software offshoring.

This does not, however, mean either that the most skilled work will shift from the United States or that American software employment will decline. More than one type of outcome is possible. First, the capacity of other countries may be constrained by the quality of their educational systems or other factors that hinder labor supply; by their infrastructure, such as telecommunications; or by institutional barriers, such as weak intellectual property laws. Second, the history of technological change suggests that new opportunities will emerge. The software industry in the United States might discover higher value-added opportunities, even as existing operations are increasingly offshored. In an era of high rates of technological change, both offshore and domestic software work can become more highly skilled. Third, assuming that the first and second outcomes are both true, developed countries other than the United States could capture the new opportunities. This possibility is not investigated in this paper.

The actual outcomes of offshoring will, therefore, depend on the evolving capabilities of developed countries vis-à-vis the capabilities of developing countries. New opportunities will depend on the pace and location of innovation, which could be affected by the development of clusters of technical excellence offshore, such as those in Bangalore, Beijing, and Shanghai. Or, perhaps, as Apte and Mason (1995) and many others have argued, the need for proximity to consumers to determine their needs will be the determining factor in the location of innovation. Perhaps the open economy and excellent educational system in the United States will enable American firms to innovate at a pace that keeps them ahead of China and India. If so, the American software industry, though possibly not specific groups of workers, may thrive by keeping the innovative, highest value-added work onshore and offshoring the rest.

Predicting the outcome of offshoring requires an understanding of (1) the software industry and its evolving supply chain and (2) the ecosystem for innovation in the United States vis-à-vis other countries. To simplify our task, we have focused on India and the United States. Data on other countries are used primarily to illustrate the challenges and opportunities in these two countries. We have chosen India as the alternative to the United States for the following reasons: first, because of its position as the largest exporter of software after the United States; second, it has the size of labor force that can pose the most significant threat to U.S. employment; third, its current stage of overall economic development is likely to keep labor costs low for several years, thus adding to its attractiveness as an offshore software destination; and, finally, because as our case studies, presented below, show, Indians have the capability of doing highly skilled work.

HISTORICAL OVERVIEW OF THE SOFTWARE INDUSTRY

Product and Custom Software

Software is usually classified either by its uses or its degree of customization. We use the North American Industry Classification System (NAICS) definitions to differentiate product software and custom software. The attributes, the size of the market, and the market shares of the two key players other than the United States, India, and Israel are summarized in Tables 1 through 3.

Types of software defined by usage are listed below:

- system-level software (i.e., programs that manage the internal operations of the computer, such as operating-system software, driver software, virus-scan software, and utilities)
- tools software (i.e., programs that make applications work better, such as database-management software)
- applications (i.e., programs that deliver solutions to the end user, such as word-processing software, search-engine software and financial-accounting software)

We define two categories of software by their degree of customization: (1) publishers of packaged software (NAICS 5112) and (2) computer systems design and related services (NAICS 5415). Software publishers such as Microsoft fit under the NAICS 5112 description of publishers of packaged software, "establishments primarily engaged in computer software publishing or publishing and reproduction. Establishments in this industry carry out operations necessary for producing and distributing computer software, such as designing, providing documentation, assisting in installation, and providing support services to software purchasers. These establishments may design, develop, and publish, or publish only."[4] Similar in some respects to mass manufacturers, enterprises in this category create software products or packages for the general consumer market and capitalize on economies of scale. Software products may be shrink-wrapped and transported physically or made available for downloading over the Internet.

The second category, computer systems design and related services (NAICS 54151), comprises "establishments primarily engaged in providing expertise in the field of information technologies through one or more of the following activities: (1) writing, modifying, testing, and supporting software to meet the needs of a particular customer; (2) planning and designing computer systems that integrate computer hardware, software, and communication technologies; (3) on-site management and operation of clients' computer systems and/or data processing facilities; and (4) other professional and technical computer-related advice and services."[5]

In contrast to the one-size-fits-all software products in the first category, custom software is used when no packaged software products are available, as in highly specialized processes, or to integrate disparate software products into a cohesive system. The latter process is common when large software products, such as Enterprise Resource Planning (ERP) or Customer Relationship Management (CRM) suites, must be integrated into already existing enterprise systems. Custom software may be constructed by using traditional programming languages and tools or proprietary scripting or configuration languages. Because custom software is made-

TABLE 1 Uses of Product and Custom Software

	Product Software	Custom Software
Operating system	All users	None
Tools	Most users	Some users
Applications	Small and large users	Large users

TABLE 2 Global Spending on Software Products by Categories of Work and Israel's Market Share, 2004

Revenue Category	Global Spending on Software Products ($ billions)	Israel's Share of the Global Product Software Market (percentage)
Systems and tools software	$93.7	1.1
Application software	$120.0	1.3
Total	$213.7	1.2

Sources: U.S. and global data: *http://www.siia.net/software/resources.asp#stats*. Data for Israel *http://www.iash.org.il/content/SoftwareInds/IsraeliSectors.asp*. Israel's share of global markets are estimated from data for Israel for 2000 and comparable data for the United States for 2001.

[4]See *http://www.census.gov/epcd/naics02*.

[5]ibid.

TABLE 3 Spending on Global Software Services by Categories of Work and India's Market Share, 2003

	Global Spending on Software Services ($ billions)	India's Global Market Share (percentage)	U.S. Wage Rate ($/hour)
Consulting	41.5	< 1	80–120
Applications development	18.4	16.4	25
System integration: hardware and software deployment and support	91.7	< 1	18–25
System integration: applications, tools, and operating systems	62.4	< 1	40
IT education and training	18.5	0	40
Managed services	124.9	1.6	60–120
Total	357.4		

Definitions:

Consulting includes IT strategy, system conceptualization, information systems (IS) consulting, architecture, design, and network consulting and integration. These services require the highest level of skills, including system design and understanding of clients' requirements.

Applications development includes creating applications programs. These require programming skills.

System integration: hardware and software deployment and support includes making software and hardware components compatible and interoperable, hardware deployment and support, and software deployment and support. The skills required vary, but are not as high-level as programming or consulting skills.

System integration: applications, tools, and operating systems includes the integration of software components (both products and custom software) in a software project. The required skills include understanding clients' requirements and programming skills.

Managed services include managing applications either on site or remotely over the Web, managing networks, applications management, IS outsourcing, network and desktop outsourcing, applications service provision, and systems-infrastructure service provision. The skills required vary greatly.

Sources: Nasscom, 2004 (pp. 19, 36, 106) for columns 1 and 2; Nasscom, 2001 (p. 24) and authors' interviews for column 3.

to-order, it is more geographically constrained than product software. Proximity to the stakeholder is often crucial, especially if tacit (uncodified) knowledge is involved. Thus, software products are more readily exportable than custom software.

Nearly every computer needs systems software, and the mass market provides very favorable conditions for creating systems software as packaged products. Hence, systems software is now marketed almost exclusively as packaged products. And, over time, the need for compatibility among operating systems has become a critical requirement of both enterprise and retail users; this need has increased with the advent of the Internet. As a result, a few operating systems now dominate the computing landscape and have considerable pricing power. Compared to the demand for applications software, the demand for systems software has relatively little "give" in terms of pricing. Consumers of systems software, such as high-availability server-operating systems and real-time embedded operating systems, are willing to pay high prices for quality and interoperability. Consequently, the producers of systems software are less sensitive to production costs than product quality and the need for people with highly specialized skills.

Although product software is designed to meet a wide range of customer requirements, it can incorporate only a limited number of variations. Beyond this limit, software must be written to a customer's specifications. Industries such as banking, in which customer requirements vary significantly, need custom software. In general, the more varied the needs of different end-users, the more likely software is to be customized. And, because needs vary most at the applications stage, most customized software is applications software. Table 1 compares the uses of product and custom software.

The United States is the market leader in software product development, accounting for 41 percent of the total.[6] The U.S. share of exported software products is probably even higher because many countries only produce software products for protected local markets. For instance, data on Brazil and Japan (Table 6) show that while Brazil's annual output of product software earns revenue of about $3 billion and Japan's annual output earns about $21 billion, these products are only available to domestic markets. Western Europe and Israel, like the United States, develop product software for global markets.

Custom software is part of a larger category called software services, as defined in NAICS 54151. Software services are described by type and size in Table 3.

Independent Software Vendors

The independent software-vendor (ISV) industry was created by two events, both related to market leader IBM. First, in 1956, IBM settled a long-standing antitrust suit by the federal government by agreeing, as part of a consent decree, to stop offering computer-consulting advice (McKenna, 2006).[7] With IBM out of the picture, leading accounting firms, such as Arthur Andersen, then began offering computer consulting services. Second, in 1969, IBM

[6]See *www.siia.net.*

[7]When the consent decree was lifted in 1991, IBM immediately created an IT consulting group, which, within five years, had annual revenues of $11 billion (McKenna, 2006).

decided to unbundle its mainframe operating system, applications software, and hardware by creating open standards. Subsequently, some end-user firms set up in-house software development and maintenance operations and some began outsourcing work. As a result, ISV businesses were created (Table 4).

The columns in Table 4 do not describe mutually exclusive choices. For example, a firm might purchase system-level software products and develop its own applications. The columns are arranged by sequentially dominant work types over the decade, starting with the shift from external data processing and managed services (Column A) to in-house hardware at the beginning of the decade. Initially, firms developed their own software (B), but as hardware and software became more complex, in-house software development and management became increasingly difficult. This led to the outsourcing of system integration (C) and then system-level and applications products (D). The outsourcing of customized applications (E) was an indication that industry-specific products did not meet the needs of sophisticated users, particularly large banks (Steinmuller, 1996).

In the 1980s, the IBM PC was introduced, but within a decade, IBM had lost control of the operating system to Microsoft Windows, which combined with the Intel microprocessor (Wintel) to became a market-created standard by the late 1980s. The result was a decline in hardware prices and an increase in demand for applications. Unlike mainframes, PCs were made for individual users who relied on product software. PCs in the 1980s had neither the programming capacity nor the performance capabilities necessary for mid-sized and large enterprises. Hence PCs did not impact the custom software business. However, they did create a mass market for retail product software.

The workstation, which was introduced in the early 1980s, provided many end uses for enterprises but could also be used for stand-alone programming for mainframes. The adoption of Unix as the operating system for all computers, combined with the workstation (in short, the U-W standard), revolutionized the ISV industry. An ISV could now own a workstation made by any manufacturer and write programs for a client with a different brand of installed hardware (including a mainframe). In other words, software creation became modularized, or platform independent.[8]

With the simultaneous widespread adoption of Unix/C as the programming language, other functions of software creation, such as system architecture, design, and integration, could be done separately from programming, thus modularizing the programming component. Programming could now be done anywhere in the world by programmers whose only raw material, apart from a workstation, was a specified software system. Programmers did not even have to know which firm's hardware a program would work on or the type of application the program would support.

The workstation also had sophisticated graphics and enough computational capacity to satisfy the needs of small enterprises, which now shifted from outsourcing data-processing services to running their own workstations. In the early 1980s, the first workstation-based local area networks were established, increasing the demand for more sophisticated software for running these networks and for applications compatible with networked users.

In the 1990s, the success of database software packages further simplified the creation of applications software. Platform independence, combined with the rise in demand for custom software by small firms, resulted in the growth of a large custom software industry.

Also in the 1990s, PCs with more computing power were able to process programs written in Unix/C, thus making them more acceptable to small enterprises. As costs for PCs fell in the mass market, PCs superceded workstations as the hardware platform for programming. Later in the decade, PC-based networks made applications accessible to many more users in an enterprise.

The spread of the Internet beginning in the mid-1990s was accelerated by declining costs for bandwidth and storage. The Internet provided a platform for networked development of software and software installation, hosting, and maintenance. At this point, data no longer had to be on servers located on the premises of an enterprise but could be housed in remote data centers. The Internet also significantly reduced the cost of collaboration among remote teams. These factors further reduced the need for the proximity of user groups or of developers and users.

With the establishment of the Internet, several new models of preparing and delivering software appeared. These include service-oriented architecture that provides a standards-based environment for sharing services independent of development technologies and platforms; network-based access to and maintenance of software (software-as-a-service); and open-source software (i.e., software based on nonproprietary code) developed by voluntary contributions of networked developers. With the exception of the Linux open-source operating-system software, which is believed to have about a one-third share of the server market (although less than 2 percent of all operating systems), the new models described above have not impacted the spatial distribution of software development.[9]

The first three columns in Table 5 show the major changes and driving forces in the software-services industry in the United States described above. The two right-hand columns show (for later reference) developments in the Indian and

[8]Modularization is the conversion of a component of the production process with one or more proprietary inputs, design, or fulfillment techniques into a component with standardized inputs, design, and fulfillment techniques.

[9]See *www.idc.com/getdoc.jsp?containerID=202388.*

TABLE 4 Independent Software-Vendor Industry, 1970–1979

	External Data Processing	Clients That Own Hardware			
Clients' Options	Managed services	Develop and maintain software	Buy bundled software and outsource maintenance services	Buy software products from ISVs	Buy custom software services
ISV Services	Managed services Electronic data processing	None	Integration of hardware and software Software maintenance	Systems-level and applications products	Custom applications software
	A	B	C	D	E

Source: Adapted from Steinmuller, 1996.

Israeli software industries. Note that this table does not include information on the product-software industry.

Offshoring of Software Development

American IT firms began to offshore software development to India, Ireland, and Israel (the 3 I's) in the 1970s, about a decade after the offshoring of IT hardware manufacturing. Siwek and Furchgott-Roth (1993) argue that the lag between hardware and software offshoring was because software development, unlike hardware manufacturing, required close coordination with clients throughout the process.

A widespread knowledge of English and relatively low labor costs were common attractions of the 3 I's. Small domestic markets and the lack of domain knowledge (less so for Israel) were common disadvantages. Beginning in the 1990s, many other countries, including China, several countries in Eastern Europe, Brazil, Mexico, Russia, the Philippines, and Vietnam, began exporting software to developed countries (Table 6).

As Table 6 shows, China and Brazil sell software services mostly to their domestic markets. Ireland develops software products and services for Europe, mostly by customizing U.S. software products. This should properly be included in the category of software services. Russia, the Philippines, and Vietnam, like India, primarily export software services. Countries in Eastern Europe and Russia export mostly to Europe. Other countries export mostly to the United States. Israel is the only significant non-American producer of software products for the U.S. and other global markets.

As Table 6 shows, the most significant producers of offshored software for global markets are India and Israel. Israel focuses on software products for the global market and India on custom software for the global market. Ireland is the largest provider of localized products and services for Europe.

Ireland

Hardware offshoring began in Ireland after policy makers offered export incentives following Ireland's entry into the EU in 1973 (Enterprise Ireland, *http://www.enterprise-ireland.com*, downloaded 1/20/2007; Torrisi, 2002). Software offshoring, which began in the 1980s, followed hardware offshoring (Torrisi, 2002). The main clients initially were, and continue to be, American transnational corporations (TNCs). These use Ireland to localize their software products for European markets (Torrisi, 2002). American TNCs account

TABLE 5 New Work Types and Driving Forces in the U.S. Software-Services Industry and Their Impact on the Software Industry in India and Israel

	New Work Types in the U.S. ISV Industry	Market Change	Technology Change	New Work Types in the Indian ISV Industry	New Work Types in the Israeli ISV Industry
1960–1970	Software maintenance Electronic data processing		Mini-computers	Electronic data processing	Software maintenance exports Electronic data processing
1971–1980	Custom applications	IBM unbundles software and hardware		Programmers exported	No change
1981–1990	Software system integration	Increased complexity of applications	Unix-Workstation (U-W) standard adopted	Custom applications exports	Custom applications for domestic market
1991–2004	Managed services		Internet, database management systems, PC-based networks	Managed services, contract R&D exports	Contract R&D exports, products for global markets

Source: Adapted from Steinmuller, 1996; Mowery, 1996; and *http://www.siia.net/software/resources.asp#stats* for columns 1–4. Columns 5 and 6 are the authors' analyses.

TABLE 6 Software Exports from Developing Countries, 2001

Country	Sales ($ billions)	Exports ($ billions)	Labor Force (2000)	Sales per Employee ($)	Primary Work Type
Brazil	7.7	0.1	220	35	P/S = 40/60[b]
China	7.4	0.4	186	40	Domestic
	(15.0)[a]	*(2.0)*	*(750)*	*(20)*	
EE5 (Bulgaria, Czech Republic, Hungary, Poland, Romania)	0.6	0.5	75	8	Services to Western Europe
India	8.2	6.2	350	23	Services to U.S.
	(22.3)[a]	*(17.1)*	*(878)*	*(25)*	P/S = 25/75
Ireland	7.7	6.5	24	160	Localization of U.S. product software for Western Europe
Israel (2000)	3.7	2.6	35	106	P/S = 70/30
Japan	85.0	0.07	535	159	P/S = 25/75
Philippines	0.2	0.15	0.05	12	Services to U.S.
Russia	0.2	0.1	0.1	13	P/S = 30/70
United States (2002)	200.0	n/a	2,600	77	P/S = 40/60

Notes:
[a]Figures in italics are for 2005.
[b]P/S = the ratio between revenue from software products and revenue from software services.
Sources: Arora and Gambardella, 2005 (pp. 45, 77, 101); Sahay et al., 2003 (p. 17); Nasscom, 2006 (pp. 46, 47).

TABLE 7 Software Exports from India, Ireland, and Israel (in $ millions, except where otherwise noted)

	India	Ireland	Israel
1990	105	2,132	90
2000	6,200	8,865	2,600
2002	7,500	12,192	3,000
2003	8,600	11,819	3,000
2005	17,100	18,631	3,000
Number employed (2003)	260,000	23,930	15,000
Revenue/employee (2003)	33,076	493,988[a]	273,000
Number employed (2005)	513,000	24,000	n/a
Revenue/employee (2005)	33,333	776,000[a]	n/a

[a]Note: Sands (2005, p.45) argues that the revenue/employee for Ireland is overstated because of in-country transfers and should be about $160,000. If so, total exports in Table 7 are overstated by a factor of three.
Source: Data for India are from Heeks (1996) and Nasscom (2003–2006). Data for Ireland are from *http://www.nsd.ie/htm/ssii/stat.htm*, downloaded September 26, 2006. Data for Israel are from *http://www.iash.org.il/Content/SoftwareInds/SoftwareInds.asp*, downloaded August 31, 2003, and *http://www.israel21c.org/bin/en.jsp?enDispWho=InThePress&enPage=BlankPage&enDisplay=view&enDispWhat=Zone&enZone=InThePress&Date=08/11/05*, downloaded September 26, 2006. Data for Ireland prior to 2003 are in euros (converted at 1 euro = $1.043, rate on January 5, 2003). From 2003 on, data are converted at 1 euro to $1.26, the rate in January 2004. Most recent figures for Israel are for 2001.

for about 90 percent of Ireland's software exports (Arora and Gambardella, 2005).[10] Since the 1990s, an indigenous software sector has developed in Ireland, initially providing support services for TNCs but subsequently developing products for the European telecom and financial sectors. In 2003, the indigenous sector in Ireland employed 40 percent of the total software workforce (Sands, 2005).

Israel

As in Ireland, though a decade earlier, hardware firms were established in Israel during the 1960s first in response to export incentives.[11] Software TNCs followed in the early 1970s (Torrisi, 2002). These initially undertook software product maintenance and, later, R&D.

In the 1980s, domestic firms were established, funded by government research contracts. They initially provided software services to the defense industry. Key labor was drawn from the Israeli defense industry. In the 1990s, with support from global venture capitalists, security product firms were established. These offered products for global markets (Teubal, 2002, see also Table 5). TNCs currently account for about 25 percent of total employment in the Israeli IT industry and focus on R&D, but growth is being driven by local firms producing software products for export markets (Torrisi, 2002). The three largest software firms in Israel are product firms that jointly account for 60 percent of the industry's revenue (Bresnitz, 2005).

India

From Tables 6 and 7 above, we note that the most significant increase in offshoring to global markets is in India. Unlike in Ireland and Israel, where fiscal incentives were

[10]By contrast, in India, only 15 to 20 percent of the work since 1990 is estimated to be done by TNCs. According to Enterprise Ireland, the official state website, *http://www.nsd.ie/htm/ssii/stat.htm*, Irish-owned companies generated about 11 percent of software exports in 2002, with the rest coming from TNCs.

[11]For example, Motorola's first offshore manufacturing subsidiary was set up in Israel in 1964 (Ariav and Goodman, 1994; Sahay et al., 2003).

critical for private-sector entry, the software industry in India began when government policy was hostile to all private industry. State policy at that time was appropriately described as "statist, protectionist and regulatory" (Rubin, 1985). An industrial licensing regime and state-owned banks strictly regulated private-sector activity. In IT, the state was the main producer of products and services. The strategy was to create "national champion" state-owned enterprises, which were granted monopolies (Sridharan, 2004).

A key protectionist policy was the Foreign Exchange Regulation Act of 1973 (FERA-1973), under which a foreign firm could only have a minority interest (up to 40 percent) in a company operating in India. Many foreign firms, including IBM, closed their Indian operations, citing concerns about the protection of intellectual property (IP). FERA-1973 effectively closed the door to software development by TNCs in India.

Domestic firms found an innovative way to benefit from global opportunities for ISVs. Because software development could not come to India, Indian programmers were sent to developed countries. This began in 1974 when Burroughs, an American mainframe manufacturer, asked its Indian sales agent, Tata Consultancy Services, to supply programmers for installing system software for a U.S. client (Ramadorai, 2002). Other firms followed suit, including foreign firms in joint ventures with Indian firms.[12]

Initially, the exported Indian programmers worked for global IT firms. Later in the decade, as IBM gained a larger share of the total global market, end-users such as banks hired Indian firms to convert existing applications software to IBM-compatible versions.

The state remained hostile or, at best, indifferent to the software industry throughout the 1970s. Import tariffs were high (135 percent on hardware and 100 percent on software). Software was not considered an "industry," which meant that exporters were not eligible for bank financing. Even overseas sales offices were disallowed until 1979 (Ramadorai, 2002).

Such protectionism interfered with learning and prevented Indian-based programmers from moving up the value chain. Programmers returning from overseas assignments were the main source of learning about new opportunities, but because of their short assignments overseas—typically less than a year—their learning was also limited (Ramadorai, 2002). In addition, many chose to remain overseas after completing their assignments. As a result, the software industry during its first decade was mostly limited to the recruitment of engineers.

It being easier for established private conglomerates than for small firms to navigate anti-private-sector policies, large firms became the dominant players in the industry. Mumbai, the country's commercial and industrial capital, became the center of the business. In 1980, five of the top eight exporters (including the top four) had large-firm pedigrees. Seven of the eight, all headquartered in Mumbai, had a 90 percent market share (Table 8).

The industry changed when the global industry adopted the U-W standard in the 1980s and, as we discussed earlier, software creation and, within it, programming were modularized. Beginning at that time, coincidentally, the state gradually abandoned its protectionist, anti-TNC stance. The New Computer Policy of 1984 (NCP-1984) reduced import tariffs on hardware and software to 60 percent; reclassified software exports as a "delicensed industry" eligible for bank financing and not subject to the intrusive licensing regime (Heeks, 1996); gave foreign firms permission to set up wholly owned, export-dedicated units; and initiated a project to set up a chain of software parks that would offer infrastructure at below-market costs. In 1985, all export revenue (including software exports) was exempted from income tax.

The new policies encouraged TNCs to introduce new businesses and new business models. Some TNCs (e.g., Texas Instruments and Hewlett Packard) did R&D and wrote product software using cross-country teams; others (e.g., ANZ Bank and Citigroup) wrote custom software for in-house use, again using cross-country teams. Thus TNCs used approaches that had been successful in other environments, such as Ireland and Israel.

Although the initial entrants, such as Texas Instruments, persuaded the government to improve the infrastructure,[13] TNCs still faced daunting communications costs and intrusive regulation (Parthasarathy, 2000). Thus product-focused TNCs remained small. Domestic firms (e.g., Wipro) that tried to imitate the TNC product-software model also failed because (1) the domestic markets could not supply adequate domain expertise (Athreye, 2005), and (2) there was no venture capital industry to speak of.[14] By 1990, product development accounted for less than 5 percent of exports (Heeks, 1996), and, by 1999, it had only increased to 8 percent (Nasscom, 2002).

However, the combination of the U-W standard and lower costs engendered a successful new business model, pioneered by TCS. Domestic firms began to supply software programs coded entirely in India, while relying on foreign co-vendors for program design and specification. This approach succeeded because it matched the expertise of Indian firms (programming) with the expertise of overseas vendors (client understanding, design, and integration) and because it reduced costs by keeping programmers at home—although

[12]These included Datamatics (a joint venture between Wang, the U.S. minicomputer maker, and ex-employees of TCS), Digital, and Data General.

[13]According to Naidu (2002), Texas Instruments' decision to enter India was conditional on the state providing adequate power and telecommunications bandwidth.

[14]Through the 1980s, domestic venture capital was concentrated in state-run firms. Two of today's leading IT firms, Wipro and Infosys, were both turned down by state-run venture capital firms in the 1980s.

TABLE 8 Top Eight Indian Software Exporters

Rank	Firm, HQ 1980	Firm, HQ 1990	Firm, HQ 2004	Founder, Education, Experience
1	TCS, Mumbai	TCS, Mumbai	TCS, Mumbai	Kanodia (MIT)
2	Tata Infotech, Mumbai	Tata Infotech, Mumbai	Infosys, Bangalore	Murthy (U. Mysore, IIT Kanpur)
3	Computronics, Mumbai	Citibank, Mumbai	Wipro, Bangalore	Premji (Stanford) and Soota (IISc)
4	Shaw Wallace, Kolkata	Datamatics, Mumbai	Satyam, Hyderabad	Raju (Loyola College, Chennai; Ohio U)
5	Hinditron, Mumbai	Texas Instruments, Bangalore	HCL, Delhi	Nadar (PSG College, Coimbatore)
6	Indicos Systems, Mumbai	Dell, Mumbai	PCS, Mumbai	Patni (MIT)
7	ORG, Mumbai	PCS, Mumbai	i-Flex, Mumbai	Hukku (BITS, Pilani) (TCS, Citicorp)
8	Systime, Mumbai	Mahindra-BT, Mumbai	Mahindra-BT, Mumbai	Mahindra (Harvard)
Total Market Share	90%	65%	38%	

Notes:
1. IBM was probably in the top eight firms in 2004 (it was ranked 6th in 2002), but the company has not given permission for its name to be displayed in subsequent Nasscom rankings: *http://www.nasscom.org/artdisplay.asp?art_id=4413#top20* (downloaded August 26, 2005).
2. Column 5 data is for firms listed in Column 4.

Sources: Heeks, 1996 (p. 89), for columns 2 and 3; Nasscom, 2005 (p. 76), for column 4; company websites and authors' interviews for column 5.

TABLE 9 Exports of Indian Software

Year	Total Exports ($ millions)	Number of Firms	Average Revenue per Firm ($)	Average Revenue per Employee ($)	Exports/Total Revenue (percentage)
1980	4.0	21	190,476	16,000	50.0
1984	25.3	35	722,857	18,741	50.0
1990	105.4	700	150,571	16,215	n/a
2000	5,287.0	816	7,598,039	32,635	71.8
2004	12,200.0	3170	7,003,154	35,362	73.9

Notes:
1. Data for 1980, 1984, and 1990 are from Heeks, 1996 (pp. 72, 73, 87, and 88).
2. Data for 2000 (financial year ended March 2001) are from Nasscom, 2002, and Nasscom, 2004 (pp. 23, 26, and 64).
3. Data for 2004 (fiscal year ended March 2005) are from Nasscom, 2005 (pp. 75–76). 2004 data for number of firms and average revenues are based on figures for software, software services, and IT-enabled services combined because disaggregated data are not available.
4. Number of employees for 1980, 1984, 1990, 2000, and 2004 was 250, 1,350, 6,500, 162,000, 260,000, and 345,000, respectively. Data for 1980–1990 are from Heeks, 1996. Data for 2000 and 2004 are from Nassscom, 2004 and 2005.

the number of personnel dispatched overseas declined slowly at first.[15]

Thus Indian firms gradually shifted from exporting programmers to programming outsourced custom software in India. The shift, though gradual, induced many domestic firms to enter the market. The number of software firms increased from 35 in 1984 to 700 in 1990, and the share of smaller firms also rose (Table 9).

This shift raised the standards required for physical infrastructure in India. It also marked a turning point in the role of Bangalore, where real estate was cheaper than in Mumbai, as a center for software development. Several new firms, including Infosys and Wipro decided to locate their facilities in Bangalore (Premji, 2003). The first software technology park under NCP-1984, with a reliable supply of electricity and telecommunications bandwidth, was also located in Bangalore. Another advantage of Bangalore over competing locations was low labor costs. Unlike Mumbai and Delhi, which had histories of large firms and militant labor unions, small companies in Bangalore had relatively few problems with unions (Heitzman, 1999).

In addition, Bangalore, the capital of Karnataka, is located at the center of the four southern states, Karnataka, Tamil Nadu, Andhra Pradesh, and Kerala, which together produce 52 percent of India's engineering graduates. Bangalore's best known academic institution, the elite Indian Institute of

[15]By 1988, 10 percent of the Indian software industry's labor force was located in India; this had risen to 41 percent by 2000 and 71 percent by 2004 (Nasscom, 1999, 2002 [p.28], 2005 [p.58]).

Science (IIS), was established in 1909. Most IIS graduates and most research were directed toward the public sector, but some indirectly supported Bangalore's development in software. This was because the government had decided to locate several high-technology state-owned enterprises there, thus creating a trained labor force (Balasubramanyam and Balasubramanyam, 2000). However, according to some industry observers, the quality of that labor force was dubious and could meet only a small part of the software industry's needs (Ramadorai, 2002). The biggest success related to IIS, Wipro Technologies (India's third largest software exporter), was founded at IIS by a group of engineers working under Ashok Soota (Parthasarathy, 2003).

Policy reforms in the 1990s and 2000s reduced import tariffs to near zero[16] and regularized foreign ownership, intellectual property protection, venture capital, stock market listing, and telecommunications policies to global best practices. In addition, technological changes during this period, particularly the Internet, led to a sharp decline in data storage and transmission costs. These changes attracted a new round of TNCs, particularly foreign outsourcers and U.S.-based start-ups, and provided new opportunities for existing firms in remote software services, such as e-mail management and remote software maintenance (Table 4).

Interestingly, TNCs initially focused on programming only, which was the approach adopted by domestic firms. The TCS remote-programming method was used for in-house product development by Texas Instruments, Agilent, Hewlett Packard, Oracle, and General Electric, as well as for services by ANZ Bank, ABN Amro Bank, Accenture, IBM, and Dell. During this phase, TNCs and foreign start-ups overwhelmingly chose Bangalore for their IT operations (Naidu, 2002).

Over time, the level of sophistication of work done in India rose. As Table 10 shows, routine programming work and maintenance accounted for 68.9 percent of total export revenue in 2001, but fell to 58.5 percent by 2005. During this period, foreign firms earned 14.5 percent of total revenues in 2001 and 31 percent in 2005. We believe that there was a causal relationship between the declining share of routine work and the entry of foreign firms doing more sophisticated work.[17] Data provided by Sridharan (2004) supports this inference; he notes the presence of 230 TNCs in Bangalore employing about 25,000 engineers in R&D work by 2001 and an estimated 30 to 40 chip-design start-up firms all over India between 1999 and 2002.

Of course, several domestic firms also do high-end work. Wipro, the third largest domestic firm, with 14,000 employees, provides contract R&D services and filed 68 U.S. patents on behalf of overseas clients in 2005 (Premji, 2006).

As the share of routine programming work declined, the share of engineering services, R&D, and product development rose from 8 percent in 1999 to 23 percent in 2005 (Nasscom, 2002, 2006).

TABLE 10 Share of Foreign-Firms' Revenue and Share of Custom Programming and Applications Management Work in Indian Software Exports

Financial Year	2001	2002	2003	2004	2005	2006 (E)
CAD and AM ($ billions)	3.65	4.40	4.87	5.98	7.67	10.16
Total software exports ($ billions)	5.3	6.16	7.1	9.8	13.1	17.1
Share of CAD/AM (percentage)	68.9	71.4	68.6	61.0	58.5	59.6
Share of foreign firms' revenue (percentage)	14.5	22.0	26.0	31.0	31.0	n/a

Notes: CAD = custom application development. AM = applications management.

Sources: Nasscom, 2006 (pp. 47, 59, 60, 70); 2005 (pp. 50, 51); 2004 (pp. 36, 40); 2003 (p. 39); 2002 (pp. 29, 30).

Case Studies of Software Products Offshoring

Although a comprehensive study of value-added work in offshored software development is not presented here, evidence from case studies is provided to support the sectoral shift discussed above. In this section we present some examples based on our interviews. From these descriptions, the key constraints in performing higher value-added work appear to be the recruitment and retention of qualified persons and the small size of domestic markets.

Problems with recruitment and retention derive from earlier problems with educational policy and minimal interactions between universities and industry (Parthasarathi and Joseph, 2002). Until recently, faculty at even the best engineering institutions, almost all of which are public universities, were not required to conduct research. Those who chose to do so faced, according to the government's own reckoning, severe problems: "obsolescence of facilities and infrastructure are experienced in many institutions . . . the IT infrastructure and the use of IT in technical institutions is woefully inadequate . . . the barest minimum laboratory facilities are available in many of the institutions and very little research activity is undertaken . . . engineering institutes have not succeeded in developing strong linkages with indus-

[16]The reduction of import tariffs was a key feature of the 1990s reforms. These tariffs had risen to 110 percent by 1991 but were reduced to 85 percent in 1993, 20 percent in 1994 for applications software and 65 percent for systems software, and to 10 percent for all software in 1995 (Heeks, 1996). Duties on hardware ranged from 40 percent to 55 percent in 1995, but by 2000 they had come down to 15 percent for finished goods, such as computers, and had been eliminated for components (microprocessors, storage devices, ICs, and subassemblies, display screens, and tubes, etc) (Indian Ministry of Finance, 2000).

[17]Unfortunately, data on employment in foreign firms is not available, so causality cannot be proved. In 2001, the only year for which data are available, foreign firms employed 13 percent of the workforce (Nasscom, 2002).

try . . . the curriculum offered is outdated and does not meet the needs of the labor market" (Indian Ministry of Human Resource Development, 2001). Until very recently, nearly all of the best students migrated (Siwek and Furchtgott-Roth, 1993), although this may already be changing as opportunities at home increase.

Small domestic markets have also limited the ability of Indian engineers to move up the value chain. As Rosenberg and Mowery (1979) have argued (in a more general context), vendors become technologically sophisticated through understanding customer preferences. D'Costa (2002) has criticized the dependence of the Indian software industry on exports. He argues that international outsourcing of software, although lucrative, discouraged domestic firms from doing more complex projects at home because "excessive dependence on outsourcing limits the synergy between vibrant domestic and foreign markets."

For purposes of this discussion, we consider software product development by two types of firms, start-ups and established firms. The former are dependent on venture capital and tend to be staffed very tightly. For start-ups, coordination costs are a large share of total costs. Established firms have sources of revenue, a more reliable labor pool, and, perhaps, an interest in establishing a base in China or India for accessing domestic markets. In consequence, established firms may use offshoring as a non-integral part of product development, for purposes such as product upgrades and second-generation product maintenance.

Both types of firms also are known to use outsourcing as a strategy rather than doing work in house, despite concerns about the protection of intellectual property, labor force control, and management efficiency (Mukerji, 2006). Offshoring of product development (including engineering and R&D services), whether outsourced or done in house was estimated to be an $8 billion industry in 2005 (Nasscom, 2006), about 4 percent of the software product industry. In 2005, India was the largest participant, generating revenue of $3.9 billion in this segment. Israel came next, with $750 million.[18]

Case Study: Agilent Technologies[19]

Agilent Technologies, which produces test and measurement equipment, chose India as a base for software development in 2001. India offered a potential talent pool, a mature judicial system, favorable protections for intellectual property compared with other developing countries in Asia, and mature management talent. Nevertheless, because of some concerns about intellectual property protection and managerial control, the company decided to do most of the work in house rather than outsourcing it (although some software maintenance and programming work was outsourced). To address these concerns and concerns about reversibility in the event of failure, there was a six-month overlap in staffing between the United States and India.

The work began with simple activities and moved to more complex activities over time (see Figure 2). The engineering-services group was the first user of the Indian operations. The initial work was providing parts lists to customers worldwide and data entry for the CAD group in the United States. Over time, most support services were moved to India.

In early 2002, the second Agilent user, the communications-solutions group, established a 10-person team to automate test suites for Netexpert, one of Agilent's projects. However, a lack of coordination between the Indian and U.S. teams led to the initial failure of this experiment. The situation improved after the time allocated for coordination was increased and a quality-enhancement program was introduced in the Indian operations. By 2005, the development and maintenance of Agilent's EDA software products were being done jointly by multicountry teams located in both countries.

Case Study: Broadcom

Broadcom, a Silicon Valley-based fabless chip firm, acquired an India operation through the acquisition of Armedia Labs, another Silicon Valley-based company founded in 1997 to develop a single-chip (popularly, system-on-a-chip [SOC]) for high-definition TV. From its inception, Broadcom's work in Silicon Valley was tightly integrated with work at its Bangalore subsidiary, except for market development, for which the Silicon Valley team took responsibility (Khare, 2006). All other work, such as the design and development of embedded software and libraries was shared.

When Broadcom acquired Armedia in 1999, its 25-person Indian subsidiary became Broadcom India. Broadcom subsequently expanded the team and brought in complementary technology for SOC work, such as in graphics and digital conversion and processing. By 2006, the team in Bangalore had grown to 190. Employees were, as in the firm's San Jose offices, divided into functional teams, each of which was part of a global team consisting of engineers in San Jose and Irvine, California; Israel; Andover, Massachusetts; and Singapore.

As of 2006, product development was driven by the engineering director of the project, based in San Jose, and the marketing team, based in Irvine. The team might consist at any one time of more than 100 people located in various places who travel, as needed, from one location to another. The final chip-integration design (tapeout), which may take as long as two months, is always done at one location because of the need for close coordination. Tapeout was initially done either in San Jose or in Irvine, but is increasingly being done

[18]Sources: Nasscom, 2006 (p. 47), and Torrisi, 2002 (pp. 9 and 18). Torrisi's data are extrapolated for Israeli exports in 2005 and may not be entirely accurate.

[19]Based on Dossani and Manwani, 2005.

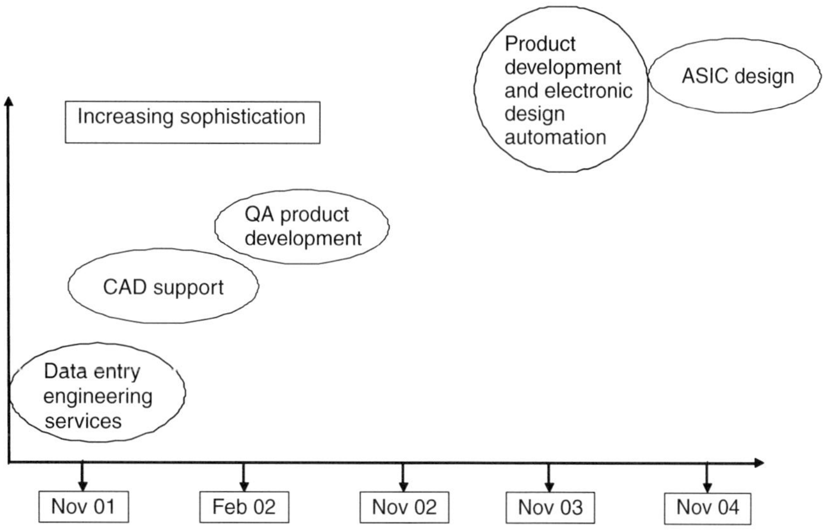

FIGURE 2 Activity transfer to Agilent's Indian operation, by date. Source: Dossani and Manwani, 2005.

in Bangalore. Early in the chip-development process, one of these three locations takes the lead.

The logic for Bangalore sharing the lead position in product development, a status not granted to other locations (such as Andover, Israel, and Singapore), is a logic of scale and capability. From 2003 to 2005, the Indian team had filed for 140 U.S. patents and been granted 10. From 2006 onward, the firm expected that the Indian team would be granted 25 to 30 patents annually. According to the CEO of Broadcom India, these numbers are comparable to U.S. patent rates (Khare, 2006).

Despite the progress of the Bangalore team, proximity still matters in some cases. Once a chip has been fully designed (after tapeout), software libraries and firmware are necessary to accommodate the specific requirements of customers, which may change considerably after the product is released. Understanding customer needs turned out to be difficult from Bangalore. Hence, in the event that the project is led by Bangalore, one member of the Bangalore team is sent to the United States for an eight-week rotation after the first release and until maturity (Khare, 2006).

The CEO also noted that the main challenges to having operations in different locations is the time it takes to establish respect among teams and to build a large enough team with the high level of skills necessary for chip development. By comparison with Silicon Valley, where putting together a 100-person skilled team of ASIC designers might take up to 18 months, putting together a similar team in India might take a good deal longer. To improve skill levels, Broadcom India recruits engineers from the United States, mostly of Indian origin, as a result of which about 5 percent of its Indian workforce is Indian expatriates. Initially, the Indian recruits were experienced engineers who were hired away from competitors. Because of low attrition rates, however, the average work experience of engineers at Broadcom India is now more than nine years. Thus the company can now recruit from universities and offer internships to university students.

This hybrid approach has two major payoffs. First, despite the recruitment of expatriates, costs in India average one-third of costs in the United States. Second, the center of expertise is growing not only in Broadcom India, but also in Bangalore generally, in embedded software and very large chip development.

Case Study: Hellosoft

Hellosoft is a Silicon Valley start-up established in 2000 and funded by U.S., Taiwanese, and Indian venture capitalists. The company provides high-performance communications intellectual property for VoIP and wireless devices. From the beginning, the firm intended to use Indian engineers to create its intellectual property. All R&D is conducted by a subsidiary located in Hyderabad, India, that employs more than 100 digital signal-processing engineers (Yarlagadda, 2005). The Hyderabad center develops software for advanced cell phones and networking technologies. Marketing and sales are located in the company's headquarters in San Jose.

Case Study: Ketera Technologies[20]

Ketera Technologies, headquartered in Santa Clara, California, provides inventory-management software on demand

[20]Information based on a case study compiled by Shah (2005).

(i.e., software-as-a-service). As of 2005, the company had 150 employees worldwide. Its objectives for having subsidiaries in India was to cut costs and speed up time-to-market. In 2002, the company established a relationship with an Indian vendor, which had a peak of 105 workers in June 2004. The engineers in the India operations worked on software development and mundane tasks, such as configuring software for customers and other support services.

The relationship with the vendor turned out to be unsatisfactory because the engineers there were relatively unproductive and attrition rates were high. In addition, the U.S. operation was understaffed as a result of the 2001–2003 downturn. For example, there was only one architect for about 80 engineers, less than half the norm.

In late 2004, the firm created its Indian subsidiary and transferred the work in phases, beginning with software programming. The company also decided to shift its product management to India. To ease coordination problems, staff was added in the United States.

It took about nine months for Ketera to hire 75 engineers in Bangalore. Close coordination was essential to the company's success; product management was divided between the U.S. and Indian teams, with the U.S. team taking responsibility for market requirements and the Indian teams converting those into product specifications.

A key challenge in new-product development is measuring team productivity. Unlike well specified software, for which productivity can be measured by error rates or lines of code, a "new level of complexity" (Shah, 2005) is always associated with the release of a new product, which makes measuring productivity difficult.

Case Study: Netscaler[21]

Netscaler was founded in 1998 to redesign a specific component of infrastructure used in regulating traffic flow on the Internet. After Netscaler had developed the product, the company realized some functionality had to be added to attract customers who were wary of moving from legacy products to the Netscaler product. Because Netscaler was constrained financially and needed to cut costs, in 2001 it hired an Indian outsourcing firm, NodeInfoTech, to help develop the new features.

The success of this contracting arrangement convinced the company to establish Netscaler India, which was staffed by many of the developers from NodeInfoTech (Tillman and Blasgen 2005). In 2004, Netscaler India employed approximately 60 engineers to develop other features and planned to grow to 200 employees by 2005 (Hindu Business Line, *http://www.thehindubusinessline.com/*, downloaded 1/13/2006). At that point, however, it was purchased by Citrix Systems for $300 million.

The reason Netscaler formed a subsidiary rather than continuing to outsource was to increase the number and sophistication of projects done in India and encourage tighter engineering integration (Tillman and Blasgen, 2005). After its initial foray into India, Netscaler offshored high-value work to its subsidiary and outsourced some lower level engineering support to local Indian vendors. Having Indian and U.S. internal engineering teams made it possible for Netscaler to provide all levels of support 24 hours a day. As the Indian team grew, it became feasible to add a technical writer in India to provide software documentation.

Case Study: Tensilica[22]

Tensilica is a Silicon Valley start-up established in 1997. The company, which has 120 employees worldwide, develops and licenses its embedded processor technology to SOC suppliers. The downturn of 2001 affected demand for Tensilica's products and led the firm to consider shifting second-generation work, such as adding features and improving product reliability, to India, thus freeing up expensive U.S.-based engineers for new-product development. To save on initial setup costs, and because the firm did not have a brand name in India to help recruit the best talent, Tensilica decided to begin working with a vendor, eInfochips, and then transfer to a subsidiary over time.

The initial work involved adding features to an existing product, such as improving the graphical user interface. An experiment with quality assurance was unsuccessful because it required too much U.S. management time. In general, coordination costs were much higher than expected. e-Infochips agreed to let Tensilica handle recruitment, but this turned out to be much more difficult than expected because the level of skills available was too low. In addition, some qualified engineers were unwilling to work for an outsourcer.

In January 2006, Tensilica transferred engineers from e-Infochips to its own subsidiary, which, as of September 2006, employed 15 persons, or 12 percent of Tensilica's workforce. Without the veil of an outsourcer, recruitment became much easier, and attrition rates have fallen. After working with the India team for a year, the company has also greatly reduced coordination costs. The company now does work that involves much more complexity in India.

Case Study: SAP

SAP, a large German applications software firm, began its offshoring operations to Bangalore in 2000. Initially, a CRM project was supported from India. About 40 percent of the programming work for the project was done in Bangalore. The work was done on an ad hoc basis. Project managers based in SAP's German offices would request programming

[21]This discussion of Netscaler is based on Tillman and Blasgen (2005) and Jagadeesh (2006).

[22]Based on Dixit (2005, 2006).

support from the Bangalore operations when needed, on a short-term basis.

Despite the success of this approach, SAP found that attrition rates in its Bangalore operations rose to over 30 percent. A workforce analysis revealed that its Bangalore team would have to be given more responsible and long-term work in order to induce them to stay on with SAP. The firm responded in 2003 by shifting all the programming work for selected projects to Bangalore, while retaining the management of the project in Germany. This approach enabled Bangalore-based engineers to offer all the programming support for a project through the life of the project.

While this approach led to a reduction in attrition, the coordination required to manage complete projects globally was proving to be very high. In 2004, SAP shifted the work of some project and sub-project (component) managers to Bangalore in order to ensure that engineers only reported locally. This approach proved to be so successful that, by 2006, SAP had grown to 3,200 persons in Bangalore. The Bangalore operations were given the status of a "Global Development Center" (i.e., it had achieved across-the-board capabilities to support any of SAP's projects globally). This is a status hitherto granted by SAP only to its operations in Germany, Palo Alto in the United States, and Tel Aviv, Israel. SAP Bangalore was also designated as SAP's center of excellence for several verticals, including oil and gas, steel and telecommunications. Attrition rates by the end of 2006 were at industry-standard rates of 12 percent.

Lessons from the Case Studies

Extrapolating from this admittedly small base of information, we found two basic models: (1) offshoring as a supplement to onshore operations (i.e., the purpose of the offshore facility is to lower costs and/or accelerate product or product-line extensions); and (2) offshore operations as an integral part of the business model. The ultimate goal in both models is for the India business to become an integral part of the company.

Interestingly, both start-ups and established firms often begin by using an outsourced provider rather than establishing their own facilities. One advantage of outsourcing is that operations can be ramped up quickly. In addition, the company may learn about the Indian environment through the operation of the outsourcer, thereby facilitating the later establishment of a subsidiary.

There are also risks to this approach. First, as a company cedes control over the labor force to an outside vendor, it risks losing control of its intellectual property and also its ability to respond directly to attrition. Second, because the ultimate goal for both new and established firms appears to be that the India operations become integral to the business, a subsidiary must be established at some point. Integration into the company may sound like an irrevocable end point, but we have observed cases of firms that later contracted out routine in-house work. Established firms have less critical cost concerns and are, therefore, more likely to create a subsidiary and begin in-house work right away. Third, in all cases, coordination costs have been surprisingly high, not because of inadequate communications facilities, but because of the complex nature of the work. Fourth, finding and retaining qualified persons for higher value-added work is difficult, most likely because of the small size of India's domestic markets and its inadequate educational system.

Table 11 provides a summary of the stages of offshoring described in the case studies. Undoubtedly, evolution will continue. For example, Agilent India plans to increase outsourcing once the offshoring process is stabilized.

THEORETICAL FRAMEWORK

A framework for offshoring of software services in international trade requires some definitions, some as basic as a definition of "service." Most people agree that "manu-

TABLE 11 Stages of Software Offshoring to India by U.S. Firms

Firm	Type of Work	Initial Stage Onshore	Offshoring Stage 1	Reason for Stage 1 Offshoring[a]	Offshoring Stage 2	Reason for Stage 2 Offshoring
Agilent	Embedded software	In house	In house, not integral	Control	In house, integral	Coordination stabilized in Stage 1
Broadcom	Chip design	In house	In house, integral	Scale		
Hellosoft	IP development	Offshoring operations from the start	Integral			
Ketera	Software-as-a-service	In house	Outsource, not integral		In house, integral	To improve coordination and resolve labor-quality issues
Netscaler	Router software	In house	Outsource, not integral		In house, integral	To undertake more complex product development
Tensilica	Embedded processor	In house	Outsource, not integral	Rapid ramp-up	In house, not integral	To improve coordination and resolve labor-quality issues
SAP	Applications development	In house	In house, not integral	Cost and scale	In house, integral	To improve coordination and resolve labor-quality issues

[a]in addition to labor cost arbitrage

facturing" is a process that involves the transformation of a tangible good. Most people also agree that, in many cases, manufacturing does not require face-to-face contact between the buyer and seller. Usually, manufacturing creates a good that can be stored, thereby allowing a physical separation of the buyer and the seller.

"Services" have been defined as the opposite of manufacturing in many respects. Services are transactions that involve intangible, non-storable goods, and client and vendor must be face-to-face when the service is being delivered. For example, Gadfrey and Gallouj (1998) define services as goods that are "intangible, cosubstantial (i.e., they cannot be held in stock) and coproduced (i.e., their production/consumption requires cooperation between users and producers)." This is obviously true when the service requires customization, such as receiving a haircut, but is also true when the "service experience" does not require customization, such as when a bank client wants to check the bank's home loan offering, or even proximity, as when a customer wants to check a bank balance.

Thus certain services are intrinsically more difficult to offshore than manufactured goods. When a service activity is considered as a totality, it indeed appears to resist relocation. In fact, very few service operations can be done only on the computer (the modern form of "mundane work"). Most services require at least some level of face-to-face interaction, either among coworkers or with persons outside the organization, such as vendors and clients.

Following Bhagwati's (1985) framework, we divide services that require proximity between user and provider into three categories:

1. Mobile user-immobile provider (e.g., a cell-phone user who visits a service center for a software upgrade).
2. Immobile user-mobile provider (e.g., a software consultant who visits a client prior to designing an IT system to understand the information flows in the client's business).
3. Mobile user-mobile provider (e.g., two delegates at a conference who exchange information through Bluetooth-enabled laptops).

For software services, the required interaction between seller and consumer has been substantially reduced. Advances in information technology have made possible the parsing of the provision of certain services into components requiring different levels of skill and interactivity. Besides the standardization of hardware and software platforms and the reduced cost of computing power, new language-structuring mechanisms, such as object orientation, have been developed. In addition, the Internet allows for the standardization of data-transmission platforms. As a result, certain portions of serviced activities—that might or might not be skill-intensive and that require little face-to-face interaction—can now be relocated offshore. Digital technology has made this possible.

The first fundamental change with digitization was that service flows could be converted into stocks of information, making it possible to store a service. For example, a consultant's assessment that once had to be delivered to a client in person could now be prepared as a computer document and transmitted via e-mail or, better yet, encoded into software. Easy storage and transmission allowed for the physical separation of client and vendor, as well as their separation in time. In addition, services could be separated into components that were standardized and could be prepared in advance (such as a template for the assessment) and components that were customized for the client (such as the assessment itself), which were non-storable. By taking advantage of the subdivision of tasks and the economies of the division of labor, costs could be reduced by having lower cost laborers prepare the standardized components, possibly at another location.

The second fundamental change was the conversion of non-information service flows into information service flows. For example, the assessment of information-technology needs for an automobile assembly line, which had required a site visit to make the assessment, can now be made through virtualization models of the assembly line delivered over the Internet. Once converted to an information flow, the service may then be converted into a stock of information, which can reduce costs through the standardization of components and remote production.

Third, by enabling the low-cost transmission of digitized material, digitization accelerated the offshoring of services. Early on, services, such as the writing of software programs, which were offshored to India in the early 1970s, were enabled by digitized storage, and, in the 1980s, by the standardization of programming languages. Later, in the 1990s, as the cost of digital transmission fell, even non-storable services, such as customer care, could be offshored.

The events that enabled software offshoring did not happen all at once and may not even have happened in the same way in every country. Israel, for example, was able to move quickly to product development for global markets by domestic firms. India, by contrast, until a few years ago, had offered only routine programming work for more than two decades. As of 2006, there was no evidence of successful product development that originated in India, although work to support product development conceived in developed countries was being done.

Thus moving to higher stages of work is not automatic, sequential, or time bound. Based on the available evidence, we cannot specify the conditions for movement to higher stages or predict that an exporter will capture a rising share of the economic rents (income in excess of cost).

At the very least, our case studies suggest that one factor that can hinder movement to higher stages is the cost of global coordination, whether it be between a developing country vendor and a developed country consumer or a team of vendors located across the world. For this reason, the developed-country firm can be compensated for being the middleman. Much of the market-related coordination and networking requires developed-country institutions, enabling the capture of value by the developed-country firm. However, competition is likely to force price compression on developed-country firms, especially if it comes from developing countries. This is happening now with major Indian software services firms, which are evolving into systems integrators as they develop the requisite skills and customer confidence.

The inference is that certain aspects, such as deciding on a product and its specification, design, marketing, and sales, are usually retained by the importer. But there is no guarantee that developed-nation firms will continue to maintain this privileged position. For the time being, however, the exporter's ability to rise to new stages of growth is limited, and developed-country buyers will continue to reap the rewards.

CONCLUDING THOUGHTS

It is tempting to view software offshoring as the cause of unmitigated job losses for U.S. workers. Software offshoring raises fears that, as a result of digitization, skilled jobs will rapidly disappear from U.S. shores. This would not only leave the United States digitally divided from other countries, but would centralize demand for U.S. workers in non-offshorable jobs. In software, the argument is often made that U.S. workers will ultimately do only those jobs that are impossible to offshore, a few of which will undoubtedly be highly skilled but most of which will require lesser skills, such as information-technology training and hardware and software systems integration.

Our analysis of the software industry shows that the effects of offshoring on employment in developed nations vary, even though the impact of software offshoring on developing countries is to generate increasingly high levels of employment. The kinds of work initially offshored typically have low entry barriers and are subject to automation. Thus services exported from developing countries initially lack brand value and thus are very different from services exported from developed countries. In consequence, there is likely to be competition and price compression in these sectors.

However, over time, the level of sophistication of work being done offshore has risen rapidly. This can be a subtle process. As Shah (2005) notes in his discussion of Ketera's offshoring, "The primary challenge [of offshoring most of the head count to India] was the lack of informal communication in our Silicon Valley office. We missed the informal hallway and coffee station side chats. We missed going to the white-board and brainstorming an idea." After observing the progress of the Indian operation, he concluded, "We then realized that the hallway discussions and white-board brainstorming are still happening [in our firm], but in India."

In summary, there is little doubt that work that is modularized and standardized and does not require regular customer contact is more likely to be moved offshore. This was evidenced by the rapid offshoring of the programming function. As our case studies show, the digital revolution (a catch-all term for a series of changes) has increased the scope of work in the software supply chain that can be spatially disaggregated and outsourced. Even when a customer interface is necessary, it is possible (as the case study of Broadcom India showed) to manage customer interfaces remotely through "body-shopping" that focuses on understanding customers rather than, as in the old days, accessing customers' software and hardware. In the case of Broadcom India, offshore workers are substitutes for U.S. workers.

Lowering the costs of some aspects of software development lowers total costs and makes a company more competitive globally. It can also make possible the creation of new firms that would otherwise not be economically viable, as the case study of Netscaler showed. Jobs created by this entrepreneurship can be counted against jobs lost to offshoring. As Rakesh Singh, Netscaler's general manager of Asia operations, said, "The cost savings through outsourcing have helped us become more competitive and experience rapid growth as a company. As a result, we have a lot more employees in the United States today than we did when we set up the India operations" (Tillman and Blasgen, 2005). In this case, offshore workers are complements to U.S. workers.

Ongoing technological development typical of the software industry can both speed up and slow down job losses. For example, prior to the establishment of the Internet as a reliable medium of digital communication, installing software or fixing a software problem required an on-site technician. In most cases, these tasks can now be done remotely, thus reducing the need for on-site work and increasing the demand for offshore maintenance. Similarly, the invention of the router led to the creation of remote data centers, thus reducing the need for on-site storage hardware and support services.

At the same time, the Internet enables access to many more software applications that are developed elsewhere, including open-source applications. Raza (2005) notes that chip designers who used to offshore components of chip development to vendors in India can now usually find some components already available in open source, thus reducing the need for offshoring (although this does not increase demand for U.S. software developers).

An alternate view of the impact of technological change is that, because the developers of new technology are mostly in developed nations, a faster rate of technological progress is advantageous to employment in developed nations because it makes it harder for developing countries to catch up. From this point of view, anything that helps developed-country engineers innovate more quickly and efficiently is a plus for the developed country. Hence, offshoring software development that is a step behind the work being done in the developed country enables engineers in developed nations to innovate even more and is good for both developing and developed nations.

As we noted in our introduction, scholars concede that the effects of offshoring on the quality of work done in developed nations are uncertain, because we do not know whether the productivity gains will be captured by the developing country or the developed country. This depends on their relative productivity gains. Hence many would concede that the jobs left for workers in developed nations will certainly include low-wage work that cannot be done remotely (such as the physical installation of a hard-wired network). Many would also agree that short-term unemployment is possible. However, they also argue that most of the new work will require higher skill levels than are available in developing countries, will pay more, and will even leverage work being done in developing countries.

Based on the experience of offshoring in the manufacturing sector, a second issue is the speed with which services offshoring takes place. The decline in manufacturing in the United States happened gradually and was accompanied by rising revenue per employee, reflecting in part that, as the more commoditized parts of manufacturing were being outsourced offshore, the more customized or specialized parts and some service components, such as design and integration, were still being done onshore (Figure 3 and Table 12). The slow pace of manufacturing offshoring also gave displaced workers time to acquire skills to shift to other occupations.

As the rate of offshoring in the Indian software industry shows, some aspects of software offshoring may be rapid, leaving little time for labor-force adjustment. The reason for the rapid rate can be attributed to digitization, which has been firmly established since the mid-1990s (the Telecom Regulation Act of 1996 is often considered a turning point). Digital technology has been crucial to the rapidity of services offshoring. Unburdened by the need for large factories, offshored services can be set up almost as rapidly as workers with the requisite skills can be hired. Certainly the growth rate of the Indian information-technology industry has been much, much faster that in manufacturing offshoring.

TABLE 12 Share of Employment in Manufacturing Employment in the United States

	1970	1980	1990	2004
Employment in manufacturing	18.9%	19.8%	18.7%	14.1%

Source: BLS statistics (*http://www.bls.gov*) accessed 10/6/05.

This raises the question of whether the digital revolution has done more than provide a one-time boost for Asian competitors. Apart from the labor-cost advantage, developing countries will continue to have a comparative advantage for two reasons: (1) economies of scale and scope, and (2) specialization.

Countries such as India have large labor pools that could offer significant economies over smaller labor pools or country-specific labor pools. In addition, by locating software developers in India, the vendor can supply services for clients in different time zones, thus making efficient use of capital and real estate. Or, vendors can manage episodic peak requirements, such as when a new upgrade of software is released, more efficiently.

Many efficient practices for offshore software development that resulted from the remote software-programming businesses were developed in India. Thus remote management is emerging as a specialized skill that is applicable in a variety of other offshoring situations, such as providing

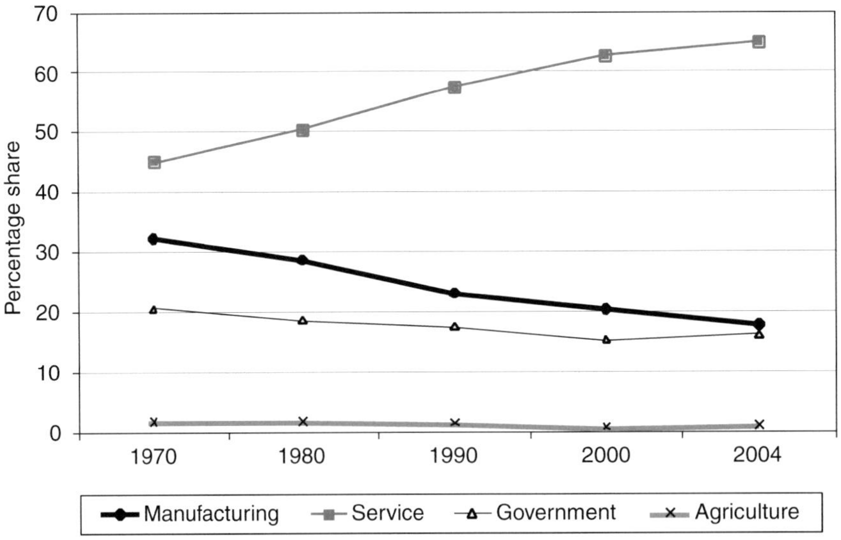

FIGURE 3 Share of employment for various economic sectors in the United States, 1970–2004. Source: BEA Statistics (*http://www.bea.gov/bea/dn/nipaweb/*) Table 6.5, accessed 10/6/05.

R&D and product-development services. Of course, Indian firms with these specialized skills must compete with the remote project-management skills developed by global firms in other environments (e.g., Accenture's skills in system integration).

At the beginning of this paper, we suggested two trajectories in offshoring that might protect employment in developed countries. The first was that constraints on capacity (both educational and infrastructural) in low-cost countries might limit the scale of offshoring. Based on the evidence we have presented, this is unlikely to happen. The second trajectory was that developed nations would reinvent themselves to a higher value-added path. It appears that the only viable strategy for developed nations is to develop the capacity to generate continuous high-value new opportunities that cannot be immediately offshored, which will require ongoing innovation. Although there is no guarantee that a developed country will have the capacity for continuous innovation, a country with an open economy that invests in education has a better chance than others. We can be hopeful that the United States will continue to demonstrate the truth of this proposition.

REFERENCES

ACM (Association for Computing Machinery). 2006. W.Aspray, F. Mayadas and M. Vardi (eds.). Globalization and Offshoring of Software: A Report of the ACM Job Migration Task Force. New York: ACM.

Apte, U., and R. Mason. 1995. Global disaggregation of information-intensive services. Management Science 41(7): 1250–1262.

Ariav, G., and S. Goodman. 1994. Israel: of swords and software plowshares. Communications of the ACM 37(6): 17–21.

Arora, A., and A. Gambardella. 2005. From Underdogs to Tigers: The Rise and Growth of the Software Industry in Brazil, China, India, Ireland and Israel. Oxford, U.K.: Oxford University Press.

Athreye, S. 2005. The Indian software industry and its evolving service capability. Industrial and Corporate Change 14: 393–418.

Balasubramanyam, V., and A. Balasubramanyam. 2000. The Software Cluster in Bangalore. In Regions, Globalization and the Knowledge-Based Economy, edited by J. Dunning. Oxford, U.K.: Oxford University Press.

Bhagwati, J. 1985. Why Are Services Cheaper in the Poor Countries? In Essays in Development Economics, Volume 1: Wealth and Poverty, edited by J. Bhagwati and G. Grossman. Cambridge, Mass.: MIT Press.

Bhagwati, J., A. Panagariya, and T. Srinivasan. 2004. The muddles over outsourcing. Journal of Economic Perspectives 18(4): 93–114.

Bresnitz, D. 2005. The Israeli Software Industry. Pp. 72–98 in From Underdogs to Tigers: The Rise and Growth of the Software Industry in Brazil, China, India, Ireland and Israel, edited by A. Arora and A. Gambardella. Oxford, U.K.: Oxford University Press.

D'Costa, A. 2002a. Software outsourcing and development policy implications: an Indian perspective. International Journal of Technology Management 24(7/8): 705–723.

Dixit, A. 2006. VP, Tensilica. Personal Interview, 9/20/06.

Dossani, R. 2006. Globalization and the Offshoring of Services: The Case of India. Pp. 241–267 in Offshoring White-Collar Work, edited by S. Collins and L. Brainard. Washington, D.C.: Brookings Institution.

Dossani, R., and A. Manwani. 2005. Agilent's Supply Chain: A Locational Analysis of its Indian Operations. Paper presented at Conference on Globalization of Services, Stanford University, June 2005.

Ellis, R., and C.L. Lowell. 1999. Core Occupations of the U.S. Information Technology Workforce. IT Workforce Data Project 1.

Gadrey, J., and F. Gallouj. 1998. The provider-customer interface in business and professional services. Services Industries Journal 18(2): 1–15.

GAO (Government Accountability Office). 2005. International Trade: U.S. and India Data on Offshoring Show Significant Differences. Report #GAO-06-116. Washington, D.C.: U.S. Government Printing Office.

Gomory, R., and W. Baumol. 2000. Global Trade and Conflicting National Interests. Cambridge, Mass.: MIT Press.

Hanson, G., R. Mataloni Jr., and M. Slaughter. 2001. Expansion Strategies of U.S. Multinational Firms. In Brookings Trade Forum, edited by S. Collins and D. Rodrik. Washington, D.C.: Brookings Institution.

Heeks, R. 1996. India's Software Industry. New Delhi: Sage Publications.

Heitzman, J. 1999. Corporate strategy and planning in the science city: Bangalore as Silicon Valley. Economic and Political Weekly, January 30, 1999.

Hira R., and A. Hira. 2005. Outsourcing America: What's Behind Our National Crisis And How We Can Reclaim American Jobs. New York: American Management Association.

Indian Ministry of Finance. 2000. Finance Minister's Budget Speech. New Delhi: Ministry of Finance.

Indian Ministry of Human Resource Development. 2001. Technical Education Quality Improvement Project of the Government of India. New Delhi: Ministry of Human Resource Development.

Jagadeesh, B. 2006. CEO, Netscaler. Personal Interview.

Khare, R. 2006. CEO, Broadcom, India. Personal Interview, 2/1/06.

Landefeld, S., and R. Mataloni. 2004. Offshore Outsourcing and Multinational Companies. Working Paper. Washington, D.C.: Brookings Institution.

Mankiw, G., and P. Swagel. 2006. The Politics and Economics of Offshore Outsourcing. Working Paper. Washington, D.C.: American Enterprise Institute.

Mann, C. 2006. Accelerating the Globalization of America: The Role for Information Technology. Washington. D.C.: Institute for International Economics.

McKenna, C. 2006. The World's Newest Profession. Cambridge, U.K.: Cambridge University Press.

Mowery, D., ed. 1996. The International Computer Software Industry: A Comparative Study of Industry Evolution and Structure. New York: Oxford University Press.

Mukherji, A. 2006. Senior VP, Wipro. Personal Interview, 8/1/06.

Naidu, B. 2002. Director, Software Technology Parks of India, Bangalore. Personal Interview, 2/15/02.

NAPA (National Academy of Public Administration). 2005. Department of Commerce Offshoring Study. Available online at *http://www.napawash.org/pc_management_studies/index.html*, accessed 5/12/07.

Nasscom. 1999, 2001–2006. The IT Industry in India: A Strategic Review. New Delhi: Nasscom.

Parthasarathi, A., and K. Joseph. 2002. Limits to Innovation in India's ICT Sector. Science, Technology and Society 7(1): 13–50.

Parthasarathy, B. 2000. Globalization and Agglomeration in Newly Industrializing Countries: The State and the Information Technology Industry in Bangalore, India. Ph.D. thesis, University of California, Berkeley.

Parthasarathy, S. 2003. CEO of Aztec Software. Personal Interview, 3/2003.

Premji, A. 2003. CEO of Wipro. Personal Interview, 12/2/2003.

Premji, A. 2006. CEO of Wipro. Personal Interview, 2/10/2006.

Ramadorai, S. 2002. CEO of TCS. Personal Interview, 11/29/02.

Raza, A. 2005. CEO of Raza Foundries. Personal Interview, 12/15/2005.

Rubin, B. 1985. Economic liberalization and the Indian state. Third World Quarterly 7(4): 942–957.

Sahay, S., B. Nicholson, and S. Krishna. 2003. Global IT Outsourcing. Cambridge, U.K.: Cambridge University Press.

Samuelson, P. 2004. Where Ricardo and Mill rebut and confirm arguments of mainstream economists supporting globalization. Journal of Economic Perspectives 18(3): 135–146.

Sands, A. 2005. The Irish Software Industry. Pp. 41–71 in From Underdogs to Tigers: The Rise and Growth of the Software Industry in Brazil, China, India, Ireland and Israel, edited by A. Arora and A. Gambardella. Oxford, U.K.: Oxford University Press.

Shah, R. 2005. Ketera India Case Study. Paper presented at Conference on Globalization of Services, Stanford University, June 2005.

Siwek, S., and H. Furchtgott-Roth. 1993. International Trade in Computer Software. Westport, Conn.: Quorum Books.

Sridharan, E. 2004. Evolving Towards Innovation?: The Recent Evolution and Future Trajectory of the Indian Software Industry. Pp. 27–50 in India in the Global Software Industry: Innovation, Firm Strategies and Development, edited by A. D'costa and E. Sridharan. New York: Palgrave Macmillan.

Steinmuller, W. 1996. The U.S. Software Industry: An Analysis and Interpretive History. Pp. 15–52 in The International Computer Software Industry: A Comparative Study of Industry Evolution and Structure, edited by D. Mowery. Oxford, U.K.: Oxford University Press.

Teubal, M. 2002. The Indian software industry from an Israeli perspective: a microeconomic and policy analysis. Science, Technology and Society 7(1): 151–187.

Tillman, J., and N. Blasgen. 2005. Case Study of Netscaler. Paper written for CRD 199 Special Study Course, University of California, Davis, June 16, 2005.

Torrisi, S. 2002. Software Clusters in Emerging Regions. Working Paper. University of Camerino. Italy.

Yarlagadda, K. 2005. Founder, President, and CEO, Hellosoft. Personal Interview, 2005.

BIBLIOGRAPHY

Arora, A., and S. Athreye. 2002. The software industry and India's economic development. Information Economics and Policy 14(2): 253–273.

Collins, S., and L. Brainard. 2006. Offshoring White-Collar Work. Washington D.C.: Brookings Institution.

Correa, C. 1996. Strategies for software exports from developing countries. World Development 24(1): 171–182.

D'Costa, A. 2000. Technology Leapfrogging: The Software Challenge in India. Pp. 183–200 in Knowledge for Inclusive Development, edited by P. Conceicao, D. Gibson, M. Heitor, G. Sirilli, and F. Veloso. Westport, Conn.: Quorum Books.

D'Costa, A. 2002. Export growth and path dependence: the locking-in of innovations in the software industry. Science, Technology and Society 7(1): 51–87.

Dixit, A. 2005. Tensilica's India Operations: The First Seven Months of Embedded Processor Engineering Offshore. Paper presented at Conference on Globalization of Services, Stanford University, June 2005.

Feller, J., B. Fitzgerald, S. Hissam, and K. Lakhani. 2005. Perspectives on Free and Open Source Software. Cambridge, Mass.: MIT Press.

NSF (National Science Foundation). 1999. Assessing the demand for information technology workers. IT Workforce Data Project: Report 1. Washington. D.C.: NSF.

Scholte, J. 2000. Globalization: A Critical Introduction. Basingstoke. U.K.: Palgrave Publishers.

Schware, R. 1992. Software industry entry strategies for developing countries: a "walking on two legs" proposition. World Development 20(2): 143–164.

The Changing Nature of Engineering in the Automotive Industry

John Moavenzadeh
Executive Director
International Motor Vehicle Program
Massachusetts Institute of Technology

Engineering has always been essential to the global automotive industry, which spends more on research and development (R&D) than any other industry except the pharmaceutical industry (Figure 1).[1] Ranked by R&D spending, four of the top 10 global firms are automotive companies (Figure 2). The vast majority of the $55 billion spent on R&D in the automotive industry is on development, rather than basic or applied research,[2] and most steps in the vehicle-development process require engineers and technicians. A typical new-vehicle development program costs between $500 million and $1 billion and takes two to three years from concept to customer. A new-engine development program costs roughly $100 million to $500 million, and a new-transmission development program costs roughly $50 million to $250 million. Thus corporate engineering capability is a key competitive differentiator for vehicle manufacturers.

PRODUCT ENGINEERS

There are two basic types of automotive engineers—product engineers and manufacturing engineers. In general, product engineers design cars and trucks and their components. Individual product engineers focus on specific systems (e.g., braking, steering, or interiors) or specific components within those systems (e.g., antilock braking controllers, steering columns, or instrument clusters). Product engineers can also be development engineers who evaluate prototype vehicles and tune vehicles in the preproduction phase (e.g., calibrating the power train to meet the customer profile for a vehicle). Product engineers can also be test engineers responsible for performing durability, stress, thermal, or noise and vibration testing.

Although product engineers have traditionally been grounded in mechanical and industrial engineering, as the software content of vehicles has increased, the industry has increasingly hired electrical, electronics, and software product engineers. Many vehicle manufacturers also operate advanced engineering departments to search for new ideas and develop new technologies for future vehicles.

MANUFACTURING ENGINEERS

Manufacturing engineers, who tend to be trained as industrial and mechanical engineers, are responsible for determining the most efficient way to produce vehicles. Some manufacturing engineers are part of a central engineering staff dedicated to production. However, most are located in offices at production facilities, such as vehicle-assembly plants and component-manufacturing plants.

Most firms encourage close coordination between product and manufacturing engineers. Design for assembly, design for manufacturing, and value engineering require that product and manufacturing engineers work together to engineer excess cost and waste out of a vehicle.

SUPPLIERS

The importance of the supply base cannot be overstated. A typical automobile is made of 20,000 to 30,000 indi-

[1]If information and telecommunications technology industries are lumped together, the automotive industry ranks third in R&D spending.

[2]Not all of the companies could estimate the precise split, but the three that provided data spent less than 10 percent for research and more than 90 percent for development.

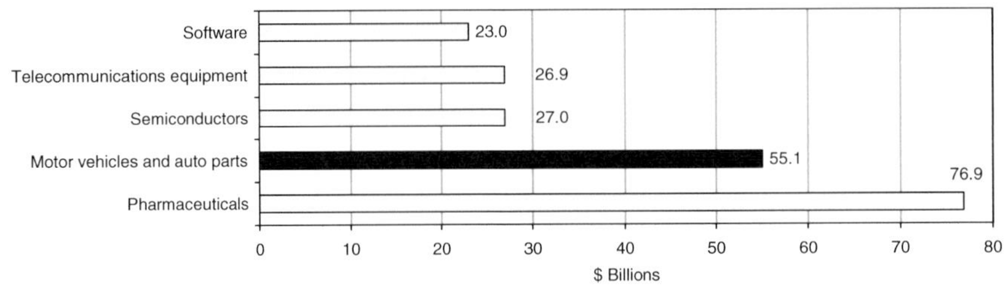

FIGURE 1 Estimated R&D spending for top industries, 2006. Source: Schonfeld & Associates, 2006. Reprinted with permission of Schonfeld & Associates. Note: Industry SIC Codes are: Software: 7372; Telecom Equipment: 3663 and 4812; Semiconductor: 3674; Automotive: 3711 and 3714; Pharmaceutical: 2834.

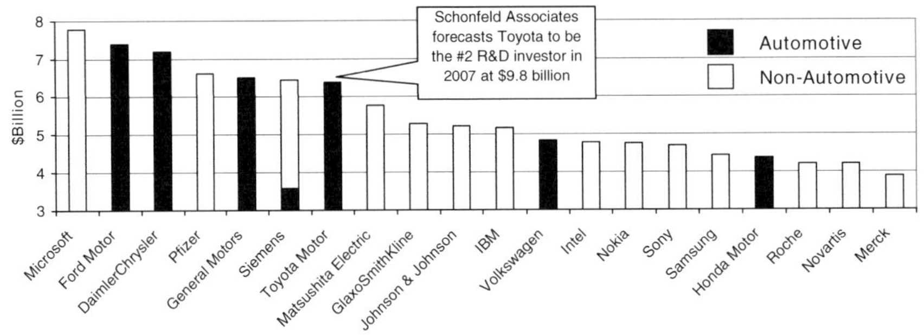

FIGURE 2 R&D spending for top 20 global companies, 2004. Sources: Corporate R&D Scorecard, Technology Review, 2005; Industrial Research Institute, 2005; company annual reports. Note: Siemens includes Siemens VDO automotive business, which accounted for 12.7 percent of 2005 revenue.

vidual parts engineered into hundreds of components and subsystems. Vehicle manufacturers purchase one-half to three-quarters of these parts from their suppliers. All of the major vehicle manufacturers spend at least 50 percent of their revenue on components from suppliers.[3] Vehicle manufacturers increasingly specify overall system requirements and give suppliers free rein to engineer and design a component or vehicle subsystem to meet those requirements. This contrasts with the traditional business model (which still exists for some components),[4] in which vehicle manufacturers give suppliers detailed technical specifications for components. Supplier engineers, who frequently work closely with engineers at the vehicle manufacturers, play a critical role in introducing technology into vehicles.

Many of the hundreds of firms that primarily supply the automotive industry have consolidated into global enterprises that employ thousands of people in facilities spread across the planet. In theory, the industry supply base is divided into tiers. A tier-one supplier sells directly to the vehicle manufacturer (e.g., BorgWarner may sell a transmission to General Motors [GM]). Tier-two suppliers sell to tier-one suppliers (e.g., Timken may sell roller bearings to BorgWarner). In practice, however, the distinctions are often blurred, and some very small firms may sell directly to vehicle manufacturers (although these should not be considered tier-one suppliers for the purposes of analysis). Some firms, such as Freescale (formed when Motorola spun off its automotive semiconductor business), Siemens, Sumitomo Electric, DuPont, and even Microsoft), are not thought of as automotive supply firms, although they have large automotive businesses. In addition, many firms supply production equipment to the automotive industry (e.g., stamping presses or robotics systems) or test equipment (e.g., dynamometers and road simulators). All of these firms employ product and manufacturing engineers.

PRODUCT ARCHITECTURE

Product architecture, the relationship between the functions and structures of the vehicle, greatly influences how a vehicle is engineered. The terminology developed by Clark and Fujimoto (1991) provides helpful distinctions:

[3]Some vehicle manufacturers and suppliers have significant equity relationships. In the Japanese *keiretsu* system, for example, Denso and Aisin Seiki, two large Japanese suppliers, are partially owned by Toyota. In France, PSA Peugeot Citroën and Faurecia have an equity relationship; and Hyundai-Kia and Mobis in South Korea have a similar relationship.

[4]For more on the rise of the "black-box parts ratio" in automotive product development, see Clark and Fujimoto, 1991.

- Modular architecture is based on a one-to-one correspondence between functional and structural elements.
- Integral architecture is based on a many-to-many correspondence between functional and structural elements.
- Open architecture is based on a mix and match of component designs across firms.
- Closed architecture is based on a mix and match of component designs within one firm.

Figure 3 illustrates where some typical products fall in a product-architecture matrix based on this terminology. Lego, the children's toy, is an example of a perfectly modular, closed architecture. The bicycle and PC system are examples of products with modular, open architectures. PC components, such as printers, displays, and other devices, are interchangeable among many manufacturers and are mapped closely to specific features (e.g., printers are used for printing).

Automobiles have traditionally had integral, closed architectures (although in the past few years, vehicle manufacturers have attempted to reduce costs through modularization). The many internal parts of a vehicle are not interchangeable among manufacturers, even though the same suppliers may make very similar parts for different vehicle manufacturers. The integral architecture of the vehicle often forces close, coordinated interaction among teams of engineers from vehicle manufacturers and suppliers.

The product architecture for heavy trucks is significantly more modular and open than for cars (e.g., trucks can be ordered with engines from different engine manufacturers).

ENGINEERING EFFICIENCY AS A DRIVER OF CHANGE

From a financial perspective, most vehicle manufacturers and many tier-one suppliers destroy value, meaning that their real market value is lower than the real value of capital put into the firm by investors. Most American and European automotive firms have lost value in recent years, while most Japanese automotive firms have returned value to their investors (Marcionne, 2006).

Although some original equipment manufacturers (OEMs) (e.g., Toyota, Honda, Nissan, BMW, and more recently Hyundai) are profitable and create value, the rest have not created value for several years. In addition, the fortunes of the winning firms and losing firms are diverging. For example, in 2006 the value of Toyota, the most valuable automotive firm in terms of market capitalization, was more than 10 times that of GM. Almost every manager and executive in the industry—even at profitable firms—reports tremendous pressure to reduce costs and improve performance, reflecting the fiercely competitive nature of the current automotive market.

In light of the extraordinary R&D costs for a typical vehicle manufacturer (Figure 2), firms that can engineer a vehicle at lower cost and bring the vehicle to market faster have an extraordinary advantage over their competitors. Fujimoto and Nobeoka (2004), who have studied automotive product development for many years, found significant differences in efficiency among vehicle manufacturers. Their data show that differences in engineering efficiency—as measured by engineering hours adjusted for comparison—are actually increasing between American, European, and Japanese automakers. Figure 4 shows the product-engineering hours

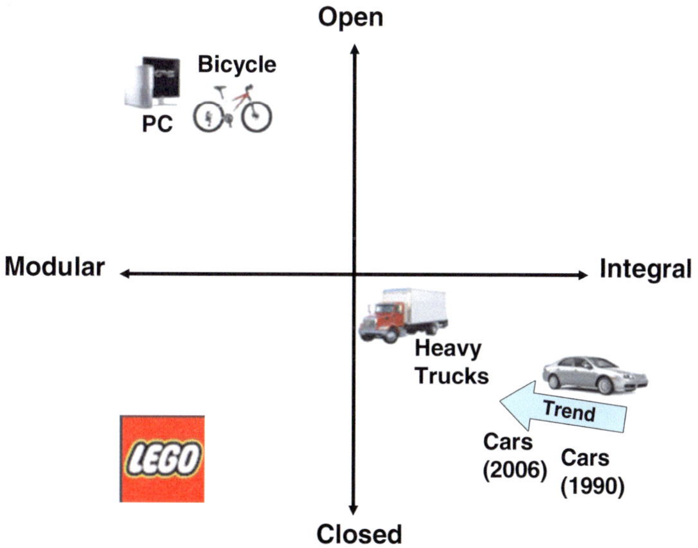

FIGURE 3 Product architecture matrix for cars, heavy trucks, and other products.

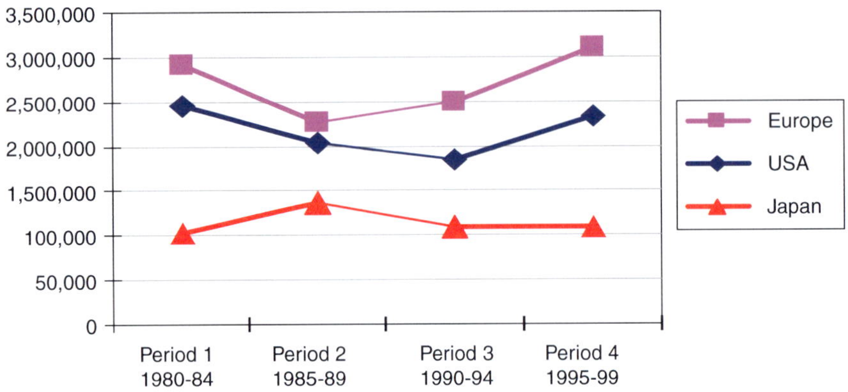

FIGURE 4 Adjusted product engineering hours for vehicle manufacturers in three regions. Source: Fujimoto and Nobeoka, 2004. Reprinted with permission.

required for a typical vehicle program averaged for vehicle manufacturers from three regions and for four time periods. (The data are presented as regional averages to mask the identity of individual firms; so, for example, an individual Japanese OEM may be less efficient than an individual American OEM.).

Note that product-engineering loads in the United States and Europe increased in the last five-year period (1995–1999) as a result of significantly more stringent regulatory requirements. Fujimoto and Nobeoka (2004) argue that in Japan, regulatory requirements cancelled out improvements in engineering efficiency; as a result, the number of engineering hours remained about the same. Indeed, returning to Figure 2, it is entirely unclear whether vehicle manufacturers that spend more on R&D than their competitors have an advantage or disadvantage. To evaluate R&D output, one must also consider the efficiency of the engineering operation.

One vice president of engineering reported that his single greatest challenge is the pressure "to do more with less." This manager had been asked to meet a corporate target of increasing engineering efficiency by 30 percent in three years—a remarkably ambitious objective. This particular manufacturer measures engineering efficiency by dividing engineering output by total engineering costs; engineering output is measured by a point system that assigns various weightings to the company's new vehicle programs, significant vehicle redesigns (known in the industry as product freshenings), and new power trains.

The drive to improve efficiency (i.e., to increase engineering output while lowering engineering costs) has led to several interrelated developments:

- pressure to manage a firm's global footprint more effectively across the enterprise
- changes in the working relationship between vehicle manufacturers and their suppliers
- a shift toward a more open model to accelerate innovation

The first item, managing the global engineering footprint, is the subject of this paper. Items two and three are discussed below.

Relationship between Vehicle Manufacturers and Suppliers

One of the most significant trends in the automotive industry in the past two decades has been the emergence of mega-suppliers capable of designing and developing large portions of the vehicle and, in some cases, manufacturing entire vehicles. The focus of the largest tier-one suppliers has been shifting from components to full-vehicle systems, or "modules." Their customers, the vehicle manufacturers, have granted them greater engineering responsibility and have announced plans to work more closely with fewer suppliers.

Contract Manufacturing

The increasing importance of suppliers in the global automotive industry is reflected in the emergence of contract manufacturers. For example, Magna Steyr, a wholly owned subsidiary of Magna International, builds complete vehicles for several OEMs. In 2005, Magna International declared more than $20 billion in automotive sales, making it the third largest automotive supplier in the world.[5] Magna Steyr's production volumes have increased steadily; in 2005, the company sold 230,505 units representing $4.1 billion in sales to OEMs. The company's manufacturing complex in

[5]2005 revenue of the top three automotive suppliers: Robert Bosch GmbH, $28.4 billion; Denso Corporation, $22.9 billion; Magna International, $22.8 billion (Automotive News, 2005).

Graz, Austria, includes two assembly plants that build about 1,000 vehicles a day, including the BMW X3, Mercedes E-class and G-class cars, Saab 9-3 convertible, Jeep Grand Cherokee, Chrysler 300, and Chrysler Voyager.

Magna has also moved into the upstream business of contract engineering for automakers, and the company now employs 2,300 engineers in 10 locations around the world. The largest engineering center, in the Graz complex, employs 1,000 people. Magna Steyr says it not only completely engineered the 9-3 Cabriolet, G-class; BMW X3; and Audi TT coupe and roadster, but also performed engineering projects for Alfa Romeo, Audi, Iveco, Lancia, Lincoln, Pontiac, Smart, and VW. These projects range from adding a body derivative to creating a four-wheel-drive version.

The blurring of the lines between OEMs and suppliers is reflected in DaimlerChrysler's Toledo Supplier Park in Toledo, Ohio. The 2007 Jeep Wrangler is manufactured at this facility with the significant involvement of a variety of suppliers. Kuka Flexible Systems, a German company, runs the body shop; Magna-Steyr runs the paint shop; and Mobis, a Korean company, supplies chassis modules. This arrangement is in sharp contrast to traditional assembly plants, where vehicle manufacturers are responsible for all of these functions.

A More Open Innovation Process

Another result of the tremendous pressure to engineer vehicles more efficiently is a migration toward openness in the innovation process. Vehicle manufacturers have historically looked inward for new ideas and better ways to engineer vehicles. In the previous section, we described how vehicle manufacturers are working more closely with suppliers. They are also turning to their competitors, universities, and even customers to improve their products through joint programs, technology alliances, online technology brokers, and university research programs.

Vehicle manufacturers have always shared programs among their internal brands; for example, a Buick and Oldsmobile product from GM might have been given different names although they were nearly identical. In addition, manufacturers with an equity relationship, such as Ford and Mazda, have shared vehicle platforms. However, in the past 10 years collaborations on vehicle programs have increased among manufacturers that do not have an equity relationship and that are otherwise fierce competitors in the marketplace; examples include the Toyota Aygo and the Peugeot 107, or the Pontiac Vibe and the Toyota Matrix.

Vehicle manufacturers that do not have equity relationships are also increasingly entering into technology alliances. The alliance of most interest in the industry currently is an agreement announced in September 2005 among GM, DaimlerChrysler, and BMW to develop a new hybrid electric power train to surpass the one developed by Toyota for its Prius vehicle. GM and BMW have been collaborating on the development of hydrogen refueling systems since May 2003, and Ford and PSA Peugeot Citroën have been working on small diesel engines since March 2000.

Vehicle manufacturers and suppliers have increasingly leveraged the Internet to solicit new ideas and technical solutions to specific problems. Online technology brokers, such as NINΣ, Yet2com, and InnoCentive, are like eBay for technology. Automakers and suppliers describe a problem in detail and request proposals (sometimes anonymously). Researchers from all over the world can offer solutions at various stages of development, from vague ideas to well tested technology. BMW has taken the search for outside solutions directly to its own website, where anyone can point out a problem or need and offer a solution.

Automakers have reached out to universities for decades, but the volume of research funding and depth of collaboration seem to be increasing. GM's collaborative research laboratories (CRLs) program, which was established in 2002, includes 10 long-term strategic relationships with professors or teams of professors at specific universities to focus research on specific technical areas. An electronics and controls CRL, with Carnegie Mellon University, is one of the largest; others include an engine technology CRL at the University of Aachen and a lightweight-materials CRL at the Indian Institute of Science. Ford and MIT have also established a multiyear, multimillion dollar research relationship. Toyota has pledged as much as $50 million to the Stanford University Global Climate and Energy Program.

THE ENGINEER'S PERSPECTIVE

At the working level, most automotive engineers interviewed reported that the single greatest change since 1990 has been the introduction of remarkable new tools that have changed their daily work routines. Most of these tools were enabled by tremendous advances in information and communications technologies. At first, in 1990, computer-aided design (CAD), which enables engineers to fit components together in a virtual three-dimensional space, and computer-aided engineering were specialty areas, and just a few engineers were taught to understand the software. Since then, design engineers have had far more exposure to these powerful systems. Today, every Ford product engineer either has a dedicated UNIX workstation at his or her desk or shares a UNIX machine with a neighboring engineer.

Access to information has also greatly improved. From the company intranet, engineers can access assembly plant quality data in real time and call up engineering prints, engineering specifications, and engineering test procedures. They can also assess critical data from suppliers.

The changing knowledge boundary between OEMs and suppliers has had a significant impact on both OEM engineers and supplier engineers. The role of engineers at vehicle manufacturers and suppliers has changed as the structure of the industry has changed. When Ford spun off many of its automotive-parts businesses to form Visteon, engineering work that had been done in house (e.g., axle engineering) was moved to the new company. The same thing happened when GM spun off Delphi. Several OEM engineers described the change as shifting from a designer of components and subsystems to a systems integrator. Several supplier engineers noted that their customers now grant them greater autonomy to design components (or even full-vehicle systems)—although the degree of autonomy varies by vehicle manufacturer.

Finally, some engineers stated that they are much more aware of potential legal liabilities related to their daily work than they were 10 years ago, which has changed the way they document information. Many engineers also mentioned that they feel pressured to work more efficiently today than they did 15 years ago, because fewer engineers seem to be doing more of the work.

Requirements for Entry-Level Engineers

The general requirement for entry-level engineers in the United States is a bachelor's degree in engineering or physics. However, some interviewees noted that the number of entry-level hires with master's degrees has increased.

The Supply of Qualified Engineers

Several press reports have suggested that the United States is losing its technological lead by graduating fewer engineers than India and China. Typical reports state that the United States graduated roughly 70,000 undergraduate engineers in 2004, while China graduated 600,000 and India graduated 350,000 (Figure 5). However, these numbers may be misleading. Duke University researchers determined that the data were not comparable. The numbers for China and India include graduates of three-year training programs and diploma holders, whereas the numbers for the United States include only graduates from four-year accredited engineering programs.

GLOBALIZATION

Historical Context

The automotive industry has been international since its earliest days. Daimler vehicles were produced under license in France in 1891, England in 1896, and America (New York City) in 1907.[6] Proximity to customers—wealthy individuals in the early days of craft production and mass markets in the days of mass production—has always been a key determinant for the location of vehicle-production facilities. The development of Henry Ford's system of mass production around 1910 was a key enabler of offshoring of vehicle-production facilities. Mass production, with its interchangeable parts, greatly reduced the amount of labor required to assemble a motor vehicle (and reliance on craft assembly skills). This led to a proliferation of automotive assembly plants around the world to gain access to new markets.

American automotive firms were pioneers in the early age of globalization. Both Ford and GM established their first production facilities outside the United States only one year after each company was founded. The early development of the "build where you sell" philosophy was driven by the high costs of shipping finished vehicles and later by increases in trade tariffs in the 1930s. To reduce transport costs, most early offshore assembly plants were based on the assembly of completely knocked down (CKD) kits. Ford could ship eight unassembled Model T CKD kits in the same amount of space that it could ship one completed vehicle. Table 1 shows the tremendous investment in offshore assembly plants made by Ford, GM, and Chrysler prior to 1929.

The appeal of CKD kits gained traction during the 1930s when higher tariffs and other trade restrictions were implemented by governments around the world. CKD kits were assessed at a lower tariff rate in exchange for the investment and employment provided by local CKD facilities. Eventually, offshore CKD plants began to procure components locally, especially in Europe where tariffs were high and markets were large.

Ford and GM followed different paths in Europe. Ford established wholly owned subsidiaries that were initially tightly controlled by Detroit. GM increased its European operations through acquisitions. In 1926, GM bought Vauxhall in England, and in 1929 the company bought Adam Opel AG in Germany; Opel was seized by the German government in 1940 and reclaimed by GM in 1948.

By the 1950s, both Ford and GM's European operations were largely autonomous; each had its own engineers who designed vehicles specifically for the European markets (and, in the case of GM, its own European brands). Each had developed extensive local supply chains and no longer relied on CKD units shipped from America. In fact, Ford and GM's operations in the United Kingdom and Germany were largely autonomous and organizationally distinct. The creation of Ford of Europe in 1967 by Henry Ford II, which forced the integration of Ford's German and British units, is considered one of the most significant reorganizations in the company's history.

The automotive industry in the mid-1960s was dominated by two large markets—America and Europe—and one emerging market—Japan. At the time, interregional trade in vehicles was insignificant. For the most part, Americans purchased vehicles manufactured by GM, Ford, Chrysler, and American Motors. In Europe, where national markets were far more distinct than they are today, the French bought French vehicles, the British bought British vehicles, and so on. A firm like Adam Opel, although it was owned by GM,

[6]For an excellent historical account of globalization in the automotive industry, see Sturgeon and Florida, 2000.

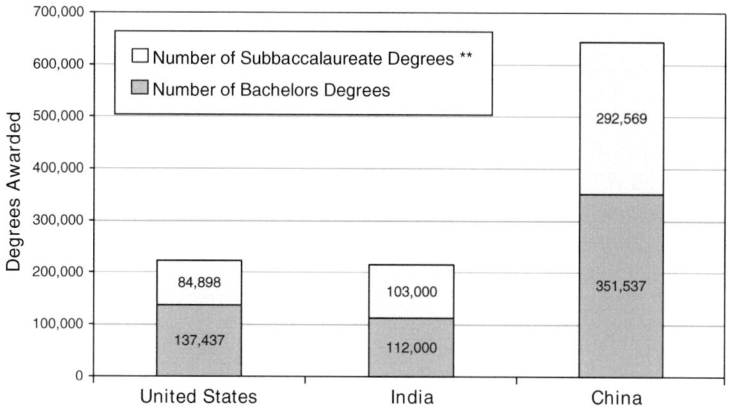

FIGURE 5 Engineering, IT, and computer science degrees awarded in the United States, India, and China (2004). Note: Subbacculaureate degrees refer to associate degrees in the United States, short-cycle degrees in China, and three-year diplomas in India. Source: Gereffi and Wadhwa, 2005.

TABLE 1 Ford, GM, and Chrysler Offshore CKD Assembly Plants as of 1928

Company	Number of Plants	Location of Plants (Year Opened)
Ford Motor Company	24	Canada (1904); England (1911); France (1913); Argentina (1915); Argentina (1919); Spain (1919); Denmark (1919); Brazil (1919); Belgium (1919); Sweden (1922); Italy (1922); South Africa (1923); Chile (1924); Japan (1924); Spain (1925); Germany (1925); France (1925); Australia (1925); Brazil (3 locations, 1926): Mexico (1926); India (1926); Malaysia (1926)
General Motors	19	Canada (1907); England (1908; not a CKD plant); Australia (1923); Denmark (1923); Belgium (1924); England (1924); Argentina (1925); England (1925); Spain (1925); Brazil (1925); Germany (1926); New Zealand (1926); South Africa (1926); Uruguay (1926); Indonesia (1926); Japan (1927); India (1928); Poland (1928); Sweden (1928)
Chrysler	3	Germany (1927); Belgium (1928); England (1928)

Sources: Rhys, 1972; Maxcy, 1981.

was largely managed and operated like a German company. The next big automotive production powerhouse—South Korea—had not yet appeared on the scene; Hyundai Motor Corporation was founded in 1967.

The automotive industry underwent a second wave of globalization starting around 1970, when international trade in motor vehicles—especially fuel-efficient Japanese vehicles—increased in response to the oil shocks of the 1970s. In the 1980s, foreign direct investment in manufacturing facilities increased. Honda opened the first transplant[7] in Ohio in 1982, beginning a wave of investment that continues today. Japanese manufacturers followed a similar pattern of investment in transplant production facilities in Europe a few years later. Beginning in the late 1980s, but greatly accelerating throughout the 1990s and the first few years of the 2000s, the world's automotive firms—both OEMs and suppliers—underwent a wave of mergers, acquisitions, and various kinds of strategic alliances.

Today, the level of business integration among vehicle manufacturers varies greatly. The list below is organized from the most integrated to the least integrated:

- **Merger/Acquisition:** Daimler Benz and Chrysler Corp. (until August 2007); Ford and Jaguar; Ford and Volvo; Volkswagen and Seat; Volkswagen and Skoda
- **Controlling Equity Stake:** Ford and Mazda; DaimlerChrysler and Mitsubishi Motors (until July 2005)
- **Non-controlling Equity Stake:** GM and Fiat Auto (until February 2005); GM and Fuji Heavy (until October 2005); DaimlerChrysler and Hyundai (until July 2005)
- **Product-Development Agreements/Shared Platforms:** GM Pontiac Vibe and Toyota Corolla (shared platform); Peugeot 107 and Toyota Aygo (small-car program)
- **Technology Alliances:** Ford and PSA on diesel engines; GM, BMW, and DaimlerChrysler on dual-stage hybrid vehicles; PSA and BMW on small gasoline engines

This evolution has blurred the distinction between domestic and foreign automakers in all countries, including the United States. Ford owns Jaguar, Volvo, and Land Rover and a controlling stake in Mazda. GM owns Saab and Daewoo and has only recently divested itself of equity stakes in several Japanese manufacturers. At the time this was written in

[7]A transplant is a foreign-owned manufacturing facility, such as a Toyota or BMW assembly plant, located in the United States.

2006, Chrysler was owned by DaimlerChrysler AG, a company based in Germany; 74 percent of DaimlerChrlysler's capital stock was owned by European investors, and the single largest shareholder was the Kuwait Investment Authority (DaimlerChrysler, 2005). Some of these international relationships are considered great successes (e.g., Renault-Nissan), but many are considered failures that have destroyed shareholder value (e.g., GM-Fiat, Ford-Jaguar).

Current Level

Although traditional global business relationships in the industry are breaking down (e.g., the GM-Fiat relationship has been terminated), the automotive industry today is more globally integrated than ever. Figure 6 shows the percentages of employment, sales, and production outside the home country for the top 10 vehicle manufacturers (in terms of 2005 global sales). Because these 10 vehicle manufacturers account for about 83 percent of global sales, we can draw some conclusions from these data:

- All 10 automakers sold more vehicles outside their home markets than in their home markets. In 2005, for the first time, GM sold more than half of its vehicles outside the United States; the average for both U.S.-based automakers, Ford and GM, is slightly more than half. For the other eight manufacturers, the percentages range from about 70 to 80 percent.
- Among these 10, the lowest percentage of sales, production, or employment outside the home country was about 38 percent, but the percentages for all of them are increasing. While GM and Ford sales are declining in their home market (USA), their competitors' share in the U.S. market is growing.

We can also look at globalization from the market perspective—how open major national and regional automotive markets are to foreign-brand or foreign-made products. Figure 7 shows 2005 sales in the U.S. market divided into four-categories: foreign-owned foreign-brands (e.g., Honda); foreign-owned domestic brands (e.g., Chrysler); domestic-owned foreign brands (e.g., Volvo); and domestic-owned domestic brands (e.g., Chevrolet). In 2005, 54 percent of the vehicles sold in the United States were sold by foreign-owned firms.

Table 2, which compares U.S. data with data from Western Europe, Japan, and Korea, shows that the U.S. market is the most open, but penetration of foreign brands and foreign-owned domestic brands in other developed markets is increasing. Japanese automakers are following a similar pattern of building transplants in Europe.[8] The 26.6 percent penetration of foreign brands in Western Europe includes Chrysler vehicles, but not Opel vehicles (owned by GM). The 38.2 percent penetration of foreign-owned vehicles includes Opel vehicles, but not Chrysler vehicles. The 9.0 percent figure for Japan includes Mazda vehicles (controlled by Ford), and the 26.2 percent for South Korea includes Daewoo vehicles (controlled by GM).

The U.S. Market

Competition from foreign automakers in the United States has steadily increased providing more choices for U.S. consumers:

- Since 1980, several foreign brands have entered the U.S. market or dramatically increased their share. Foreign automakers have attacked their U.S. competitors on all fronts. In 1986, Honda made a strong move to attract upscale consumers when it introduced the Acura brand in the United States. Toyota followed suit with the introduction of the Lexus brand in 1989, the same year Nissan launched the Infiniti brand.
- New market segments are being created. Toyota moved toward the downscale/hip-youth segment with the introduction of the Scion brand in 2004. DaimlerChrysler introduced the Maybach, a new super luxury car that costs more than $300,000.
- Manufacturers are offering more models to cover all market segments. Low-end producer VW tested the U.S. market with the high-end Phaeton, while high-end producers Audi and BMW have introduced lower cost models, such as the Audi A3 and the BMW 1-series.
- The threat of reentries also looms large. Speculation is rampant that both French automakers—Renault and PSA Peugeot Citroën—will soon reenter the U.S. market.
- The Koreans have also entered the fray. In 1986, Hyundai entered the U.S. market but retreated in the early 1990s because of problems with quality. Over the past five years, however, U.S. sales of Hyundai vehicles have come roaring back as quality has greatly improved. Hyundai also acquired majority ownership in Kia Motors in 1998, and by 2005, Hyundai/Kia U.S. market share had increased to 4.3 percent.

Figure 8 shows the increases in sales of foreign-brand vehicles, at the expense of domestic brands, in the United States in the past 25 years. The combined U.S. market share of the traditional Big 3 automakers since the mid-1980s steadily declined to 58.5 percent in 2005. In 1985, GM's market share was slightly more than 40 percent; that figure had dropped to 25.8 percent in 2005. In 1985, Ford was number two with about 22 percent of the market. Ford's share crept up to about 26 percent in the mid-1990s but had dropped back to 18.2 percent by 2005. DaimlerChrysler's 2005 U.S. market share of 14.5 percent is nearly identical to the 1985 market share

[8]Japanese automakers operated 16 transplants (assembly plants) in European Union member countries in 2006, producing over 1.5 million vehicles (more than double the production for 1995). Japanese automakers operated 13 R&D centers in European Union member countries in 2006 (JAMA, 2007).

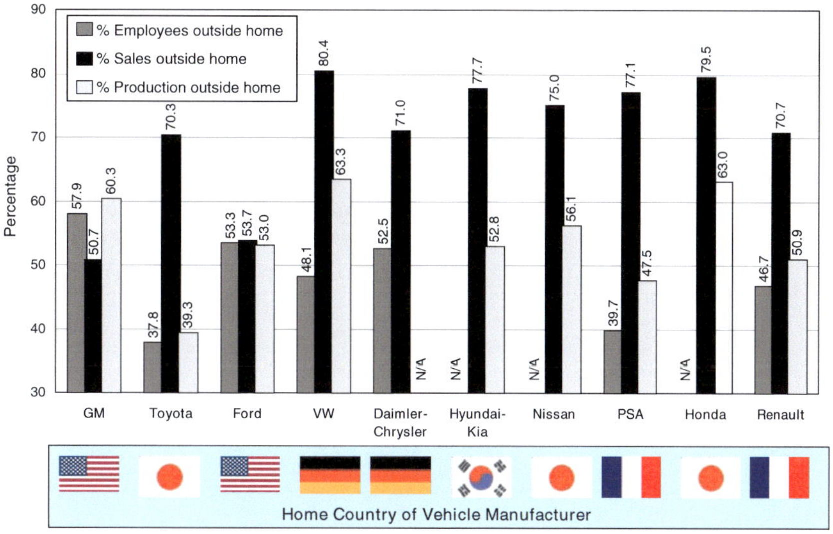

FIGURE 6 Percentage of employees, sales, and production outside home country for the top 10 global automakers. Source: Compiled from annual reports and market literature and Automotive News, 2005.

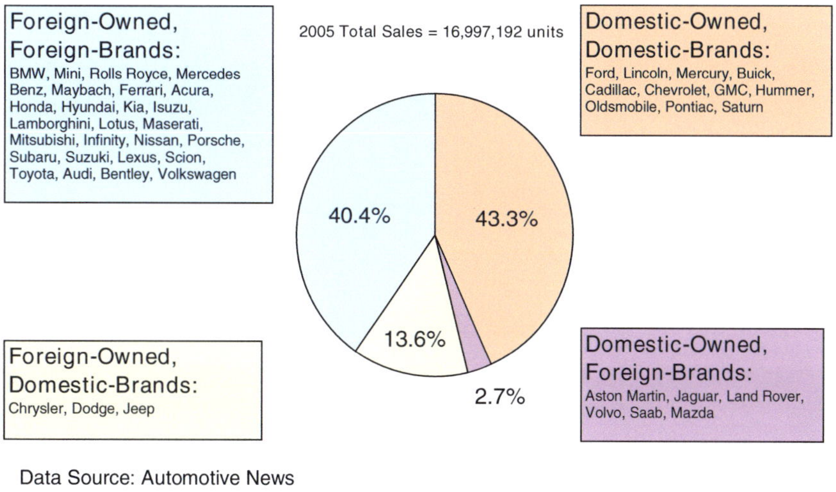

FIGURE 7 U.S. vehicle sales by category, 2005. Source: Automotive News, 2005.

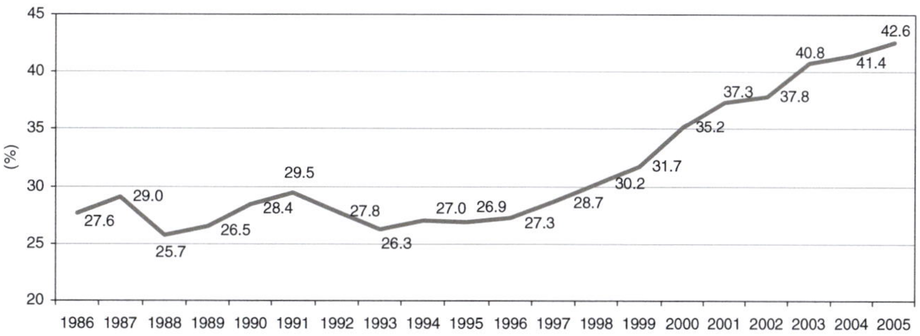

FIGURE 8 Foreign-brand market share in the United States, 1986–2005. Note: Includes domestic-owned foreign-brands, such as Volvo (Ford) and Saab (GM). Source: Automotive News data, 2005.

TABLE 2 Foreign Penetration in Four Developed Markets, 2004

Country or Region	Penetration by Foreign Brand (%)	Penetration by Foreign Ownership (%)
United States	41.3	51.2
Western Europe	26.6	38.2
Japan	4.2	9.0
South Korea	2.3	26.2

Data sources: ACEA, 2004; JAMA, 2004; KAMA, 2004.

for Chrysler Corporation. The combined share for Japanese brands steadily increased from about 20 percent in 1985 to almost 34 percent in 2005.

As shown in Figure 9, U.S. sales of foreign-brand vehicles were driven by imports through the mid-1980s, when they were supplemented by transplant-produced vehicles. Figure 10 shows the 17 transplants now sold in the United States—14 from Japanese OEMs, one Korean OEM (Hyundai), and two German OEMs (Mercedes Benz and BMW).

As of early 2005, transplants employed about 65,000

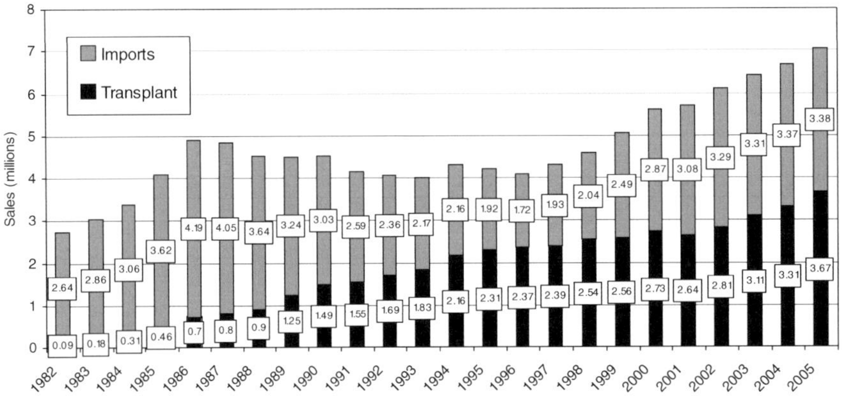

FIGURE 9 U.S. sales of foreign-brand vehicles transplant-produced and imports, 1982–2005. Source: Adapted from Center for Automotive Research study prepared for Association of International Automobile Manufacturers Inc.; Automotive News data; U.S. Department of Commerce; IMVP.

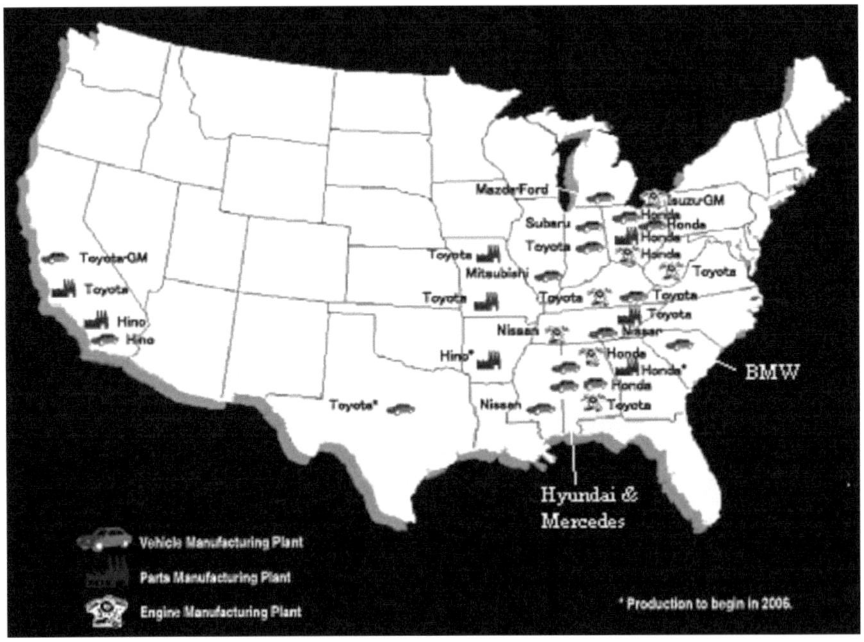

FIGURE 10 Transplants in the United States. Source: IMVP, 2004; JAMA, 2004.

TABLE 3 North American Assembly Plant Footprint as of October 2006

Manufacturer	United States	Canada	Mexico	North America Total
GM	17	1	3	21
Ford	10	2	2	14
DaimlerChrysler	8	2	2	12
Other OEMs	14	3	7	24
Totals	49	8	14	71

Notes: Locations that include two assembly plants, such as Honda in Lincoln, Alabama, and Toyota in Princeton, Indiana, counted only once above. Mercedes plant in Alabama included with DCX USA. This accounts for the difference between 14 U.S. transplants shown above and 17 cited previously. Other OEMs USA includes NUMMI Toyota-GM facility and AutoAlliance Ford-Mazda facility. Other OEMs Canada includes CAMI GM-Suzuki facility.

Sources: Automotive News, 2005, and company reports.

people and accounted for a cumulative investment of more than $27 billion, and these figures have rapidly increased since then. In April 2006, Toyota announced a major expansion of its Indiana plant. In June 2006, Honda announced it would build a new assembly plant in Indiana to begin production in 2008. Kia (a brand of Hyundai) broke ground for a second assembly plant in Georgia in October 2006.

During that same period, Ford closed its St. Louis and Atlanta assembly plants, and GM closed its Oklahoma City plant. The assembly plant footprint in North America as of October 2006 is shown in Table 3.

Figure 11 shows light-vehicle production for domestic plants and transplants in the United States since 1982. Overall U.S. production has hovered around 12 million vehicles since 1994, so in a sense, the industry remains relatively healthy. However, Figure 11 shows a gradual, but relentless shift from domestic plants to transplants, which produced a record 3.58 million vehicles in the United States in 2005. By 2006, when the new Hyundai plant in Alabama and the new Toyota plant in San Antonio had ramped up production, the figure had risen to almost 4 million units. Thus roughly one of every three vehicles built in the United States is from a foreign company.

Following the "power train is core business" mantra, all major vehicle manufacturers engineer and manufacture engines and transmissions. However, OEMs are increasingly sharing engine and transmission programs or obtaining them from other manufacturers. A report by the Center for Automotive Research estimated that the engine-production capacity of foreign-brand automakers in 2003 was 3.5 million units, 30.5 percent of the overall capacity in the United States (Center for Automotive Research, 2005). Honda has major engine-manufacturing facilities in Anna, Ohio, and Lincoln, Alabama; Nissan has an engine plant in Decherd, Tennessee; and Toyota has engine plants in Georgetown, Kentucky; Huntsville, Alabama; and Buffalo, West Virginia. In 1996, a similar report had estimated the total engine-production capacity of foreign-brand automakers at 1.5 million units. Hence, over an eight-year period, foreign engine-production capacity increased by 133 percent.

Although globalization in the United States has been disruptive for automakers and parts suppliers, it has generated tremendous benefits for U.S. consumers: (1) Americans have more vehicle-model choices than ever before; (2) manufacturing productivity and quality levels have improved and converged among all automakers; (3) vehicle prices have fallen in real terms; and (4) significant product enhancements in safety, environmental impact, and performance have been made.

The Automotive-Supplier Industry

Since the 1990s, suppliers of components—a critical link in the automotive value chain—have also undergone relentless globalization. Nowadays, vehicle manufacturers "shop at the global mall"—that is, they purchase components

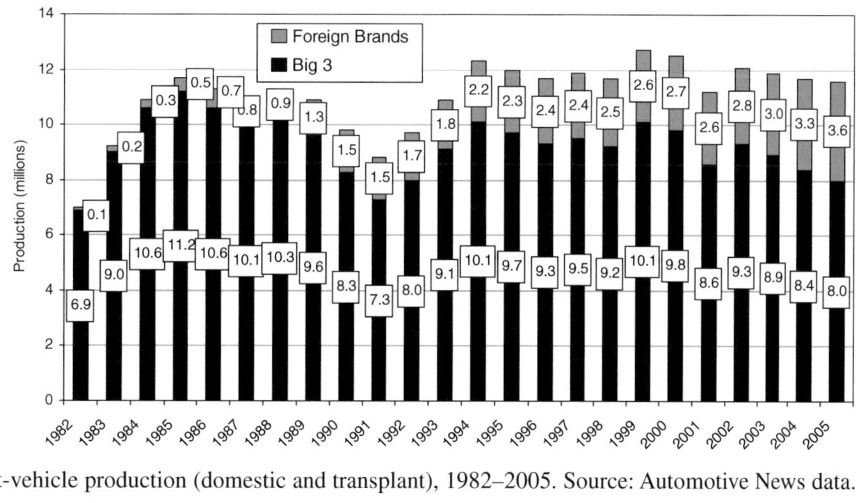

FIGURE 11 U.S. light-vehicle production (domestic and transplant), 1982–2005. Source: Automotive News data.

TABLE 4 Top 20 Global Automotive Suppliers by Sales to Automotive OEMs, 2005

Company	Home Region	2005 Sales to Auto OEMs (US$ billion)	North America (%)	Europe (%)	Asia (%)	Totals (%)
Robert Bosch GmBH	EU	28.4	17	69	14	
Denso Corp.	JP	22.9	21	14	64	1
Magna International Inc.	NA	22.8	56	43		1
Delphi Corp.	NA	22.6	71	21	7	1
Johnson Controls Inc.	NA	19.4	46	47	7	
Aisin Seiki Co. Ltd.	JP	17.9	18	8	73	1
Lear Corp.	NA	17.1	54	38		8
Visteon Corp.	NA	15.9	61	24	12	3
Faurecia	EU	14.0	11	81	4	4
TRW Automotive Inc.	NA	11.7	38	54		8

Source: Automotive News, 2005.

from locations around the globe, regardless of where suppliers' headquarters are located. Globalization has advanced on two levels: (1) suppliers have followed their traditional home-market customers to other parts of the world (e.g., Denso, a large Japanese supplier, followed Toyota to the United States); and (2) suppliers have focused on winning business from OEMs based in other parts of the world. Mergers, acquisitions, and spin-offs have led to the creation of "mega-suppliers."

Table 4 shows that the top 10 suppliers have significant sales volumes outside their home regions. An analysis of the top 100 global suppliers (based on 2004 data) reveals that 38.3 percent of total sales were to customers outside their home markets, with North American suppliers topping the list:

- 41.2 percent of sales by North American suppliers in the top 100 were to customers outside North America
- 38.2 percent of sales by Japanese suppliers in the top 100 were to customers outside Japan
- 35.2 percent of sales by European suppliers in the top 100 were to customers outside Europe

A study by the Federal Reserve Bank of Chicago analyzed the ELM Guide database, which tracks suppliers to the U.S. auto industry (Klier, 1999). As shown in Table 5, in 1997, at least 60 percent of suppliers to transplants were domestic. The study also sheds light on the importance of geographic proximity to customers.

Table 6 is based on a study in 2004 by McKinsey and the Original Equipment Suppliers Association of 57 large suppliers operating in North America. The study showed that both European and North American suppliers wanted to reduce their reliance on North American OEMs, from a high of about 80 percent in 2003 to about 60 percent in 2008. Japanese suppliers indicated that they were content with their customer mix, which includes about 40 percent North American OEMs. In general, diversification of a supplier's customer base away from its traditional home region seems to make good financial sense. Some reports indicate that OEM financial performance improves as reliance on business with the Detroit 3 decreases (e.g., Casesa et al., 2005).

The complexity of the global supply base makes measuring the local content of most modern automobiles nearly impossible. U.S. vehicles contain thousands of components from European and Japanese suppliers, each of which is

TABLE 5 1997 Proximity of Suppliers to Transplants: Number of Suppliers, Median Distance to Assembly Plant, and Percentage Domestic

Company	Location	Start-Up Year	Number of Suppliers	Median Distance (miles)	Domestic (%)
Honda	Marysville and East Liberty, OH	1982	507	251	65
Toyota	Georgetown, KY	1988	452	285	69
Subaru-Isuzu	Lafayette, IN	1987	292	245	60
Diamond-Star (Mitsubishi-Chrysler JV)	Normal, IL	1988	286	309	63
AutoAlliance (Ford-Mazda JV)	Flat Rock, MI	1987	360	242	71
Nissan	Smyrna, TN	1983	460	423	70
BMW	Spartanburg, SC	1994	119	477	75
Mercedes Benz	Vance, AL	1997	77	610	68
NUMMI (Toyota-GM Joint Venture)	Freemont, CA	1984	178	1,966	60
Saturn (GM)	Spring Hill, TN	1990	300	462	81
Ford (1970–1980)	Dearborn, MI	n/a	222	405	89
Ford (1983–1993)	Dearborn, MI	n/a	301	200	77

Source: Klier, 1999.

TABLE 6 Customer Mix for 2003 and 2008 (projected)

	European Suppliers		Japanese Suppliers		North American Suppliers	
Customers	2003	2008 (Projection)	2003	2008 (Projection)	2003	2008 (Projection)
Korean OEMs	0	4	0	2	1	5
Japanese OEMs	6	17	60	55	8	14
European OEMs	14	18	2	5	11	24
North American OEMs	79	61	38	38	80	57

Source: McKinsey and OESA, 2004.

built from smaller components and materials from around the world. Consider, for example, the 2005 Dodge Dakota shown in Figure 12.

Foreign-brand automakers are major customers of U.S. suppliers. The report by the Center for Automotive Research estimated that, in 2003, foreign-brand automakers purchased $66.7 billion worth of goods and services from suppliers in the United States. Of this total, $49.1 billion was for manufacturing/production purposes, and $17.6 billion was for non-production purposes (e.g., engineering and design, sales, distribution, finance, and port services). As Figure 13 shows, purchases by foreign automakers from U.S. suppliers increased rapidly from 1986 to 2005.

U.S. ENGINEERING WORKFORCE

Overall Employment in the U.S. Auto Industry

The automotive industry is one of the biggest employers in the United States. According to data from the Bureau of Labor Statistics (BLS), the automotive manufacturers and suppliers directly employ roughly 1.1 million people (not including sales, service, etc.). Figure 14 shows total employment for vehicle manufacturers and vehicle and parts manufacturers. Overall employment in the parts sector has declined slightly more (17.4 percent) than in the vehicle manufacturing sector (15.5 percent).

FIGURE 12 Some non-U.S. suppliers to the 2005 Dodge Dakota. Source: Automotive News, 2005.

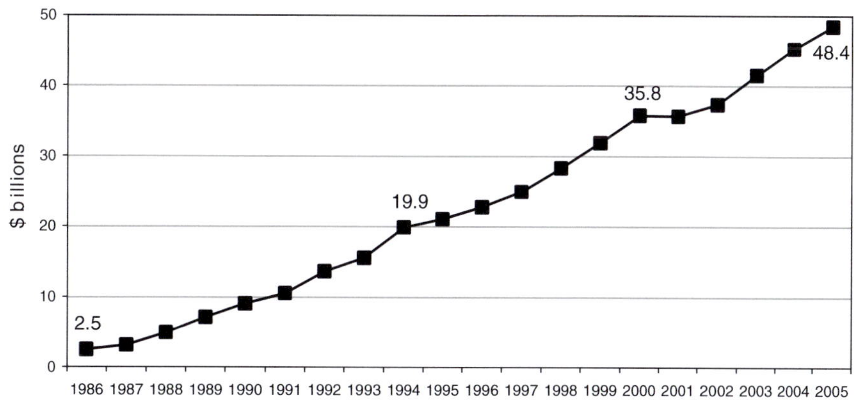

FIGURE 13 Purchases of U.S. parts by Japanese automakers, 1986–2005. Source: JAMA, 2006.

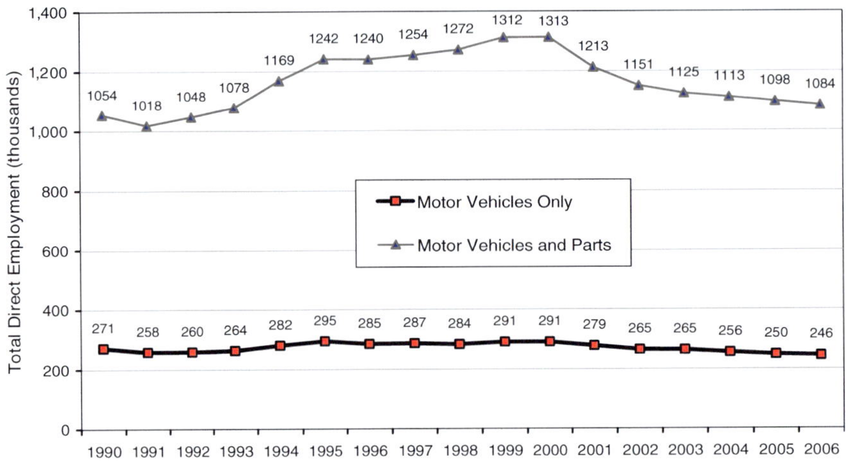

FIGURE 14 Employment in the U.S. automotive industry, 1990–2006. Note: 2006 data are for first half 2006 for NAICS code 3361 for motor vehicles and 3361, 3362, and 3363 for motor vehicles and parts. Source: BLS, 2006.

However, according to GM and Ford annual reports (GM, 2005; Ford Motor Company, 2005a), the U.S. automotive workforce has undergone dramatic changes:

- In January 2006, Ford announced its Way Forward plan, which included idling 14 manufacturing facilities and reducing employment by 25,000 to 30,000. The plan calls for reducing North American production capacity by 1.2 million units, or 26 percent, by 2008. The company also announced a 10 percent reduction in salaried costs in North America and a related reduction in head count of 4,000. Ford recently announced plans to accelerate the Way Forward restructuring plan by slashing its North American workforce by 44,000 and reducing fourth-quarter 2006 production by 21 percent.
- In 2005, GM announced plans to close 12 U.S. assembly plants by 2008 and reduce its manufacturing workforce by 30,000. This will reduce GM's U.S. manufacturing capacity by about one million units. GM had already reduced its U.S. manufacturing capacity by about one million units between 2002 and 2005. GM's U.S. salaried workforce (including contract staff) had been reduced by 33 percent since 2000.
- At the same time, Hyundai, Toyota, and Honda were building new plants and increasing their total employment and R&D workforce in the United States.

U.S. automotive suppliers are under tremendous financial pressure as a result of these production cuts by domestic manufacturers and downward price pressure by their OEM customers coupled with rising costs for steel, aluminum, resins, and other materials used to make automotive components. Several U.S. automotive suppliers have filed for bankruptcy in the past few years (Table 7), many accompanied by substantial reductions in employment.

Engineering Employment

It is extremely difficult to estimate the number of automotive engineers in the United States using BLS data. There are three NAICS codes for the automotive industry: 3361

TABLE 7 Recent Bankruptcies of U.S. Automotive Suppliers

Company	Date of Filing	Total Assets	Number of Employees
Delphi Corporation, Troy, Michigan	October 8, 2005	$17.1 billion	185,000
Federal-Mogul, Southfield, Michigan	October 1, 2001	$10.1 billion	50,000
Dana Corporation, Toledo, Ohio	March 3, 2006	$7.9 billion	46,000
Collins & Aikman Corporation, Troy, Michigan	May 17, 2005	$3.2 billion	23,900
Hayes Lemmerz, Northville, Michigan	December 5, 2001	$2.8 billion	15,000
Tower Automotive, Novi, Michigan	February 2, 2005	$2.6 billion	12,891
Dura Automotive Systems, Rochester Hills, Michigan	October 30, 2006	$2.0 billion	15,200
Venture Holdings, Fraser, Michigan	March 28, 2003	$1.4 billion	12,980
Oxford Automotive,[a] Troy, Michigan	December 7, 2004	$1.0 billion	3,800

[a]Oxford Automotive also filed for bankruptcy on January 18, 2002.
Sources: Automotive News, 2005; *BankruptcyData.com*; company reports.

TABLE 8 U.S. Employment of Automotive Engineers (excluding R&D engineers)

Occupational Code	NAICS 3361: Motor Vehicle Manufacturing	NAICS: Motor Vehicle Body and Trailer Manufacturing	NAICS 3363: Motor Vehicle Parts Manufacturing	Total of All Three NAICS Codes
Engineering Managers	610	570	3,960	5,140
Industrial Engineers	3,390	1,240	14,460	19,090
Mechanical Engineers	1,920	1,360	9,300	12,580
Electrical Engineers	150	110	910	1,170
All Other Engineers	n/a	180	7,200	7,380
Total	6,070	3,460	35,830	45,360
All Occupations	256,700	168,840	693,120	1,118,600

Source: BLS, 2005.

(motor-vehicle manufacturing), 3362 (motor-vehicle body and trailer manufacturing), and 3363 (motor-vehicle parts manufacturing). Table 8 shows the U.S. employment levels for various types of engineers for all three codes. However, engineers whose primary function is R&D (i.e., all product engineers) are not included. R&D engineers in the automotive industry fall under NAICS 5417 (scientific research and development services). Therefore, numbers in Table 8 are mostly for manufacturing engineers.

A BLS career brief, *Motor Vehicle and Parts Manufacturing*, compiled using May 2004 data, estimates other engineering employment in the same three NAICS codes at 18,000, which is significantly higher than the roughly 8,600 shown in Table 8. Using this figure, we can estimate the total number of manufacturing engineers in the automotive industry in 2005 at 55,000.

To estimate the number of product engineers, we can use a bottom-up approach. Table 9 shows that an estimated 34,000 engineers and technicians work for vehicle manufacturers (not parts makers) in the United States. Assuming the same ratio of supplier engineers to vehicle manufacturer engineers as the ratio of supplier employees to vehicle manufacturer employees (roughly three to one), we can estimate that at least 100,000 engineers and technicians support the automotive supply base in the United States. In addition, many firms (e.g., Motorola, Siemens, IBM, et al.) that supply the auto industry do not fall under an auto industry SIC or NAICS code. Therefore, the estimate of 100,000 is likely to the lower boundary.

If we combine the figures in Tables 7 and 8, we can estimate that 189,000 product and manufacturing engineers are employed by the automotive industry in the United States (Table 10). This estimate is also probably on the low side because many engineers work for the automotive businesses of large firms that also serve other industries (e.g., DuPont and Siemens).

TABLE 9 U.S. Engineering/Technical Employment for Major Vehicle Manufacturers, 2006

Company	Current Number of Engineers and Technicians	Projection
General Motors	11,500	Decreasing
Ford Motor Company	12,000	Decreasing to 10,000
DaimlerChrysler	6,500	Steady
Japanese companies	3,593	Increasing rapidly
Korean (Hyundai-Kia)	200	Increasing rapidly
German (BMW)	150	Increasing
Total	About 34,000	

Notes: Technicians may be included. Japanese data includes designers.
Sources: Company reports and interviews; JAMA, 2004.

Engineering Wages

Based on the BLS database, the annual mean salaries for the weighted average of NAICS codes 3361, 3362, and 3363 for May 2005 are $95,872 for engineering managers; $66,284 for industrial engineers, and $65,861 for mechanical engineers. In a National Science Foundation report, *Scientists, Engineers and Technicians in the United States: 2001*, estimates of mean annual wages were $90,086 for managers of science, engineering, and technical (SET) personnel, $61,637 for scientists, $63,107 for engineers, and $46,947 for technicians.

Mean annual wages for engineering managers, industrial engineers, and mechanical engineers based on BLS occupational wage and employment estimates for the past eight years (Figure 15), shows that wages for U.S. engineers have been gradually increasing. However, this figure should be viewed with some caution because the survey was designed for cross-sectional analysis rather than time-series analysis.

TABLE 10 Estimate of Overall U.S. Engineering Employment

Industry Sector	Product Engineers	Manufacturing Engineers	Total
OEMs	34,000	10,000	44,000
Suppliers	100,000	45,000	145,000
Total	134,000	55,000	189,000

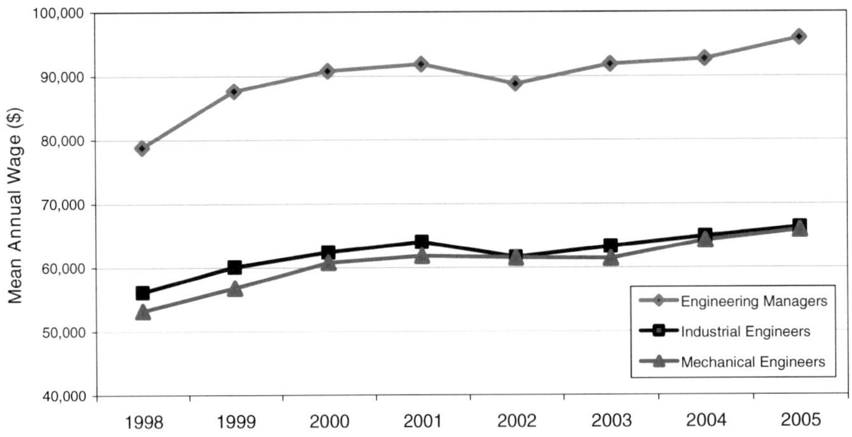

FIGURE 15 Annual mean wages for engineering occupations in the U.S. automotive industry. Note: 1998–2001 data are for SIC 3710, Motor Vehicles and Equipment. 2002–2005 data are a weighted average of NAICS 3361, Motor Vehicle Manufacturing; NAICS 3362, Motor Vehicle Body and Trailer Manufacturing; and NAICS 3363, Motor Vehicle Parts Manufacturing. Source: BLS, 2005.

THE GLOBAL FOOTPRINT OF AUTOMOTIVE ENGINEERING

Definition of Offshoring

The word *offshoring* is ambiguous and is frequently used interchangeably with *outsourcing*. Figure 16 is a matrix that can help distinguish between offshoring and outsourcing. For this example, we can adopt the perspective of GM. GM's Technical Center in Warren, Michigan, can outsource certain technical functions to one of many Detroit-area contract-engineering firms, such as MSX International or Kelly Services. These contract engineers frequently work side by side with GM engineers in the Warren tech center; this is called local outsourcing. GM can also share engineering functions with one of the 12 GM engineering centers outside the United States and can move work among those centers as it wishes. GM can also "source" engineering work to an overseas outside firm, such as Wipro Technologies in India. Finally, GM can share engineering functions with a joint-venture partner, such as the Pan-Asian Technical Automotive Center (PATAC) in Shanghai, a joint venture between GM and Shanghai Automotive Industry Corporation (SAIC).

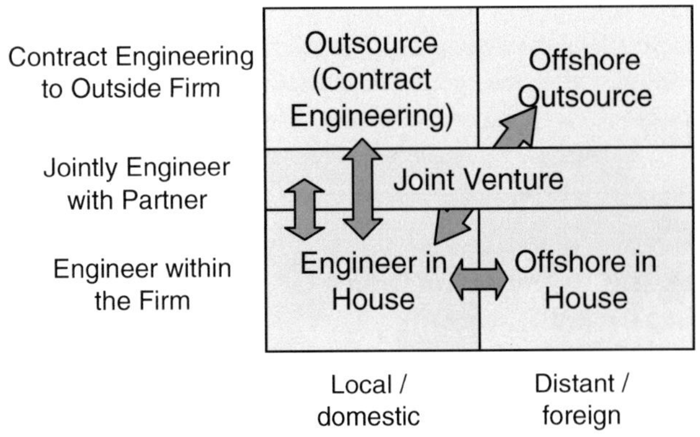

FIGURE 16 Matrix for defining offshoring and outsourcing.

The point is that all vehicle manufacturers employ engineers in all four quadrants of the matrix. To optimize overall engineering efficiency (i.e., to make the most efficient and effective use of engineering resources, both inside and outside the firm, local and distant), management shifts functions to any of the four quadrants in the matrix. Thus the arrows point in both directions. GM can choose to bring an engineering function back to Warren just as easily as it can send it out of Warren.

The term *offshoring* is frequently used to imply the *replacement* of U.S. workers with foreign workers. As Figure 16 shows, replacement is possible, but it is only one part of a much bigger story.

Management of the Footprint

Many factors are involved in deciding where to locate product engineers and manufacturing engineers for a given firm. Industry managers identified four critical factors considered by both vehicle manufacturers and suppliers in determining their footprint strategies: customer, cost, capability, and government policy. This is called the 3C+G Footprint Model (Figure 17).

Because manufacturing engineers are typically located near the production site, the production footprint can be used as a proxy for the manufacturing-engineering footprint. In other words, the factors that determine the location

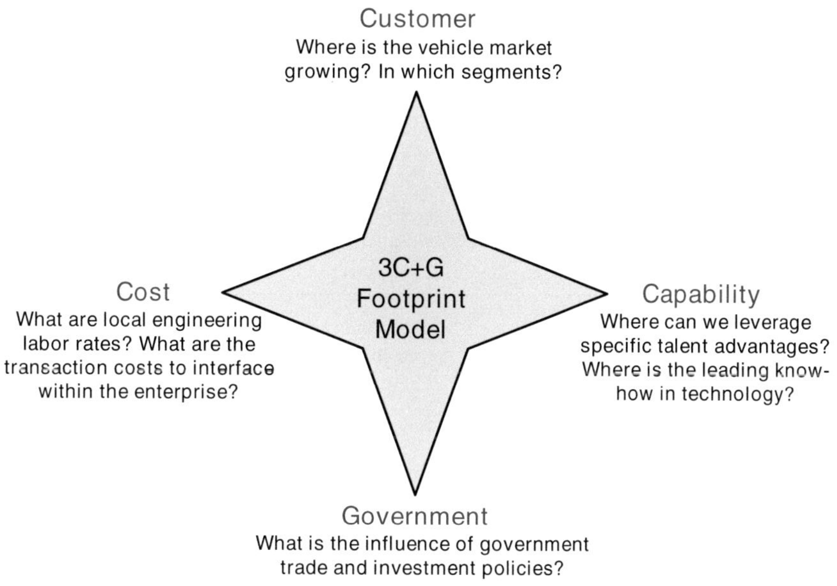

FIGURE 17 3C+G global footprint model for the automotive industry.

of production facilities will also determine the location of manufacturing engineers.

For determining the location of product engineers, the same factors are involved, but they are weighted differently (Figure 18). For example, government policy has a stronger influence on an automotive firm's production footprint than on its R&D footprint. Trade policy in particular is a major factor in the "build where you sell" strategy described in the section on globalization.

THE VALUE OF PROXIMITY

Before assessing each of the 3C+G factors, it is important to understand the value of proximity in the automotive-engineering world. Why is it valuable for vehicle manufacturers to "build where they sell"? Is it also valuable to "engineer where they sell"?

Factor	Influence on Manufacturing Engineering Footprint	Influence on Product Engineering Footprint
Customer	High	Medium
Cost	Medium	Medium
Capability	Low	High
Government	High	Low

FIGURE 18 Impact of 3C+G factors.

Vehicle Manufacturing

Transport Costs

As described in the section on globalization, minimizing transport costs is a key motivator for localizing vehicle-production facilities. Although transport costs have declined relative to the average cost of vehicles, transporting automobiles is still expensive.

Trade Policy

Trade policy is always a key factor in the localization of production. U.S. trade policy has contributed to the rise of transplants in the United States. In the early 1980s, the Big 3 automakers and the United Auto Workers (UAW) Union pressured the U.S. government to limit the import of Japanese vehicles. In response to this pressure, the Japanese Ministry of International Trade and Industry announced a Voluntary Restraint Agreement (VRA) that limited Japanese exports of vehicles to the United States.

VRA backfired, however, for several reasons. First, the U.S. limits applied only to imported vehicles and not to vehicles built in the United States. Thus they provided strong incentives for the development of transplants. Second, VRA was based on the volume of vehicles rather their value, thus providing an incentive for Japanese manufacturers to develop upscale and luxury vehicles for export to the United States (e.g., Acura, Lexus, and Infiniti). Third, Japanese manufacturers realized enormous profits (estimated from $4 billion to $7 billion per year for 1981 to 1985) on their high-demand, VRA-limited vehicles (Ries, 1993; Smitka, 1999). Thus the net effect of VRA was completely contrary to the protections sought by domestic manufacturers.

Currency Risk

Localizing production is a hedge against currency risk. Recent declines in the U.S. dollar versus the Euro have hurt European manufacturers that export large numbers of vehicles to the United States, such as BMW and Mercedes. Currency risk was a factor in the decisions of both BMW and Mercedes to build U.S. transplants in the mid-1990s.

Foreign-exchange rates were also a major factor in the increase in production by Japanese transplants in the United States in the late 1980s. In February 1985, the yen traded at an average daily rate of 260.5 to the dollar. By May 1987, the yen was trading at an average daily rate of 140.5 to the dollar, which greatly reduced the purchasing power of American consumers for imported Japanese products. This dramatic shift in the exchange rate provided an additional incentive for Japanese companies to invest in U.S. transplants.

Company Reputation and Political Influence

Localized production also improves a company's reputation in the local community and increases its local political influence. Honda has been manufacturing cars in Ohio for more than two decades, and many locals say that it now "feels like" an American company. Some buyers are more likely to buy a foreign-brand vehicle if it is built in America.

Bringing an automotive assembly plant to a community is a politician's dream. Many automotive assembly plants, even modern ones, employ an average of 3,800 people directly, and even more indirectly. In addition, assembly plants tend to be located at the confluence of major highways where they are visible to voters in the local community. Most recent transplants in the United States have also been offered tax breaks or other fiscal incentives to attract investment from state and local governments. In some cases, local governments compete against each other, offering increasingly lucrative fiscal incentives to bring a plant to their community. These factors combined heighten the impact of customer location on production location—and, therefore, on manufacturing-engineering location (Table 11).

Supplier Location

Suppliers have many incentives to locate their production facilities close to their OEM customers. The Lean Location Logic Project (of the International Motor Vehicle Program [IMVP]) was developed to interview managers to assess how suppliers make location decisions. The primary focus

TABLE 11 Value of Proximity

	Production (Manufacturing Engineering)	R&D (Product Engineering)
OEMs close to customer/ market (value of proximity)	Lower transport costs Lower trade barriers (trade policy) Improved political position/reputation Lower currency risk	Localization Engineering of regional-specific vehicles
Suppliers close to OEMs (components with high proximity value)	Components that are bulky and relatively expensive to ship (e.g., fuel tanks) Components that require sequenced just-in-time delivery to assembly plant (e.g., seat sets) Components that require careful production coordination with assembly plant (e.g., bumpers require careful color matching with assembly plant paint shop)	Components that are highly integral to the vehicle architecture

of the study was on production-location decisions, but the interviews also revealed a good deal about location decisions for engineering and design processes. During the interviews, it became clear that suppliers tend to follow their OEM customers to achieve "lean flow," an underlying principle of lean production. Proximity of suppliers to vehicle manufacturers may also support an OEM's decision to build lower volume, more flexible assembly plants (e.g., Womack and Jones, 2003).

Proximity to a vehicle-assembly plant is especially important for certain types of components, such as bulky items that are expensive to transport, fuel tanks, and built-up exhaust systems. In addition, components that must be delivered to the assembly plant in a precise sequence, such as seat sets (i.e., the combination of driver, passenger, and rear seats for a particular vehicle), are almost always produced close to the final assembly plant. Seat suppliers are given the build sequence—the particular vehicles that will be built in a given time period—usually with only a few hours notice. The seat sets are loaded into trucks and delivered to the assembly plant just-in-time for installation in the vehicle. Production facilities for bumpers, side mirrors, door cladding, and other components that require precise color matching are also located close to assembly plants (Table 11).

With the *insourcing* of foreign manufacturers to the United States, foreign OEMs have "pulled" their suppliers to the locations of new assembly plants. In 2005, when Hyundai invested more than $1 billion to build a new plant in Montgomery, Alabama, the company brought along several of its suppliers. Some, such as the large Mobis plant just down the road from the Hyundai assembly plant, had been part of Hyundai's traditional supply base. The Mobis plant produces front-end modules, chassis modules, cockpit modules, and so on—large, built-up chunks of vehicles. Hyundai also attracted U.S. suppliers to its new assembly plant in Alabama. For example, Lear, a tier-one American supplier, built a "state-of-the-art" seating factory to supply seat sets exclusively for the Hyundai assembly plant located just minutes away by truck.

Engineering and Proximity

Market Factors

Although proximity to customers is significantly less important for product engineering than for manufacturing engineering, customer proximity can be important under certain circumstances. First, "engineer where you build" makes sense if the engineered product is primarily consumed in the local market. "Localization engineering," for example, is the adaptation of a vehicle engineered in country A to meet the unique regulatory and customer requirements of country B. For example, a Buick engineered predominantly at GM's Warren, Michigan, engineering center but manufactured and sold by Shanghai GM (GM's joint-venture operation in China), may undergo local engineering changes that require different components (e.g., Chinese customers may prefer more chrome in the vehicle interior). These changes can most efficiently be made by GM's Chinese engineers in China.

Second, it makes sense to "engineer where you sell" if a vehicle will be sold predominantly in the local market. For example, the United States is the largest market for pickup trucks in the world, so it is very unlikely that GM or Ford will ever shift their engineering centers for pickup trucks outside the United States (Table 11).

Honda, however, whose sales of pickup trucks (e.g., the Ridgeline vehicle) and other light trucks (e.g., Acura MDX sport utility vehicle) for the U.S. market are increasing, is shifting the engineering for those vehicles to the United States. Honda now has 10,000 engineers in Japan and about 1,300 engineers in Ohio. The engineers in Japan work on most Honda vehicle programs, almost all power-train programs, and almost all of Honda's advanced R&D, such as hybrid power-train systems and fuel-cell vehicles. However, the 1,300 engineers in Ohio have been taking on responsibility for entire vehicle programs, especially for vehicles sold predominantly in the U.S. market.

The Engineering of Components

It makes sense to engineer some components, especially if they are highly integral to the vehicle architecture, close to the vehicle engineering center, where supplier engineers can benefit from being close to their OEM engineer counterparts, especially the OEM vehicle engineering team for a vehicle under development. When key supplier engineers are co-located at the vehicle manufacturer, as part of a vehicle-development program, proximity facilitates interaction, iteration, and, therefore, faster resolution of problems and concerns as they arise.

Conclusion

Proximity does have value in the automotive industry. In other words, the automotive world is not entirely flat. Customers pull vehicle production, and to a lesser degree vehicle engineering, closer to the end market, and suppliers follow their OEM customers for the production and engineering of certain components. Finally, the global footprint of customers is changing, as described in the next section.

THE 3C+G MODEL

The Customer Factor

The first and perhaps most important factor that influences the location of product and manufacturing automotive engineers is the location of the customer. The U.S. automotive market, like the Western European and Japanese markets, is a mature, replacement market—vehicle sales have been flat for nearly seven years and are projected to remain relatively flat. Figure 19 shows that U.S. sales of light vehicles (passenger cars and light trucks) have been high (about 17 million units), but roughly steady since 1999. Although historically the industry has demonstrated cyclical behavior, most analysts expect the U.S. market to remain relatively flat for the next five years, hovering between 16.5 and 17.5 million units through 2010.

Despite stagnant sales in the developed markets, global sales have consistently risen, increasing approximately 13

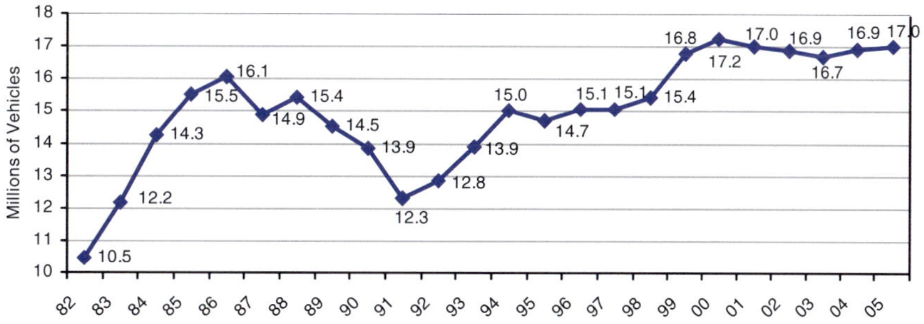

FIGURE 19 Sales of U.S. light vehicles, 1982–2005. Source: Automotive News, 2005; Wards Autoworld, 2005.

percent from 2001 to 2005. Industry sales totaled 64.7 million vehicles in 2005, representing a 3.7 percent increase over sales in 2004. Almost all of the growth was attributable to emerging markets, particularly China, India, Brazil, and Russia. Figure 20 shows vehicle sales normalized to 1999 volumes in the three developed markets, the United States, Western Europe, and Japan, and three developing markets, China, India, and Russia.

Increases in vehicle sales are primarily driven by two factors: vehicle saturation rates and income growth. Increases in vehicle sales shown in Table 12 are expected to continue in developing countries, particularly China and India. Figure 21 shows that developed markets are saturated in terms of vehicle ownership, but there is ample opportunity for motorization in the developing world.

Market growth has attracted investments in new production facilities (and the manufacturing engineers that go with them). Chinese vehicle sales exploded in 2002 and 2003, making China the third largest vehicle market (behind the United States and Japan). The promise of continued growth in China has attracted the attention of automakers around the world. According to *Automotive News*, roughly \$6 billion in automotive foreign direct investment flowed into China between 1994 and December 2002. The same amount, \$6 billion, was invested in the following 18 months. Since the mid-1990s, GM has invested heavily in China. In 2005, GM surpassed Volkswagen as the Chinese market leader with sales of 665,000 vehicles (a 35 percent increase over 2004).

In general, production increases follow market growth. Figure 22 shows actual vehicle production data for the same regions shown in Figure 20. The number of vehicles produced in China and India has increased as the number of vehicles sold in those markets increased. At a time when Ford and GM are making significant reductions in production capacity in their home market, both companies—like automakers everywhere—are vigorously pursuing growth opportunities in the developing world. Besides their pro-

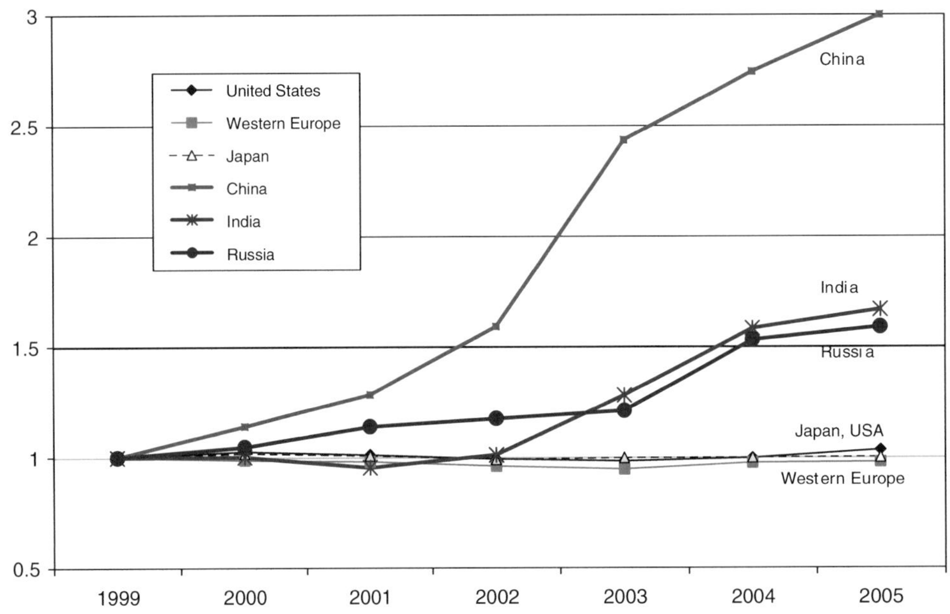

FIGURE 20 Sales rates in key markets, 1999–2005. Source: Automotive News, 2005.

TABLE 12 Vehicle Sales in Developed and Developing Markets and Growth Rates, 1999–2004

Country or Region	1999	2000	2001	2002	2003	2004	Growth in Vehicle Sales (%)	Growth in Average GDP (%)
United States	16,959	17,402	17,178	16,848	16,676	16,913	–0.3	2.78
Western Europe	17,296	17,053	16,944	16,608	16,352	16,856	–2.4	1.97
Japan	5,861	5,964	5,907	5,813	5,849	5,853	–0.1	1.66
China	1,832	2,089	2,353	2,917	4,461	5,230	185.5	8.52
India	857	859	818	867	1,098	1,298	51.5	5.74
Russia	1,102	1,154	1,256	1,295	1,334	1,654	50.1	6.87

Sources: Automotive News data; IMVP, 2004; World Bank Development Indicators, 2005.

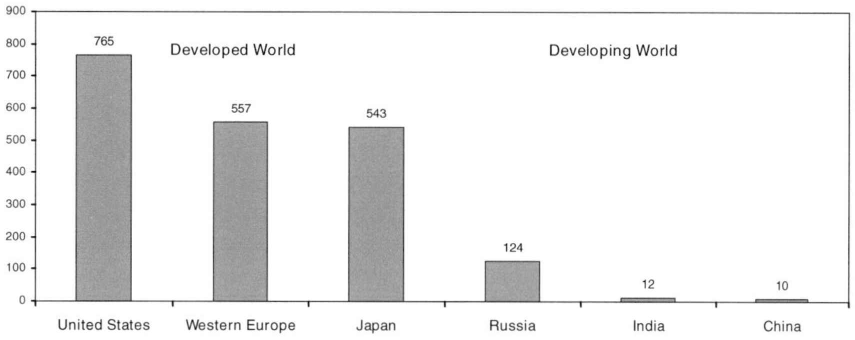

FIGURE 21 Vehicles per 1,000 people for selected countries and regions. Source: United Nations Statistical Yearbook, 2000.

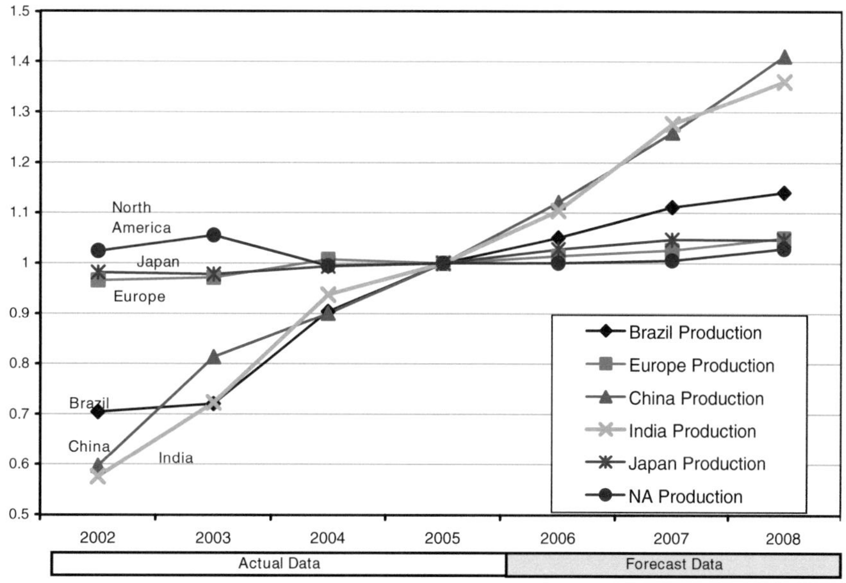

FIGURE 22 Vehicle production by country/region, normalized to 2005. Source: Automotive News, 2005; forecast data from JD Power, 2006.

duction facilities in Western Europe, Ford and GM actively invested in Latin America and Mexico during the 1980s and 1990s and in Eastern Europe and Asia since then. In the Asia-Pacific region, GM now employs more than 20,000 people at assembly and manufacturing facilities in China, India, Indonesia, South Korea, Thailand, and Australia. Ford has opened plants in the last 10 years in St. Petersburg, Russia; Chennai, India; and Chongqing, China.

The Cost Factor

Reducing costs is a critical factor in location decisions for both manufacturing engineers (production-facility locations) and product engineers (R&D locations). All automotive companies, both OEMs and suppliers, whether profitable or unprofitable, are under tremendous pressure to reduce costs. Shifting production to low-cost countries is an acceptable strategy, although there are some important caveats. One of the key findings in the interviews for the IMVP Lean Location Logic Project was that suppliers frequently underestimate the cost of ramping up production in low-cost countries.

One such cost is training workers (i.e., developing key capabilities) so they can match the quality and productivity levels of their sister plants in developed markets. In practice, this often requires that teams of production engineers and equipment technicians make extended trips to the new plant to train workers. Several managers noted that production ramp ups took longer than expected, resulting in delayed orders for customers (at the supplier's cost), and required more resources to train the new workforce than expected.

For most components, the automotive supply chain is a highly integrated system in which each component is part of an elaborate, coordinated process. Production does not take place in a vacuum. To supply something as simple as a hydraulic pump, tier-two suppliers must supply raw materials, fasteners, cast or forged components (such as pump vanes), roller bearings, and so on. The quality of the pump depends on the quality and responsiveness of the local supply base for all of those subcomponents. Even the best plant in the world will produce inadequate products if it does not have an adequate supply base.

Until very recently, China was *not* a low-cost producer of automobiles because of its highly inefficient and costly tier-two and tier-three supplier base. Local suppliers must not only produce components of requisite quality, they must also deliver those components on time—usually to meet just-in-time requirements. Therefore, the capability of the local supply base is also a function of the local transportation infrastructure and the capability of local logistics providers.

The integration of the automotive supply chain is important, but managers often analyze costs based on easily quantifiable metrics, such as wage rates, and fail to take into account broader system-level costs. The differences in manufacturing wages in different countries are striking and well documented (ILO, 2005). Automotive manufacturing wages in Germany are about $26 per hour (unloaded), compared to about $2 per hour in China.

Thus managers are strongly tempted to shift manufacturing of simple components, like the hydraulic pump, to a low-cost country where the local manufacturing wage is one-tenth the wage in the home country. However, experience has shown that the significant reduction in wages can be outweighed by lower productivity rates, higher than anticipated production training and ramp-up costs, and higher costs for quality materials and parts to build the hydraulic pump. As the CEO of a North American tier-one supplier said, "We've learned that chasing [ever lower] labor rates is not a sustainable business strategy. We invested heavily to build up production volumes in Mexico, only to discover that it was difficult to retain workers after investing in their training. We also discovered unanticipated costs much higher than we expected—such as the cost of customs and border-clearance processes for our supply flows along the U.S.-Mexico border."

For reasons described above, the general mantra in the automotive industry is "build where you sell." The three traditional automotive-production regions (i.e., the United States, Western Europe, and Japan), have invested in low-cost countries in the backyards of their production centers, as shown in Table 13. For Western Europe and the United States, production has increased in these low-cost countries, and a significant share of that production has been exported back to the traditional high-cost country, except in Japan, where increased production in regional low-cost countries has not led to increased exports back to Japan. Even today, almost all vehicles sold in Japan are produced in Japan.

When the North American Free Trade Agreement (NAFTA), which lowered or eliminated trade barriers among the United States, Canada, and Mexico, went into effect in January 1994, Mexico emerged as a low-cost country for automotive production in the backyard of the U.S. market. Canada and the United States already had a strong trade in vehicles and vehicle components, and Canada had traditionally been a lower cost base for manufacturing.[9] Both Canada and Mexico currently export more than half their vehicle production volumes to the United States, which imports more passenger vehicles from Canada than from Japan. Each of the Detroit 3 manufacturers has four production facilities in Canada and Mexico combined.

Of the $124.1 billion in imported passenger vehicles to the United States in 2005, the countries of origin were: Canada ($36.6 billion), Japan ($35.2 billion), Germany ($20.4 billion), Mexico ($10.8 billion), and South Korea ($8.8 billion) (DOC, 2006). The volume of U.S.-imported vehicles from Mexico, which provides the clearest cost advantage, declined steadily, from $15.8 billion in 2000 to $10.8 billion in 2005. This is partly because the plants in Mexico produce more trucks than cars, and truck sales have suffered in the past few years because of rising gasoline prices. Production losses in Mexico have been offset by production gains in Canada.

The production footprint of U.S. companies expanded in Central and Eastern Europe, where production volumes increased 31.4 percent from 2001 to 2005. Since May 2004, when Poland, Hungary, Slovenia, Slovakia, and the Czech Republic joined the European Union (EU), trade barriers among the EU-15 and the new entrants have been reduced or eliminated, and Poland, Hungary, Slovakia, and the Czech Republic have all offered substantial fiscal incentives to attract automotive investment. In addition to the much lower labor costs, these countries are accessible to the automotive supply chains of Germany and Austria and can provide a local low-cost production center for Western Europe. The increases in automotive production and exports to Western Europe from these countries are shown in Table 13. As the CEO of a European manufacturer noted, "Within the radius of a one hour flight in Europe, manufacturing wages range from €5 to €50. We [car companies] cannot ignore that; we are not in a position to negotiate with our customers."

Japanese manufacturers have increased production in Asian low-cost countries, such as Thailand, Indonesia, Malaysia, and Viet Nam. Nevertheless, Japanese OEMs have not exported vehicles produced in these countries to the Japanese market. Imports to Japan of Japanese-brand vehicles actually declined, from 90,682 in 1995 to 19,119 in 2005 (both numbers represent a negligible portion of the overall Japanese market of roughly 5.9 million units in 2005) (JAMA, 2006). Instead, the Japanese have used Thailand

[9]The Canadian cost advantage has declined, but automotive production in Canada is still estimated at a 5.1 percent cost advantage for the production of auto parts compared to the United States (KPMG, 2006). The U.S. corporate average fuel economy (CAFE) requirements also provided incentives for U.S. automakers to locate certain vehicle production in Canada.

TABLE 13 Production in Local Low-Cost Countries for Key Automotive Regions

Region	Low-Cost Countries	Units Produced in 2001	Units Produced in 2005	Percentage Change
United States	Canada, Mexico	4,396.5	4,390.3	–0.1
Western Europe (Germany, France, United Kingdom, Italy, Belgium, Spain)	Czech Republic, Poland, Slovakia, Slovenia, Hungary, Romania	1,341.1	1,762.6	+31.4
Japan	Thailand, Malaysia, Indonesia, Philippines	1,248.9	2,270.2	+81.8

Note: Production declined 9.1 percent in Mexico and increased 6.4 percent in Canada.
Source: Automotive News, 2005.

and other Asian low-cost countries to increase their vehicle production to the home markets in those countries and the Asia-Pacific region in general.

Other start-up costs for an offshore plant are associated with teaching local manufacturing engineers the company's basic production principles and procedures. Toyota's well known Toyota Production System involves creating value and eliminating waste from production processes through just-in-time production (smooth flow, minimal inventory), *jidoka* (building in quality, error-proof processes), *heijunka* (stabilizing variability in production schedules), and *kaizen* (continuous improvement). When Toyota opens a new assembly plant, be it in Kentucky, Thailand, Turkey, or France, a key challenge is ensuring that the Toyota Production System is understood and embraced at the new facility. However, Toyota has kept production of its high-end vehicles (e.g., Lexus) in Japan. Many other OEMs follow a similar strategy.

The "build where you sell" mantra, or "build close to where you sell" mantra, applies to fully assembled automobiles. For automotive components, the situation is more complex. It makes sense for suppliers of certain components, such as bulky or sequenced components, to be close to OEM assembly plants. Other components can more easily be supplied from a low-cost country. For example, components that are highly labor intensive but easily transportable might be shifted to a low-cost country. Wire harnesses, for example, are essential for connecting all electrical functions of a vehicle. Wire harnesses are built up from thousands of strands of individual wire braided into a complex product with many branches and end connectors. The work is highly labor intensive and cannot be easily automated. Wire harnesses are an early candidate for a component to be shifted to a low-cost country.

By contrast, highly capital-intensive products, such as a nozzle for a diesel fuel injector, may not be suitable for production in a low-cost country. Diesel fuel injectors are highly sophisticated products that require a clean-room production environment and sophisticated production equipment. Lower labor cost is not much of an advantage because the product is not labor intensive. In addition, some sophisticated components require manufacturing knowledge that cannot be easily transferred to a new production location.

All of the cost factors discussed above are pertinent to offshoring of the production of full vehicles and vehicle components, which can be used as a long-term proxy for the manufacturing-engineering footprint. The offshoring of product design is also driven in part by cost considerations. All of the industry managers interviewed were asked to estimate the cost to the company of employing a product engineer with 5 to 10 years of experience in six different countries/regions. The answers varied widely, both within and among regions, as shown in Figure 23. A "fully loaded" experienced engineer in the United States may cost $100,000, while an equivalent engineer in China may cost $15,000.

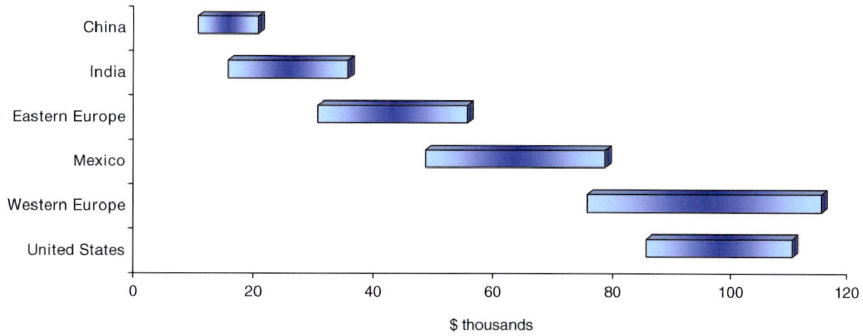

FIGURE 23 Approximate labor rates for fully loaded vehicles by country/region for an automotive engineer with 5 to 10 years experience. Note: Fully loaded was defined as total cost to the firm of employment of one full-time equivalent. Source: Interviews with industry managers.

Just as manufacturing labor costs are not the primary determinant for locating production facilities, engineering labor costs are not the primary determinant for locating the product-design function. The following factors must also be considered:

- Low labor rates may not provide a sustainable advantage, because engineering labor rates can increase over time.
- Engineering labor accounts for roughly one-third to one-half of the engineering cost of vehicle development.[10] Other major costs are for vehicle prototypes, testing equipment and laboratories, buildings/office space, software licenses, and so on. Engineering software licenses for products like CATIA are very expensive regardless of where they are used. As of June 2005, a CATIA license cost roughly $5,000 per user, regardless of the location of the user.
- Low productivity can effectively increase the cost of engineers in "low-cost countries." The same executive who estimated the annual loaded cost of an engineer in Shanghai at $10,000 per year noted that, after training and adjusting for output, the cost was easily $20,000 per year. Many interviewees cited the lack of domain knowledge as the key reason for lower productivity of engineers in countries like India and China.

In conclusion, cost is a critical factor in location decisions, and labor costs (both manufacturing labor and engineering labor) are important components of overall costs. It makes sense to manufacture certain vehicles or certain vehicle components in a low-cost country—but not all of them. It makes sense to engineer certain vehicles and vehicle components in a low-cost country—but not all of them. The dilemma facing manufacturers was summed up by one CEO of a European manufacturer, "No one has the solution to this problem. If you don't move some jobs away from your home base, you could be overwhelmed by competitors who are willing to do this. On the one hand, your family loses jobs. On the other hand, if you don't shift jobs to places like India and China, we're all dead."

The Capability Factor

Capability has little impact on the production footprint strategy for vehicles or components because, for production, the capability of the local manufacturing workforce and local manufacturing engineers is less important than customer location, government policy, and cost. However, capability has a high impact on the footprint strategy for product engineering. For a firm to shift a product engineering function offshore, there must be, at a minimum, qualified engineers available to perform the required tasks. This implies an engineering-education infrastructure that produces an adequate supply of qualified engineers.

Vehicle manufacturers can offshore product engineering in two ways: (1) offshore the full vehicle-engineering program for a specific vehicle or a family of related vehicles (e.g., large, rear-wheel-drive cars); or (2) offshore part of the vehicle-engineering process, such as a particular task or area of expertise. (Offshoring of full-vehicle programs is discussed in the next section.)

With respect to offshoring certain engineering functions, several interviewees noted that low-cost countries are best suited for certain types of engineering work:

- repetitive or routine tasks that require technical skills but not innovation or creativity, such as documenting an engineering bill of materials, performing a failure modes effects analysis (FMEA), certain types of routine stress analyses or heat-transfer calculations, and generation of a tool design from a part specification
- specialized functions that leverage local expertise or capabilities, in effect creating an offshore R&D center of excellence in a particular technology or capability, such as computational fluid dynamics
- localization tasks, that is, taking a vehicle (or component) designed in one part of the world and modifying it to comply with local regulations or customer preferences in a different part of the world

A study by Booz Allen Hamilton also concluded that higher value-added engineering tasks are more difficult to offshore. More demanding tasks, such as the full engineering responsibility for a vehicle program, are more difficult to outsource or offshore. Almost all interviewees for this report agreed that more complex engineering tasks were more difficult to offshore, although there was some disagreement about the level of complexity for some tasks. Routine tasks that require relatively low skills, such as creating a mesh for a finite-element model, are the easiest to outsource or offshore (Figure 24).

Two overarching messages emerged from interviews of automotive executives in the United States. First, many managers expressed concerns about the lack of automotive domain knowledge among engineers in low-cost countries. As the Asia-Pacific managing director of a North American tier-one supplier said, "I don't use my engineers in China for innovation. The culture is imitative, not innovative. They are great for reverse engineering, and so we use Chinese engineers for many of our aftermarket applications." Others noted that some automotive engineers in China had never even driven a car, much less owned one; thus they do not have a basic familiarity with the product.

A second concern expressed by some automotive managers was the shortage of engineers in the United States,

[10]To determine engineering labor costs, the overall engineering head count is multiplied by $100,000 per engineer and divided by overall R&D budget.

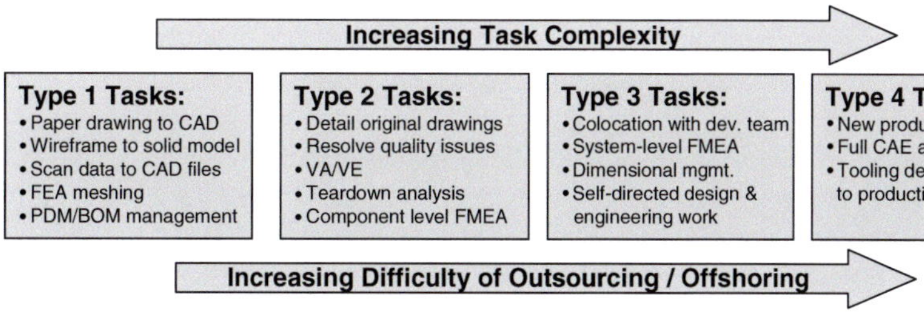

FIGURE 24 Complexity and difficulty of engineering tasks suitable for outsourcing/offshoring. Notes: CAD = computer-aided design; FEA = finite element analysis; FMEA = failure mode effects analysis; VA/VE = value analysis/value engineering; CAE = computer aided engineering. Source: Jackson et al., 2005.

particularly of engineers with certain skills. A CEO of a U.S. tier-one supplier cited this problem, "In Mexico, an engineer costs 10 times a manufacturing employee. In the United States, an engineer costs about the same as a manufacturing employee. Think about that. The issue is not cost; the issue is supply [of capable engineers]. We have a big problem with engineering in this country: it's called 'where's the talent?' My view [for my firm's engineering footprint] is that growth will occur overseas, and engineering in the U.S. will remain flat."

The value of electronics content in automobiles has increased steadily for the last two decades. Thus electrical and software engineers have become as important as the traditional mechanical engineers who have historically been associated with the automotive industry. Several interviewees indicated that electronics and software engineering functions are easier to outsource or offshore than mechanical engineering functions. Software engineers across an ocean can more easily discuss a few lines of code than mechanical engineers can discuss the modification of the design of a component. Software and electronic systems also tend to have a more modular product architecture than mechanical systems, making it easier to offshore both low- and high-value added functions. Figure 25 shows a conceptual model of transportability (i.e., ability to offshore) for the capability of performing mechanical engineering tasks compared to electrical and software engineering tasks.

GLOBALIZATION OF RESEARCH AND DEVELOPMENT

Coordinating Global R&D

Engineering managers at Ford, GM, and DaimlerChrysler report that their top priority is improving coordination among their engineering functions around the world, rather than further offshoring of engineering. Despite many attempts to improve coordination, at the beginning of this decade Ford, GM, and DaimlerChrysler each had several regional

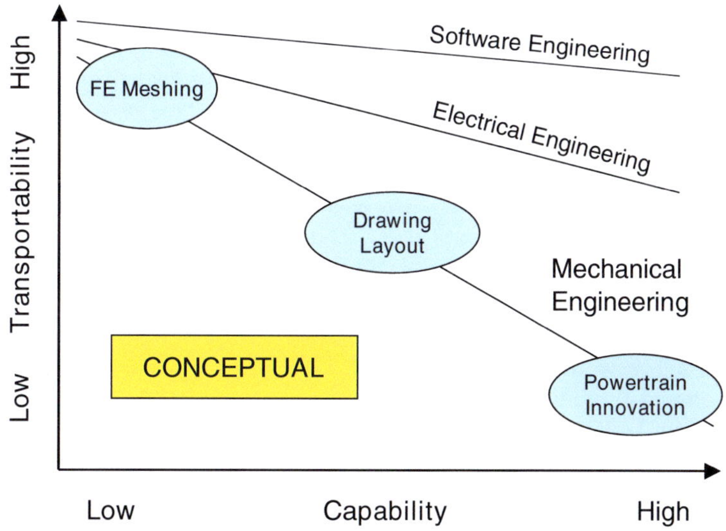

FIGURE 25 Conceptual model of trade-offs between capability and transportability versus engineering disciplines.

The Quest for the World Car

The quest for a world car has proved to be very difficult. The industry has several times tried and failed to produce a vehicle that could be sold in markets around the world with minor modifications. Ford tried to engineer the Escort of the early 1980s as a world car, but at launch time the American and European versions had little in common. The Ford CDW27 vehicle program of the early 1990s (which produced the Ford Mondeo, Contour, and Mystique) cost more than $5 billion (including new engines and transmissions) and took an agonizing seven years to bring to market. The tremendous expense of the CDW27 program ("W" indicated a "world" program) was a driving force behind the creation of a highly ambitious reorganization, the Ford 2000 program, announced in 1994, to merge Ford's European and North American vehicle development programs.

One great challenge of the world car program is that markets around the world are different. Americans have a preference for light trucks, large vehicles, and comfort-enhancing features ranging from cup holders to video displays for children. Europeans prefer smaller vehicles with better vehicle dynamics (ride and handling characteristics). Europeans have also embraced the diesel engine; nearly half the vehicles sold in Europe have diesel engines.

Many Japanese consumers prefer on-board information features, such as navigation systems, and minicars—a market segment all but unknown in the United States and still rare in Europe. Minicars are remarkably small vehicles (5 feet wide and less than 11 feet long, by Japanese law) powered by engines typically in the range of 60hp. Led by Suzuki, minicars accounted for 35 percent of new car sales in Japan from January to October 2006, compared with 24 percent a decade ago.

engineering centers that primarily supported their respective regional markets and did not work together. Product planning—making critical decisions about which vehicles are brought to market and at what level of funding—was also relatively decentralized, with regional executives exercising relative autonomy. As GM Vice Chairman Robert Lutz joked in 2004, "up until a few months ago, GM's global product plan used to be four regional plans stapled together" (Hawkins, 2004).

Ford and GM (and Volkswagen) had adopted a *multinational* business model with distributed, and (mostly) independent, regional R&D centers supporting mostly autonomous regional operations.[11] GM and Ford's highly decentralized global network of R&D centers reflected the history of their development. Both companies had developed significant European operations during the twentieth century selling distinct European vehicles engineered by European engineers built in Europe by European workers with parts supplied by European suppliers. Ford's engineering centers near Cologne, Germany, and in England supported Ford of Europe. GM's European engineering centers were aligned by brand; for example, Rüsselsheim, Germany, supported Opel, and Millbrook, UK, supported Vauxhall.

Ford and GM's acquisition of European brands during the 1980s and 1990s further complicated the picture. For example, Ford acquired Volvo's engineering center in Gothenberg, Sweden, when the company purchased Volvo in 1999, and GM acquired the engineering center in Trollhättan, Sweden, when it purchased Saab.

[11] Although Ford and GM conducted vehicle development in Europe for their European vehicle lines, both firms conducted the majority of their basic and applied research in the United States through the 1990s.

Ford and GM are now trying to integrate their regional engineering centers so that engineers across the globe can coordinate on global programs. The objective is not to engineer the same vehicle for different markets (the "world car" vision) but to engineer a family of vehicles with the same underlying structure that can be very easily modified to meet local customer and (environmental and safety) regulatory requirements. Achieving this objective will require more centralized product planning and more coordination among global product development centers. Thus both GM and Ford are changing from their *multinational* business model to a *transnational* business model.

GM has transitioned from brand-specific engineering to regional engineering and is now transitioning from regional engineering to global engineering. For example, GM headquarters declined requests from its Daewoo subsidiary to build an SUV for the Korean market rather than leverage an existing GM vehicle program already under development. GM uses the term architecture to describe a family of vehicles that may appear very different to customers but have basic engineering commonality.

For example, the Chevrolet Malibu, the Saab 9-3, and the Opel Vectra are all products of GM's midsize-vehicle architecture, developed at the Rüsselsheim engineering center, although these vehicles appear very different outwardly. GM is trying to reduce the number of vehicle architectures while making sure that the right engineers among GM's 13 global engineering centers are working to support the appropriate vehicle architecture. Table 14 shows which engineering centers have the lead responsibility for current vehicle architectures.

Toyota and Honda are also adopting a transnational business model, but from a much different starting point than

TABLE 14 GM Engineering Centers Responsible for Various Vehicle Architectures

Architecture (vehicle family)	Home Engineering Center	Architecture (vehicle family)	Home Engineering Center
Luxury RWD Car	Warren, Michigan	International Mid-Size Truck	São Paolo, Brazil
Compact Crossover	Warren, Michigan	Compact Car	Rüsselsheim, Germany
Performance Car	Warren, Michigan	Mid-size Car	Rüsselsheim, Germany
Full-Size Truck	Warren, Michigan	Small Car	Seoul, South Korea
Mid-Size Truck (regional)	Warren, Michigan	Mini Car	Seoul, South Korea
FWD Truck	Warren, Michigan	RWD Car	Melbourne, Australia
Vans / Commercial Truck	Warren, Michigan		

Source: General Motors, 2006.

Ford and GM. Toyota, established in 1937, sold almost all of its vehicles in the Japanese home market for the first two decades. Toyota Motor Sales USA was established in 1957, the Toyota Technical Center in Ann Arbor was opened in 1977, production in the United States (at the NUMMI joint venture with GM) began in 1984, and production in Europe (in the UK) began in 1987. Although Toyota has operated the Ann Arbor Technical Center for nearly 30 years, engineers in that facility have only recently been given program-level responsibilities.

Toyota has about 20,000 engineers in Japan; however, nearly 40 percent are contract employees or "guest engineers" from suppliers. Like many Japanese firms, Toyota is about to face a shortage of engineers in Japan as the first baby-boom generation there reaches the mandatory retirement age of 60. The Japanese call this the year 2007 problem.[12] Thus Toyota is being forced to look beyond its borders for engineering talent, one reason the company plans to dramatically expand employment at the Ann Arbor center in the next few years.

Honda's evolution has been similar to Toyota's, although Honda shifted more engineering responsibility to America earlier than Toyota did. Honda, founded in 1948, opened American Honda Motor as a sales operation in 1959. The company began producing the Honda Accord in Ohio in 1982, and Honda R&D Americas center in Ohio was established in 1984. The Ohio facility concentrates on product engineering, development, and testing. A newer facility in California concentrates on market research and vehicle styling. Honda R&D Americas has full-vehicle engineering responsibility for the Acura TL and MDX and the Honda Element, Pilot, and Civic Coupe.

Both Toyota and Honda started out by following an international business model with strongly centralized R&D (very little of it outside the home country) and regional operations with strong reporting lines to the home-country headquarters. As Honda migrates toward a transnational business model, the company must first shift more of its R&D to new or existing R&D centers outside Japan. Second, it must ensure that R&D is coordinated throughout its international network. Figure 26 illustrates how two groups of companies (Ford and GM; and Toyota and Honda) are migrating toward the same model.

The automotive industry was globalized first by brand (through imports and exports), then by production (through foreign direct investment in assembly and manufacturing plants), and now by changes in management of R&D operations. U.S. companies have expanded their R&D footprint outside the United States and decreased their R&D footprint in the United States. At the same time, foreign companies have increased their R&D footprint in the United States.

Offshoring of R&D by U.S. Companies

GM operates 13 engineering and design (styling) centers in 13 countries (Figure 27). While GM has maintained a strong market, production, and R&D presence in Europe and Latin America for decades, it has only recently entered into China (1997), South Korea (2002), and India (2003). Ford reports that it spent $8 billion on engineering R&D in 2005, distributed among seven engineering, research, and design centers located in Dearborn, Michigan; Dunton, U.K.; Gaydon, U.K.; Whitley, U.K.; Gothenburg, Sweden; Aachen, Germany; and Merkenich, Germany (Ford Motor Company, 2005b).

GM considers its technical center in Bangalore, India, a center of excellence for the development of math-based tools and electronic-control systems. Work in Banaglore includes the development of modules and systems; human modeling for predicting crashworthiness; development of vehicle structures; and development of control software, embedded systems, software validation and calibration tools, voice recognition and communications systems, electrical-system simulation, and electromechanical simulation. In short, the rationale for opening the Bangalore center was to develop a specialized engineering capability that might be in short supply in the United States. According to a top GM executive, "Electronics and software content will account for 40 percent of the value added in the vehicle over the next 10

[12]Toyota recently changed its re-employment system so its retirees can work up to the age of 65. The limit had been 63.

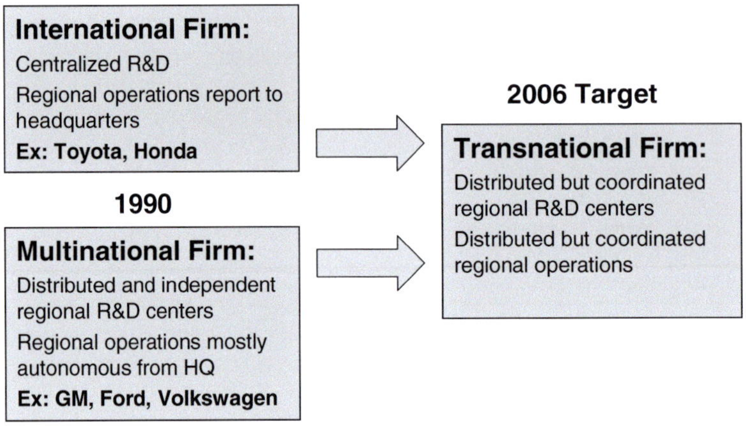

FIGURE 26 Evolution from the international and multinational models to the transnational model.

years. There's a shortage of software, electronics, and control engineers in the U.S.—that's part of why I opened our [overseas] R&D center. I think we will see a shortage of engineers in the United States."

Onshoring of R&D by Foreign Companies

Foreign-brand automakers have built product-development and design facilities in the United States, in addition to manufacturing plants. Total employment for technical and design functions by foreign-brand automakers in the United States is currently estimated at approximately 4,000 people (Table 15). This figure does not include sales and marketing staff located in the United States, which accounts for thousands more employees. Table 15 shows that foreign R&D facilities are spread across the United States; however, the majority of engineers are in Michigan and Ohio.

The number of engineers and designers employed by foreign-brand vehicle manufacturers in the United States has increased rapidly. In 1987, the Japan Automobile Manufacturers Association (JAMA) estimated that Japanese automakers employed about 200 engineers, scientists, technicians, and designers in the United States. By 2004, JAMA reported that 3,065 engineers and designers were employed at a growing number of technical R&D and design facilities. In the latest report, issued in September 2006, the number had risen to 3,593 (Figure 28).

The number of U.S. engineers employed by foreign automakers is expected to increase substantially in the next few years. Toyota plans to invest $150 million to expand its Ann Arbor, Michigan, facility and add at least 400 engineers to the current staff of roughly 950. One Toyota executive stated that Toyota plans to expand the Ann Arbor facility to 2,000 engineers in the next five years. Also in Ann Arbor, Hyundai is investing $117 million to expand its technical center from

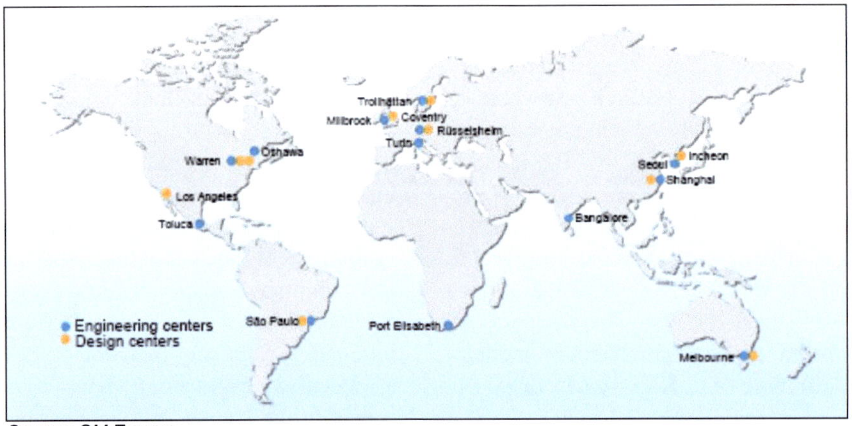

Source: GM Europe

FIGURE 27 Locations of General Motors global engineering and design facilities. Source: GM Europe, 2006.

TABLE 15 Foreign-Brand R&D and Design Facilities in the United States, 2006

Company	Location(s)	Established	Employees
BMW	Spartanburg, N.C.; Woodcliff Lake, N.J.; Oxnard, Calif.; Palo Alto, Calif.	1982	150
Honda	Torrance, Calif.; Raymond, Ohio	1975	1,300
Hyundai[a]	Ann Arbor, Mich.	1986	150
Isuzu	Cerritos, Calif.; Plymouth, Mich.	1985	100
Mazda	Irvine, Calif.; Ann Arbor, Mich.; Flat Rock, Mich.	1972	100
Mercedes-Benz	Palo Alto, Calif.; Sacramento, Calif.; Portland, Ore.	1995	50
Mitsubishi	Ann Arbor, Mich.	1983	130
Nissan	Farmington Hills, Mich.	1983	1,000
Subaru	Ann Arbor, Mich.; Lafayette, Ind.; Cypress, Calif.	1986	30
Toyota[a]	Gardena, Calif.; Berkeley, Calif.; Ann Arbor, Mich.; Plymouth, Mich.; Lexington, Ky.; Cambridge, Mass.; Wittmann, Ariz.	1977	1,000

[a]Toyota and Hyundai are currently undergoing significant expansions. BMW included approximately 50 engineers assigned to BMW-DCX-GM hybrid project in Troy, Michigan.
Sources: Automotive News, 2005; company reports and interviews; JAMA, 2004.

150 to 550 employees. The Detroit metropolitan area has an abundance of automotive engineering talent, and in the past few years, scores of engineers have left domestic OEMs to take jobs with foreign OEMs. This trend is expected to continue (Shirouzu, 2005; Vlasic, 2004).

Discussion

Many industry executives say that asking if offshoring is occurring is framing the issue the wrong way. They are quick to point out that the automotive industry has been a global industry since its inception and that the real question is how to optimize and reallocate existing resources, that is, how to develop an effective footprint strategy.

GM acknowledges that it has increased its engineering head count overseas and reduced its engineering head count in the United States. However, the company contends that *offshoring*, defined as the replacement of U.S. engineering jobs with equivalent jobs overseas, has not occurred. According to GM's executive director of global engineering processes, the main driver for increasing the engineering head count overseas is to support the growth in overseas markets in China, India, Korea, and other countries. He says the main reason for the decrease in engineering employment in the United States is a 10 percent increase in engineering productivity per year in the past five years attributable to better tools and information technology, more sharing of components among vehicles, and better coordination of R&D (Cohoon, 2006).

Many interviewees also felt that there was a great deal of hype and misunderstanding about offshoring. As one senior vice president of a North American tier-one supplier said, "I laugh about the notion of a 24/7 product-development process—the idea that engineers in Europe will hand off a project to engineers in North America, who, in turn, will pass it on to engineers in Asia. That's a myth. Handoffs don't happen for sophisticated [development] programs."

Nevertheless, some data indicate that some U.S. engineering jobs are being replaced with engineering jobs overseas. In 2003, Helper and Stanley surveyed 615 small and

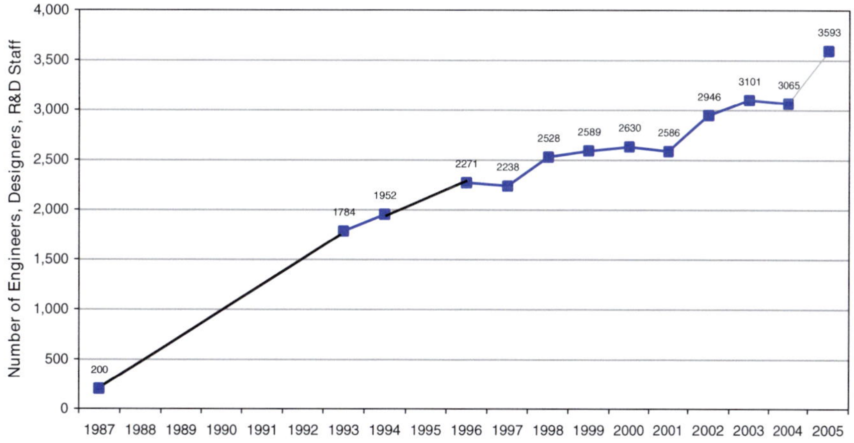

FIGURE 28 U.S. technical employment by Japanese automakers, 1982–2005. Source: JAMA, 2006.

medium-sized enterprises that produce components in the U.S. Midwest. The sample firms were second-tier suppliers that sell largely, though not exclusively, to the automotive industry. Eighty-seven percent of respondents answered "yes" to the question: "In the past three years, have any of your significant customers awarded your traditional jobs to competing suppliers in Mexico, Central or South America, Eastern Europe, or Asia?"

AUTOMOTIVE ENGINEERING IN CHINA

Automotive engineering activity is clearly increasing in India, China, and Eastern Europe for different reasons. India is seen as an emerging knowledge hub in automotive electronics, and Eastern Europe as having a low-cost, technically advanced workforce. In October 2006, Renault announced that it would invest €500 million to build a new engineering center in southern Romania. The company plans to hire 1,600 engineers and technicians by 2009.

The Rise of the Automotive Industry

As recently as 1985, the automotive industry in China was insignificant from a global perspective (total production of passenger cars was 5,200). In the early 1980s, three foreign automakers were allowed to enter the Chinese market through joint-venture agreements with Chinese partners: American Motors Corporation (subsequently bought by Chrysler), Volkswagen, and Peugeot. While Volkswagen's China partnership, based in Shanghai, proved to be very successful, the French and American partnerships were less successful. In these early joint ventures, the Chinese government limited foreign automakers to a maximum of 50 percent ownership in the joint ventures, and Chinese import duties on passenger cars in 1985 were 260 percent.

Since China's accession to the World Trade Organization (WTO) in December 2001, the industry and market have underdone a radical transformation. The WTO agreement, combined with the lure of China's huge potential market, has spurred automakers to flood China with investment. Every vehicle manufacturer has tried to find a Chinese partner to form an international joint venture. Chinese import duties on passenger cars fell from about 90 percent in 1996 to about 75 percent in 2001, and as of July 1, 2006, they had fallen to 25 percent.

Today China has a huge and growing automotive market. Last year, almost 6 million vehicles were sold in China, second in the world to the United States (about 17 million units).[13] The Chinese market exploded in 2002 and 2003 with growth rates surpassing 60 percent both years. (Remember that the sales rate in all three mature automotive markets—the United States, Western Europe, and Japan—has been essentially flat for the past five years.) After a slight slowdown in 2004, the growth rate in the Chinese market resumed. Sales of passenger cars for the first half of 2006 were 47 percent higher than in the first half of 2005.

The Chinese automotive industry is uniquely fragmented and complex. The number of vehicle manufacturers in China has remained steady—about 120—for the past 15 years, and many of these firms have insignificant sales volumes. In 2004, only 12 Chinese automakers had a production capacity of more than 100,000 units.

Leading Chinese automakers, such as Shanghai Automotive Industry Corporation (SAIC), First Automotive Works (FAW), Dongfeng, and Beijing Automobile Industrial Corporation (BAIC) have entered into a complex web of partnership arrangements with foreign manufacturers. SAIC, for example, has a joint venture with both Volkswagen and GM. In addition, a few Chinese companies, so-called independents such as Chery, Geely, and Great Wall, are developing cars without the help of joint venture partners.

Vehicles sold by joint-venture partnerships, which account for about 80 percent of the Chinese market, are sold mostly as foreign brands, such as Ford and Buick. Joint-venture facilities are clustered in six regions, Shanghai, Beijing, Changchun, Chongqing, Wuhan, and Guangzhou. There is no Chinese "Detroit," although Shanghai is the largest and fastest growing automotive center in the country.

Impact on U.S. Manufacturers and Suppliers

U.S. vehicle manufacturers have benefited from the exploding Chinese market. In 1983, Chrysler, through its acquisition of American Motors, was the first foreign player in China. Although Beijing Jeep was not a success, DaimlerChrysler has been developing an aggressive China strategy over the past few years through its joint venture with BAIC. Ford was a late entrant to the Chinese market, partnering with ChangAn, a former supplier of military equipment based in Chongqing. At the Ford-ChangAn assembly plant in Chongqing, an impressive mix of vehicles rolls down the line: Ford Focus, Ford Mondeo, Volvo S40, and Mazda 3. Ford's sales in China for the first half of 2006 were up 102 percent (U.S. sales for the same six months were down 4 percent). GM has emerged as the sales leader in China. GM sales for the first half of 2006 were up 47 percent (compared to a 12 percent decline in U.S. sales). GM made $327 million in profits from its operations in China in 2005 (Automotive News, 2006).

All of the global tier-one suppliers who followed their customers into China have also profited from the explosive growth. However, many smaller tier-two and tier-three U.S. auto suppliers have lost business to Chinese competitors. Several executives told IMVP researchers that they felt internal pressure from senior management to view investment

[13]The 2005 data were subsequently recalculated by the Chinese Association of Automotive Manufacturers (CAAM) to reveal that China had not surpassed Japan; however, China will surpass Japan in 2006 sales (Lee, 2006).

in China favorably, in order to achieve the benchmark of a "China price." This refers to the big differences in direct labor costs between the United States and China, but does not account for system-wide costs.

In general, Chinese domestic suppliers are better positioned to supply low-end parts, and foreign suppliers are better positioned to supply complex modules and sophisticated components, to Chinese joint-venture partners and vehicle manufacturers. Fourin, a Japanese-based research firm measured the percentage of foreign (i.e., non-Chinese) penetration into the production of automotive parts in China and found revealing data for chassis-related parts (Fourin China Auto Weekly, 2005). In 2003, several low-end mechanical components (e.g., wheel bolts, wheel rims, steel wheels, rear-axle housings, axle shafts) were manufactured entirely by Chinese firms. More sophisticated components (e.g., suspension systems, brake calipers, and ABS systems) had the highest degree of non-Chinese production. The data for engine-related components reveal the same trend. In 2003, 100 percent of the engine-management systems manufactured in China were produced by non-Chinese firms. These data are for components produced in China and do not include imported components.

The U.S.-China trade deficit in auto parts increased to $4.8 billion in 2005. U.S. exports of auto parts to China increased from $225 million in 2000 to $623 million in 2005. The top categories of parts flowing from the United States to China include seats, air bags, and gearboxes, which are all sophisticated components. However, U.S. exports to China are dwarfed by imports from China, which increased from $1.6 billion in 2000 to $5.4 billion in 2005. The top categories of auto parts flowing from China to the United States include radios, brake components, and aluminum wheels, which are less sophisticated or more modular components. A closer look at the data reveals that a large proportion of auto parts exported by China are produced by the Chinese operations of joint ventures with U.S. suppliers. Shanghai Delphi, for example, exports automatic door systems.

R&D Capability

Universities in China play a unique role in the automotive R&D process. Three government-funded university labs conduct applied automotive research—essentially product engineering—for Chinese vehicle manufacturers. The centers are based at Tsinghua University in Beijing (State Key Laboratory for Automotive Safety and Energy); Tianjin University in Tianjin (State Key Laboratory for Internal Combustion Engines); and Jilin University in Changchun (State Key Laboratory for Automotive Dynamic Modeling and Simulation).

At Tongji University in Shanghai, which established the nation's first College of Automotive Engineering in 2002, nearly 50 faculty members teach 730 full-time undergraduate students, 124 master's students, and 27 Ph.D. students. Congqing Lifan, China's top producer of motorcycles, recently launched its first passenger car, the Lifan 520. The vehicle was entirely engineered at Chinese university research labs using domestic R&D resources.

Until 2004, only one R&D center in China, the Pan Asia Technical Automotive Center (PATAC), was related to a foreign vehicle manufacturer. PATAC was established in 1997 as a 50-50 joint venture between GM and SAIC. PATAC currently employs more than 1,100 people, about 35 percent of whom have master's or doctorate degrees.[14] Employment is expected to increase to 1,400 in the next year to support the launch of many new products from Shanghai GM, which is now approaching a production volume of one million vehicles per year.[15] Engineers at PATAC earn approximately $12,000 per year.

PATAC is managed by an executive committee, two managers from GM and two from SAIC, but is fully integrated into GM's global engineering network. Work at PATAC includes product development, vehicle engineering, styling, and service engineering to support GM, SAIC, and Shanghai GM. PATAC also houses a GM design studio with 80 designers (out of GM's total global force of 1,200). The PATAC design studio designed all new sheet metal for the Chinese edition of the Buick Lacrosse.

Jane Zhao, an IMVP researcher at the University of Kansas, conducted extensive interviews with Chinese automakers and suppliers and complied survey data focused on R&D capability. Her studies revealed three key findings. First, domestic Chinese R&D capability is far behind the capability of non-Chinese competitors. Chinese vehicle manufacturers generally have a strong development capability for mechanical products, but have little capability for high-end electronics and software. This is consistent with the data on foreign trade cited above.

Second, R&D management is less advanced in China than in other automotive producing countries. This is consistent with media reports of shortages of management talent in certain regions and industries in China. During her interviews, the R&D manager of a well known Chinese automotive company confessed, "we don't know how to spend our R&D budget."

Recently, some Chinese companies have hired high-profile executives as R&D managers. The most notable of these was Phil Murtaugh, a talented, well respected manager who used to run GM China, who was hired by SAIC on June 18, 2006. Chery hired executives from Ford and DaimlerChrysler. Brilliance hired a former DaimlerChrysler executive to manage its R&D center, and Geely hired a former Hyundai executive to run its R&D operations. Given the remote locations of some Chinese automakers and, more importantly, the unique cultural requirements for success in China, it remains to be

[14]See *http://www.gmchina.com/english/operations/patac.htm*.

[15]Interview with Raymond Bierzynski, PATAC executive director, May 9, 2006.

seen whether Chinese companies will be able to attract and retain talented, world-class R&D managers.

Third, a great deal of R&D by international joint ventures is localization engineering, which is not nearly as sophisticated as designing a full vehicle from concept to customer. Some engineers have claimed that they had to "dumb down" to work with joint ventures where the focus was on localization rather than up-front design.

Despite the best efforts of the Chinese government to develop indigenous R&D capability, China is still heavily dependent on foreign design and technological know-how. The Chinese government's rationale for promoting international joint ventures was to develop R&D capability based on the premise that engineers from the Chinese domestic company would spend a few years working in the joint venture R&D center where they would acquire knowledge. Eventually, the domestic company would hire back the engineer and his or her acquired knowledge.

This has not happened, however. The backflow from the joint venture to the home company is much smaller than expected because of the large salary differentials, sometimes a factor of 10, between domestic companies and their joint-venture associates. In addition, the engineering infrastructure in China is very poorly developed. Take for example the lack of sophisticated test equipment—the country does not have a single automotive wind tunnel, although one is currently under construction at Tongji University.

Nevertheless, Chinese engineers working in the Chinese-foreign joint venture framework have learned a great deal about advanced automotive engineering. The Shanghai municipal government has mandated that 60,000 hybrid vehicles be sold by 2010, and Chinese engineers at PATAC are working to meet that challenge. Even though they are not leveraging the extensive research program on a dual-stage hybrid being developed by a GM-DaimlerChrysler-BMW partnership, engineers at PATAC are engaged in advanced engineering.

In addition, PATAC now also has significant design capability, such as clay modelers and CAD modelers who can design the aesthetics of a vehicle (e.g., exterior surfaces, interior materials and design, etc.). Engineering design requires not only creativity, but also highly specialized skills. Of the 1,200 people working at GM design centers around the world, 80 are in Shanghai. These are the people who designed the Buick Lacrosse sold in China by Shanghai GM, which looks significantly different from the same vehicle sold in America. As Figure 29 shows, automotive-related patent applications are on the rise in China.

Although the joint-venture model for technology transfer to Chinese engineers has largely failed, other ways of developing China's automotive R&D capability are emerging, such as strategic outsourcing to foreign knowledge centers. Chery has outsourced engineering to AVL (an Austrian firm that engineers high-tech power trains), Mira (a British firm that does special noise and vibration testing), and Pinanfarina (an Italian design, engineering, and manufacturing house). Chery and AVL successfully collaborated on a line of new advanced engines, and Chery engineers gained engine technological know-how in the process. Thus learning from collaborative outsourcing seems to be working.

China is also simply buying technology from foreigners to improve its R&D capability. The best example is SAIC buying stakes in Korean automaker SangYong and the failed British automaker MG Rover.

Recently, a debate has arisen about the possibility of China exporting vehicles to the U.S. market. Success in America and other key export markets is the ultimate test of an automaker's capabilities and would be a huge symbolic achievement, and this is a high-priority, medium-term goal for Chinese OEMs. Just as imports, followed by increased production capacity (the rise of the transplants) by Japanese, German, and Korean manufacturers, have increased in the American market, in the long term, China can be expected to develop automotive R&D capability and export significant

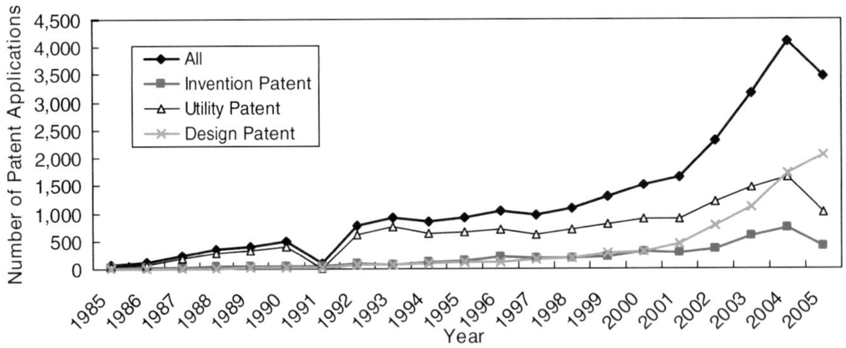

FIGURE 29 Automotive-related patent applications in China, 1985–2005. Note: Analysis by Jianxi Luo, Ph.D. Candidate, MIT. Search performed for "automotive" in the title of the patent application. Source: State Intellectual Property Office, People's Republic of China.

numbers of vehicles to the United States. However, China will first have to develop R&D capability on a par with America, Germany, Japan, and Korea.

TRENDS AND PROJECTIONS

Offshoring of automotive engineering—defined as the replacement of engineers in a high-cost country by those in a low-cost country—is just one aspect of the complex dynamics of the global automotive industry. Focusing only on the offshoring phenomenon without considering, for example, the onshoring phenomenon clearly misses the big picture. While Ford and GM are closing assembly plants in North America, Toyota, Honda, and Hyundai are building new plants in North America. While Ford and GM are reducing their engineering head counts in the Detroit area, Toyota, Honda, and Hyundai are increasing theirs in Ann Arbor, Michigan, Raymond, Ohio, and elsewhere.

Automotive engineers in the United States are legitimately concerned about offshoring, but many other issues should concern them more. Automotive engineers employed by domestic vehicle manufacturers should be more concerned that their companies are losing billions of dollars and not earning adequate returns on invested capital. They should be concerned that many of their competitors have "leaner" product-development processes, which means they can bring vehicles to market faster. They should be concerned about legacy costs, such as pension and retiree health benefit liabilities, agreements by previous managers that are no longer tenable. They should be concerned that their brands are cheapened when sales incentives campaigns essentially pay customers to buy their vehicles.

U.S. automotive engineers should keep one important fact in mind. Toyota, the benchmark of the industry and the most valuable automotive company in the world, has done the least offshoring of any large automotive company. Toyota has become the automotive MVP by focusing on value, rather than on cost. If a firm uses offshoring purely to cut costs, offshoring is unlikely to provide a sustainable competitive advantage. If a firm uses offshoring (along with onshoring) as part of an integrated footprint strategy, the firm is more likely to achieve an advantage.

Asia will continue to drive growth in the global automotive market, and the automotive production and engineering footprint in Asia will continue to expand. In the meantime, Toyota is attempting to upgrade its engineers to focus on technical areas that will be competitive differentiators in the future. For example, Toyota has invested significantly in the past few years to increase its internal capability in software development. This may indicate that Toyota believes that understanding the code that controls complex vehicle-control systems, such as the power controllers for hybrid power trains, will be one of those differentiators. Thus the most advantageous thing for U.S. engineers to do is to focus on creativity and developing cutting-edge technologies.

The global automotive industry has undergone radical changes in the past 10 years, and indications are that change will continue. Rather than stabilizing, the industry appears to be on the cusp of a significant restructuring because current business models are no longer sustainable for many firms. As vehicle manufacturers learn to engineer more with less, a company's footprint strategy will become increasingly important.

REFERENCES

ACEA (European Automobile Manufacturers Association). 2004. Data on foreign vehicle penetration of the European market provided to the author for Table 1.

Automotive News. 2005. Global 100 Supplier List.

Automotive News. 2006. 2006 Guide to China's Auto Market. May 1, 2006.

BLS (Bureau of Labor Statistics). 2005. May 2005 Occupational Employment and Wages Estimates. Available online at *http://www.bls.gov/oes/oes_2005_m.htm*.

BLS. 2006. Career Guide to Industries: Motor Vehicle and Parts Manufacturing. Available online at *http://www.bls.gov/oco/cg/cgs012.htm#emply*.

Casesa, J., et al. 2005. The Recapitalization of Detroit. Merrill Lynch. October 18, 2005.

Center for Automotive Research. 2005. The Contribution of the International Automotive Sector to the US Economy: An Update. Study prepared for the Association of International Automobile Manufacturers, Inc. March.

Clark, K., and T. Fujimoto. 1991. Product Development Performance: Strategy, Organization and Management in the World Automotive Industry. Cambridge, Mass.: Harvard Business School Press.

Cohoon, J. 2006. Presentation at the National Academy of Engineering Workshop on the Offshoring of Engineering, October 24, 2006, Washington, D.C.

DaimlerChrysler. 2005. Driven by Values: 2005 Annual Report. Available online at *http://www.daimlerchrysler.com/Projects/c2c/channel/documents/829811_DC_GB05_E_gesamt_comp.pdf*.

DOC (U.S. Department of Commerce). Accessed 2006. TradeStats Express™ Home. Available online at *http://tse.export.gov/*.

Ford Motor Company. 2005a. Ford Motor Company 2005 Annual Report: Driving Innovation. Available online at *http://www.ford.com/NR/*rdonlyres/*emwjt42turmgvsborq7bqproy43vwvnf4cztbl56l4oos4sf545rswt6jdh3gap4juvgevydld4g43xxky6y5znoote/2005_AR_full.pdf*.

Ford Motor Company. 2005b. Form 10-K filed 3/1/2006 for period ending 12/31/2005. Available online at *http://www.fordcredit.com/*investorcenter/*reports.jhtml*.

Fourin China Auto Weekly. 2005. An Analysis of China's Autopart Production: Top Manufacturers and Foreign Efforts Expand. February 7, 2005.

Fujimoto, T., and K. Nobeoka. 2004. Organizational Capabilities of Product Development: International Competitiveness of Japanese Automakers (Seishin Kaihatsu no Soshiki Noryoku: Nihon Jidosha Kigyo no Kokusai Kyosoryoku). Discussion Paper 04-J-039. Research Institute of Economy, Trade, and Industry. Available online (in Japanese) at *http://www.rieti.go.jp/en/publications/summary/04080002.html*.

Gereffi, G., and V. Wadhwa. 2005. Framing the Engineering Outsourcing Debate: Placing the United States on a Level Playing Field with China and India. December 2005. Master of Engineering Management Program, Pratt School of Engineering, Duke University, Durham, N.C.

GM (General Motors Corporation). 2005. General Motors Corporation 2005 Annual Report. Available online at *http://www.gm.com/company/investor_information/docs/fin_data/gm05ar/download/gm05ar.pdf*.

GM Europe. 2006. Information provided to the author.

Hawkins, L. Jr. 2004. Reversing 80 Years of History, GM Is Reining in Global Feifs. Wall Street Journal, October 6, 2004.

ILO (International Labour Organization). 2005. Key Indicators of the Labor Market (KILM). 4th edition. Independence, Ky.: Routledge.

IMVP (International Motor Vehicle Program). 2004. Data provided to author for Figure 10 and Table 10.

IRI (Industrial Research Institute). 2005. Data provided to the author for Figure 2.

Jackson, B., V. Sehgal, K. Dehoff, and V. Couto. 2005. Engineering Offshoring for the Automotive Industry. McLean, Va.: Booz Allen Hamilton.

JAMA (Japan Automobile Manufacturers Association). 2004. Data provided to the author for Tables 1, 8, and 13, and Figure 10.

JAMA. 2006. The Motor Industry of Japan 2006. Figures 16 and 32.

JAMA. 2007. Common Challenges, Common Future: Japanese Auto Manufacturers Contribute to the Competitiveness of Europe's Motor Industry (Brochure).

JD Power & Associates. 2006. Data provided to author for Figure 22.

KAMA (Korean Automobile Manufacturers Association). 2004. Data on foreign vehicle penetration of the Korean market provided to the author for Table 1.

Klier, T.H. 1999. Agglomeration in the U.S. Auto Supplier Industry. Economic Perspectives 23(Third Quarter): 1834.

KPMG. 2006. Competitive Alternatives: KPMG's Guide to International Business Costs.

Lee, C. 2006. In China, It's the Year of the Car. World Business, March 28, 2006.

Marcionne, S. 2006. The Re-Making of Fiat: Remodelling the Last Chance Saloon. Speech at the Automotive News Europe Congress, June 20–22.

Maxcy, G. 1981. The Multinational Automotive Industry. New York: St. Martin's Press.

McKinsey & Co. and OESA (Original Equipment Suppliers Association). 2004. Customer-Supplier Interface Study.

Rhys, D.G. 1972. The Motor Industry: An Economic Survey. London, U.K.: Butterworth.

Ries, J.C. 1993. Windfall profits and vertical relationships: who gained in the Japanese auto industry from VERs? Journal of Industrial Economics 61(3): 259–276.

Schonfeld & Associates. 2006. R&D Ratios & Budgets. Libertyville, Ill.: Schonfeld & Associates Inc.

Shirouzu, N. 2005. Foreign Car Makers Grab U.S. Resource: Automotive Engineers. Wall Street Journal, Dec. 6, 2005.

Smitka, M. 1999. Foreign policy and the US automotive industry: by virtue of necessity? Business and Economic History 28(2): 277–285.

Sturgeon, T., and R. Florida. 2000. Globalization and Jobs in the Automotive Industry. Carnegie Mellon University and Massachusetts Institute of Technology.

Technology Review. 2005. Corporate R&D Scorecard. Technology Review. September.

United Nations. 2000. United Nations Statistical Yearbook. New York: United Nations.

Vlasic, B. 2004. As Best Auto Jobs Fade: Big 3 White Collar Workers Turn to Transplants and Suppliers. The Detroit News, June 20, 2004.

Wards Autoworld. 2005. Data for Figure 19 from various issues.

Womack, J., and D. Jones. 2003. Lean Thinking. New York: Free Press.

World Bank. 2005. World Development Indicators, 2005. Washington, D.C.: World Bank.

BIBLIOGRAPHY

Boghani, A., and A. Brown. 2000. Meeting the Technology Management Challenges in the Automotive Industry. Warrendale, Pa.: Society of Automotive Engineers.

Helper, S. 2006. Testimony before the U.S.-China Economic and Security Review Commission. Hearing on China's Impact on the U.S. Auto and Auto Parts Industry, July 17, 2006. Available online at *http://www.uscc.gov/hearings/2006hearings/hr06_07_17.php*.

Helper, S., and S. Khambete. 2006. Offshoring and Value-Chain Architecture: The Case of Automotive Product Development. (Preliminary Draft) June 2006.

JAMA. 2007. Common Challenges, Common Future. Available online at *http://www.jama-english.jp/europe/auto/2007/jauto-mkrs_in_europe_2007.pdf*.

Maxton, G.P., and J. Wormald. 2004. Time for a Model Change: Re-engineering the Global Automotive Industry. New York: Cambridge University Press.

Moavenzadeh, J. 2006. Testimony before the U.S.-China Economic and Security Review Commission. Hearing on China's Impact on the U.S. Auto and Auto Parts Industry, July 17, 2006. Available online at *http://www.uscc.gov/hearings/2006hearings/hr06_07_17.php*.

NSF (National Science Foundation). 2001. Scientists, Engineers and Technicians in the United States: 2005. NSF 05-313. Arlington, Va.: NSF.

Smith, G. 2005. Mexico's Carmakers in a Ditch. Business Week, p. 58, June 13, 2005.

Womack, J., D. Jones, and D. Roos. 1990. The Machine That Changed the World: The Story of Lean Production. New York: Simon and Schuster.

Offshoring in the Pharmaceutical Industry

Mridula Pore, Yu Pu, Lakshman Pernenkil, and Charles L. Cooney
Department of Chemical Engineering
Massachusetts Institute of Technology

EXECUTIVE SUMMARY

A pharmaceutical company's competitive advantage is based on its reliance on basic science to create and develop new products. Increasing costs along the pharmaceutical value chain and an industry-wide decline in R&D productivity has placed considerable pressure on the industry to explore options for improving performance by reducing cost, increasing research productivity, and extending market penetration. Among the options are tactics for operating beyond the boundaries of companies' home countries for research, manufacturing, and sales.

A framework has been developed for investigating and assessing strategies associated with offshoring different segments of the value chain in the U.S. pharmaceutical industry. Cost, access to human capital, time to market, and market entry potential are the main drivers for offshoring. A large and expanding trained talent pool in India and China and growing infrastructure are enablers that attract multinational pharmaceutical companies to set up operations in these countries. Although government support, improvements in patent law, and growing capital markets in these countries will ultimately be the sustainers of the offshoring phenomenon in the pharmaceutical industry, the poor quality of talent, strict regulatory barriers, and cultural and economic barriers will have to be overcome for companies to maintain a competitive advantage via offshoring. The impact of offshoring on U.S. employment in the pharmaceutical industry is predicted to be minimal, and higher value-added services in the United States are expected to increase.

An interesting trend is the emergence of reverse offshoring. With the increasing success of manufacturing and research, Indian and Chinese firms are looking westward to acquire access to discovery in basic science and profitable markets by partnering or acquiring assets in the United States and Europe.

INTRODUCTION

Competitive advantage in the pharmaceutical industry first requires excellence in translating basic research and development (R&D) into new products, and then efficient manufacturing and distribution to high-margin markets in the United States, Western Europe, and Japan. Thus the traditional business model for multinational pharmaceutical companies (MNPCs) is R&D intensive, and the business is fully integrated to service key markets. R&D costs, as a percentage of sales, range from 15 to 17 percent, higher than for any other global industry. So-called "innovator" companies have invested in in-house R&D with the goal of developing a strong proprietary pipeline for new drugs. These companies, which are vertically integrated (Figure 1), are involved in everything from early-stage platform research, drug discovery, and regulatory development to the manufacturing, marketing, and distribution of their products.

In the last 10 years, MNPCs have been faced with a declining pipeline of new products, expiring patents, rising R&D costs, declining productivity, and pressure on drug pricing. From 2002 to 2004, only 58 new drugs received marketing approval from the FDA, a 47 percent drop from the peak of 110 new drugs in 1996 to 1998. In contrast, R&D spending rose from \$2 billion in 1980 to \$39 billion in 2006. Historically, only three out of 10 marketed drugs have produced revenues that matched or exceeded average

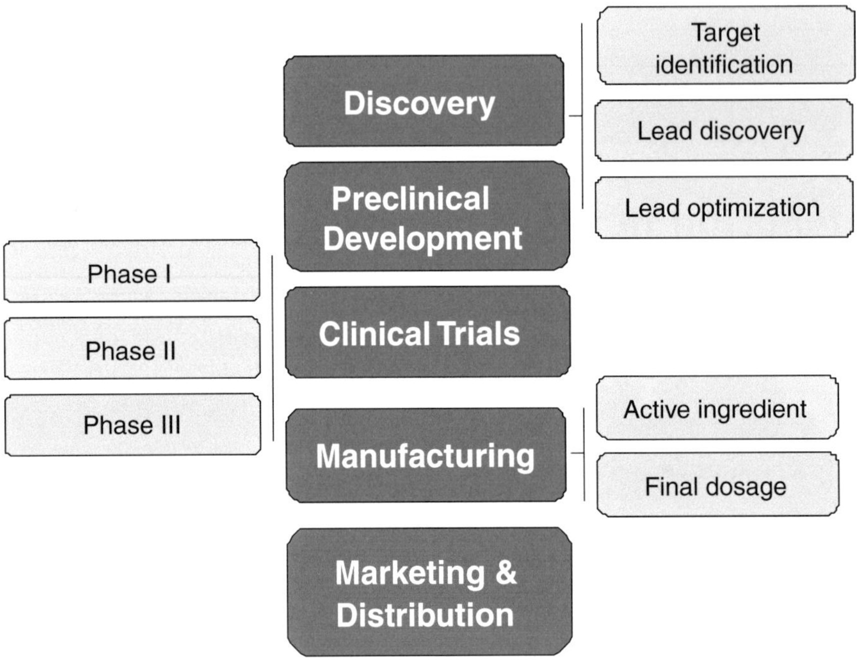

FIGURE 1 Value chain for innovator companies.

R&D costs, and it has become increasingly difficult to develop blockbuster drugs, such as Pfizer's Lipitor, which was introduced almost a decade ago in 1997. At the same time, the patents of many drug products are expiring, opening the market to competition from manufacturers of generic versions. The generic share of prescription drug units rose to 72 percent in just 18 months after generic substitutes were approved. [1] These pressures have convinced companies to consider offshoring parts of the value chain as a tactic to create competitive advantage.

This paper focuses on U.S. and European pharmaceutical companies that engage in offshore activities. Biopharmaceutical companies only recently have begun to embrace the offshoring paradigm because of complexity in the technology, their small size, and the regulatory environment surrounding their products. These factors are addressed further in later sections.

Global Employment in the Pharmaceutical Sector

Global employment in the pharmaceutical industry is estimated to account for 1.7 million full-time equivalents (FTEs). The industry is dominated by the top 20 global MNPCs, which account for 59 percent of employment. The United States, Europe, and Japan dominate the global pharmaceutical industry, and the United States has the largest single workforce by region (41% of the global workforce). [2]

Using data from the McKinsey Global Institute [2], one can see that manufacturing of the active pharmaceutical ingredient (API), or drug substance, and final dosage, or drug product, plus R&D occupy 44 percent of the workforce (Figure 2). Because these are the activities that require engineering and science expertise, they are the areas of focus in this paper. Because the industry is highly integrated throughout the value chain, these activities include engineering, such as chemical engineering, mechanical engineering, bioengineering, and materials science and engineering, as well as science, such as chemistry and biology. Furthermore, we are not aware of specific data that show the extent of involvement of each engineering and science discipline in pharmaceutical manufacturing and R&D.

Framework of Analysis

How and where a company chooses to operate its offshore activities depends on company-specific factors as well as location. Company-specific factors include the attitude of senior management and a company's regional capabilities and growth strategy. Location-specific factors fall into four categories: cost structure, business environment, workforce, and the local market. In our framework, we divide location-specific factors into drivers, enablers, sustainers, and barriers to offshoring activities at different stages of the value chain (Figure 3). We consider how these factors change over time and their impact on the global pharmaceutical industry.

The *Merriam-Webster Dictionary* defines offshoring as "The action or practice of moving or basing a business operation abroad." [7] The McKinsey Global Institute prefers the term "global resourcing," which has a more specific definition: "Decision of a company to have a location-

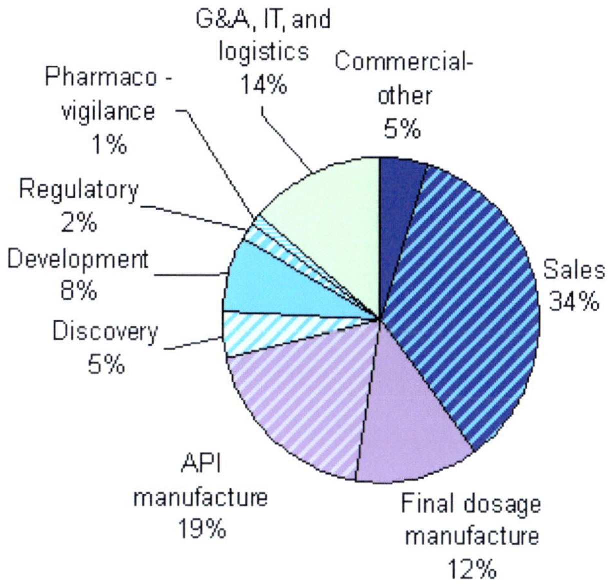

FIGURE 2 Global pharmaceutical employment, by function.

insensitive job performed in a demand market (market where the product is sold), in a border zone (near shore), or remotely (offshore)." [2] We believe it is important to explain the difference between offshoring and outsourcing. Outsourcing is defined as "procuring (as with some goods or services needed by a business or organization) under contract with an outside supplier." [8] International outsourcing is indeed one possible business model for a company offshoring some of its activities. The spectrum of operating models by which a company may operate offshore is presented in Figure 4.

Choice of Offshore Location

According to the AT Kearney Offshore Location Attractiveness Index survey in 2004, India and China are currently

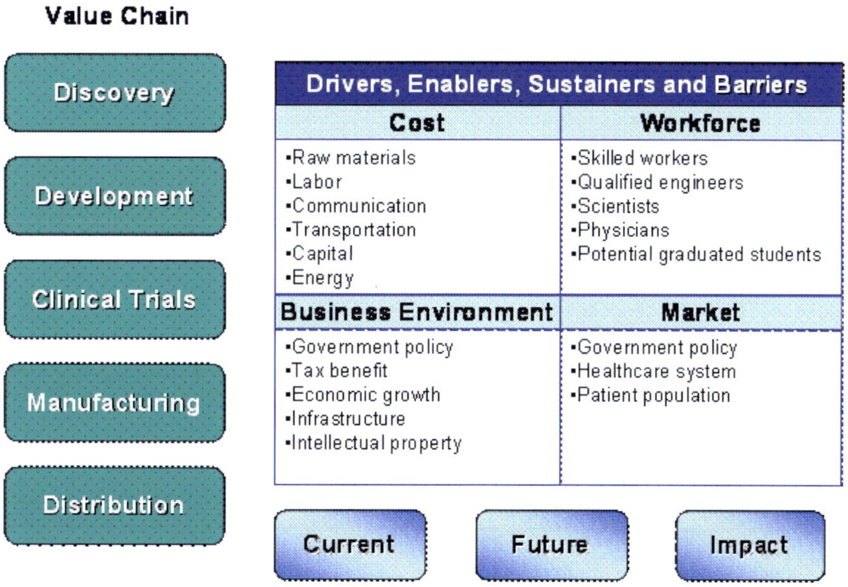

FIGURE 3 Framework of analysis.

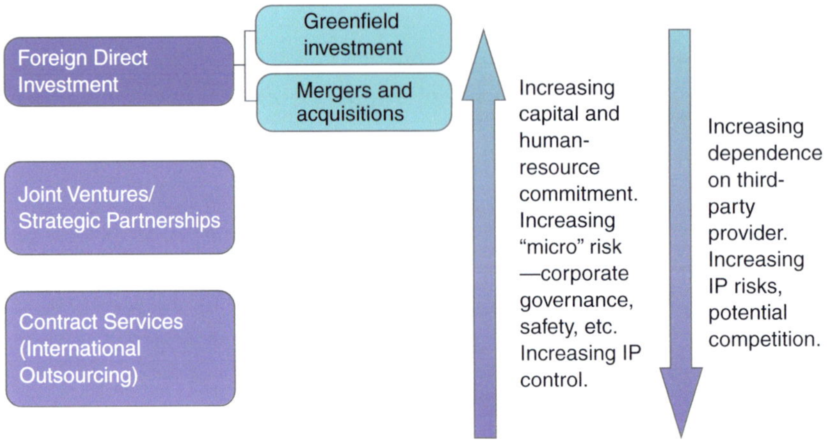

FIGURE 4 Offshoring business models.

the two most popular offshoring locations for a broad range of industry sectors because of their cost advantages and their depth and breadth of offshoring experience and people skills. [3] Malaysia, Singapore, and the Philippines also rank in the top 10, confirming the strength of Asian economies in offshoring competition. China and India are also ranked first and second in both the AT Kearney FDI Attractiveness Index and the Country Attractiveness Index for Clinical Trials. [4] Both countries have hosted offshoring activity in manufacturing and R&D in the pharmaceutical sector for several years, and this sector is still evolving and growing and is expected to have a considerable impact on the global industry. For these reasons, we focus mainly on offshore activities in these two countries.

Both China and India are net exporters of drug products. Indian exports of pharmaceuticals have been growing at a compound annual growth rate of 17 percent. China's exports are growing at a rate of 25 percent, to $5.7 billion in 2005. [5] As these economies develop, their domestic markets are also growing, involving both multinational and domestic companies. The largest players in each market are summarized in Table 1.

The exports are almost exclusively generic products, and the market is becoming increasingly competitive. As a consequence, only firms that can meet demanding pressures on manufacturing cost can compete, and margin pressure continues to erode profits. An important competitive attribute of these firms is their ability to continually improve manufacturing to reduce costs.

In the next two sections, we consider motivations for offshoring in the context of drivers, enablers, sustainers, and barriers associated with both China and India for R&D and manufacturing. Following this assessment, we address the impact of offshoring by U.S. companies to China and India and the effects on the U.S. pharmaceutical industry of a growing and increasingly aggressive domestic industry in India.

OFFSHORING IN PHARMACEUTICAL RESEARCH AND DEVELOPMENT

Overall, offshoring of R&D in pharmaceuticals is not very common but has been growing at a rapid pace in recent years. Outsourcing of drug-discovery services, such as chemistry,

TABLE 1 The Top Multinational and Domestic Pharmaceutical Companies by Market Share in China and India [5, 6]

	China	India
Multinational Companies	Pfizer Inc. AstraZeneca plc Roche AG Novartis AG GSK plc Bayer AG	GlaxoSmithKline Pharma Ltd. Pfizer Inc. Sanofi-Aventis Abbott Novartis AG Wyeth Merck Astra Zeneca plc Janssen-Cilag Infar India
Domestic Companies	Shanghai Pharmaceutical Group Co. Ltd. Guangzhou Pharmaceutical Holdings Ltd. Tianjin Pharmaceuticals Group Corp. Yangtze River Pharmaceutical Group Harbin Pharmaceutical Group Co. Ltd. Shijiahzhuang Pharmaceutical Group Co. Ltd. North China Pharmaceutical Group Corp. Beijing Double-Crane Pharmaceutical Co. Ltd. Northeast Pharmaceutical Group Co. Ltd.	Ranbaxy Laboratories Cipla Ltd Dr Reddy's Laboratories Ltd. Wockhardt Ltd. Nicholas Piramal India Ltd. Sun Pharmaceuticals India Ltd. Lupin Ltd. Aurobindo Pharma Ltd. Cadila Healthcare Ltd.

biology, screening, and lead-optimization, accounted for $4.1 billion in 2005, and is expected to approach $7.2 billion by 2009. [1] Starting with comparatively high-volume, low-value work, offshoring related to drug discovery has moved up the value chain to services ranging from preclinical chemistry to large clinical trials. Companies in India and China provide manually intensive but highly skilled outsourcing services that include nucleotide sequencing and synthesis, protein expression, and library construction. A few firms even provide chemical services in molecular biology and bioinformatics. In places with an established hospital infrastructure and support activities such as India, clinical trails with good supporting analytical work are becoming increasingly common, not only because of cost, but also because of access to skilled workers, treatment-naïve, and well stratified patient populations and the prospect of reducing development time.

Offshore Research and Development in China

Traditionally, foreign firms have shied away from investing in R&D in China because of the widespread prevalence there of generic brands and counterfeit drugs, inadequate IP protection, and Chinese consumers' inability and unwillingness to pay for expensive medicines. Even today, according to a recent report from Ernst & Young, only about 20 percent of the world's leading pharmaceutical companies have plans to invest in R&D in China. [10] According to Kalorama Information, global pharmaceutical firms will outsource about $3.5 billion in research in 2006, but less than 5 percent of that is earmarked for China. [11]

China's current pharma R&D environment (Figure 5) is reasonably advanced in clinical trials and lower complexity chemistry but less so in preclinical and biology-based drug discovery. [9] Although MNPCs have been cautious about

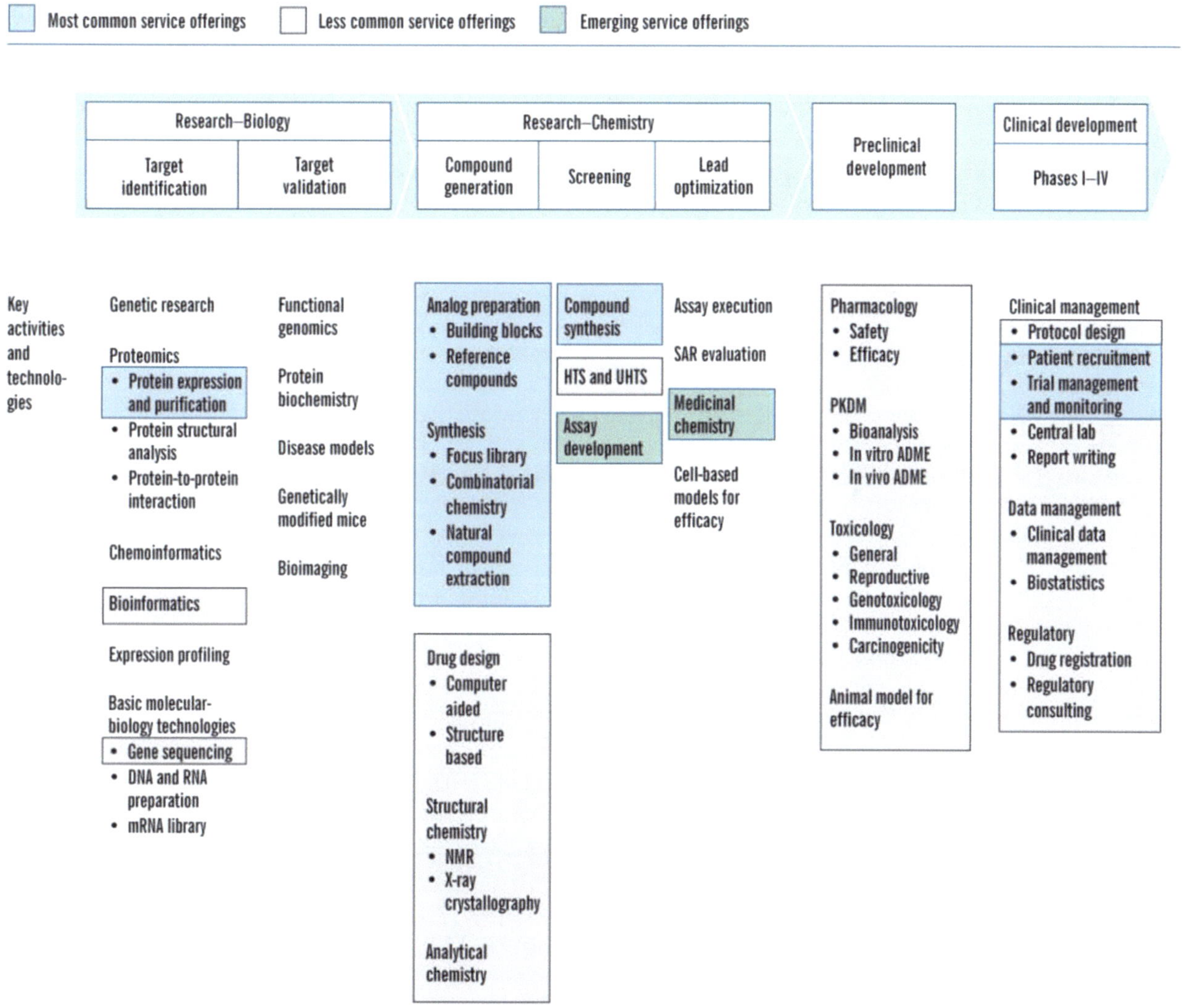

FIGURE 5 China's current environment for pharmaceutical R&D. Reprinted with permission of ©The Boston Consulting Group. All rights reserved. [9]

offshoring R&D to China, the scope and scale of these activities have risen. Today, almost all of the top 20 MNPCs are doing some form of chemistry-based work in China. Prior to 2001, 40 percent of Chinese medical enterprises had cooperative projects with foreign firms. [12]

A few leading MNPCs have built their own R&D centers in China. Novo Nordisk, the first MNPC to establish its own R&D center in China, set up a $10 million center in 1997 to conduct research in industrial biology and pharmaceuticals focused on natural products. After spinning off its industrial-enzymes divisions in 2000, the new Novo Nordisk designated China as its global center for competency in microbial protein expression. The company plans to double the staff in the next two to three years to 60 scientists and gradually transition from a highly skilled biotechnology provider to an innovator in target identification for cancer and inflammatory diseases. [9]

GlaxoSmithKline (GSK) has worked with Chinese scientific and research groups on several occasions. At the beginning of the 1990s, the company cooperated with Shanghai Institute of Material Medica (SIMM) in evaluating approximately 10,000 herbal medicines and undertook collaborative projects worth $7 million. Since the merger, GlaxoSmithKline has invested more than $10 million in R&D projects in China. [13]

Roche invested more than $10 million in a new R&D center in Shanghai's Zhangjiang High-Tech Park at the end of 2004. Currently, the center has 40 scientists working on basic chemical synthesis. The center plans to begin research in traditional Chinese medicine and is expected to gradually develop more comprehensive R&D capabilities. [9]

Twenty percent of Lilly's chemistry work is being done in China, where costs are one-quarter of what they are in the United States or Western Europe. Lilly helped start a laboratory, Chem-Explorer, in Shanghai in 2003. The start-up company works exclusively for Lilly and has a staff of 230 chemists, 20 to 25 percent of whom have Ph.D.s. In addition, Lilly does about 50 percent of its clinical research outside the United States, mostly in Western Europe. However, it has been predicted that Lilly will do 20 to 30 percent of its testing in China and India in the next few years. [14]

AstraZeneca was one of the first MNPCs to set up clinical trials in China in 2002. [15] In December 2003, the company announced a $374,000 three-year partnership with Peking University's Guanghua School of Management to fund programs at the China Centre for Pharmacoeconomics and Outcomes Research to support reform of China's health care system. [16]

Pfizer, one of the largest foreign pharmaceutical enterprises in China, has more than 1,500 employees in four state-of-the-art plants throughout the country, as well as a management center and a trade company. Pfizer China located an R&D center in Shanghai, following the lead of AstraZeneca and Roche. Part of the Shanghai center's strategic plan is R&D on biometrics, which would support the development of new drugs. [17]

Other large MNPCs, such as Servier, Novartis, and Sanofi-Aventis, are also planning to support research in China on compounds from traditional Chinese medicines as a basis for drug discovery. Novartis has announced its intention to make captive research investments (i.e., establish its own facilities) in China.

Offshore Research and Development in India

A 2004 survey of 104 senior executives in a wide range of industries, including eight pharmaceutical companies, ranked India among the top three countries where they planned to spend R&D dollars in the next three years. [65] With some of the top technical universities in Asia, a large community of entrepreneurs, Western-trained graduates, resourceful managers, and researchers who are at ease with the English language, India has a welcoming business environment for global collaboration in R&D. [18]

Global R&D companies, such as U.S.-based AMRI and Nektar, Switzerland-based Evolva, and Germany-based Taros, have already opened research facilities there. In 2002, about 40 global trials were conducted in India, and in 2005, the number rose to about 200. [18] Many leading MNPCs have invested in R&D work in India (Figure 6). For example, Pfizer doubled its investment in clinical research in India to roughly $13 million and plans to invest another $30 million in the next five years. [19] AstraZeneca made an early investment in the late 1980s in a captive R&D center in Bangalore, where its new candidate drug molecule for tuberculosis is under final development. In addition, the company has forged a partnership with Torren Pharmaceuticals to work on a drug for hypertension. [18]

Novartis has entered an agreement with Syngene International, a biopharmaceutical company based in Bangalore, to carry out R&D to support new drug development. The research teams in Syngene, with skills in synthetic chemistry and molecular biology, also conduct high-value R&D in early-stage drug discovery for other global clients. [20]

Compared with China, India has a relatively well developed R&D environment in clinical trials and basic chemistry, in contrast to biology and preclinical work (Figure 7). [21] However, China is more advanced in the field of proteomics and molecular biology for target identification, while India is better at clinical data management and lead optimization work. [22] Thus different areas of the R&D value chain are being conducted in China and India, but neither country has an environment that supports end-to-end R&D.

DRIVERS, ENABLERS, BARRIERS, AND SUSTAINERS FOR RESEARCH AND DEVELOPMENT

Drivers

There are multiple complex reasons that MNPCs are offshoring R&D work to India and China. According to a study by Thursby of R&D intensive firms, the drivers for offshoring

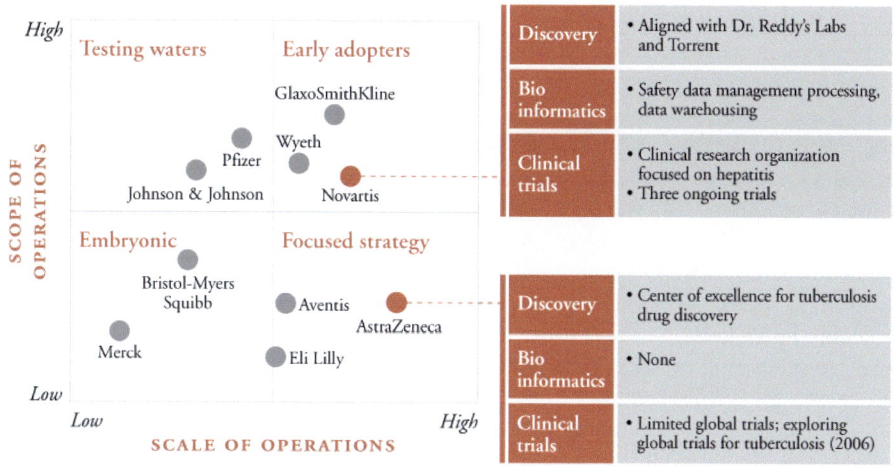

FIGURE 6 R&D activities by MNPCs in India. © 2004, A.T. Kearney, Inc. Reprinted with permission. [19]

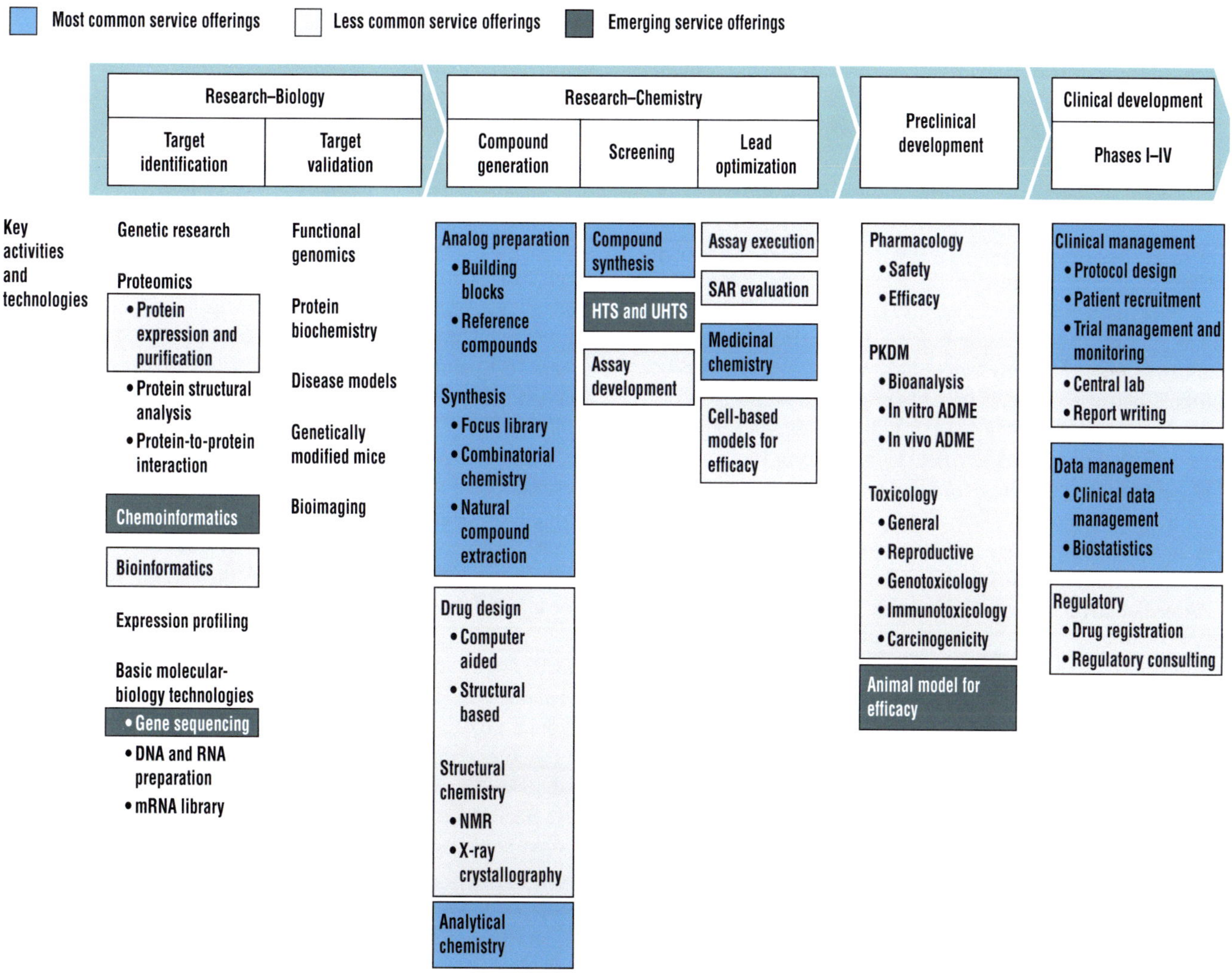

FIGURE 7 Various opportunities along the value chain. Reprinted with permission of ©The Boston Consulting Group. [21]

include low cost, market factors, the quality of R&D personnel, and collaboration with university scientists. [66]

Cost and Time

The cost of bringing a drug to market is more than $800 million and can take 8 to 12 years. Of every 5,000 drugs tested, only about five reach the clinical trial stage, and only one is approved by the FDA. [23] This high "failure" ratio adds significant risk to pharmaceutical R&D, forcing major pharmaceutical companies to focus on fewer projects to address increasingly specific indications. Cost advantage is one of the driving forces for offshoring of R&D. In general, direct cost savings can be as high as 60 percent, or even 80 percent, on salaries in the drug-discovery phase, and 60 to 70 percent per patient in clinical trials. [18]

Biologists in China are paid 20 to 33 percent of what similarly qualified biologists are paid in the United States. The average annual salary of a full-time employee with a Ph.D. in an MNPC in Shanghai is about $12,500, approximately one-fifth the salary in the United States. Because clinical trials account for 40 to 75 percent of drug-development costs, savings in this phase of R&D can be significant.

In India, clinical trials cost as little as 40 percent of those conducted in Western countries. [24] For example, in a clinical data-management center established by GSK in Bangalore, the combined salaries were barely one-third of salaries for an equivalent center in the United States; GSK had an annual cost saving of $30,000 per employee. Staff of the center has been expanded from four 10 years ago to 300 today.

Along with the lower costs, drug development time is much shorter. In low-cost countries, Phase III clinical trials can be completed six to seven months faster than in domestic markets because of faster patient enrollment and higher patient concentration. [18] For example, the German manufacturer Mucos Pharma asked SIRO Clinpharm in India to find 750 patients to test a drug for head and neck cancer. Within 18 months, the company had recruited enough volunteers in five hospitals. In Europe, it took twice as long to find just 100 volunteers in 22 hospitals. [24]

Another advantage of offshoring R&D is multi-shift work across multiple time zones. For instance, scientists in the United States can focus on more complex processes while offshore staffs perform the repetitive tasks. In this way, MNPCs gain flexibility in their pipeline management.

Market Potential

MNPCs with offshore activities and investments in China and India often seek access to the domestic markets as part of their global market strategy. (Market growth in both countries is described later.) Offshore R&D allows companies to build close relationships with local governments, research institutes, and hospitals that can help secure their positions in the local market.

In addition, offshore R&D brings companies closer to the demand and dynamics of the local market. In China, for example, the lifestyle is increasingly influenced by Western culture, leading to changes in the disease profile. As living standards rise, particularly in the cities, a number of formerly common diseases and conditions associated with poverty have been almost entirely eliminated. At the same time, higher incomes, new diet patterns, less physical exercise, and more work-related stress, including a recent decline in job security, have combined to increase the incidence of diseases new to China but common in Western countries, such as diabetes, cardiovascular disease, and other stress-related disorders. [25] In addition, the aging population in China is growing as life expectancy increases annually, and the birth rate is declining. People over 60 now account for 10 percent of the total population, and this number is expected to rise to 30 percent within five decades. [25] By 2020, people 65 or older will account for 16 percent of China's population.

These trends point toward a larger, more diversified market demand for drugs in the future. As per capita GDP rises, purchasing power will also rise, enabling sales in the pharmaceutical market to increase by 6 to 8 percent annually. In addition, Western pharmaceuticals and diagnostics are increasingly believed to be more effective than domestic versions or traditional Chinese medicines.

A Large Talent Pool

Finding qualified scientists, engineers, and physicians is essential to offshoring R&D. China and India, which have large talent pools, make it possible for R&D work to be carried out at lower cost. However, there are still questions about quality, such as whether there are enough well qualified researchers to maintain or even improve the quality of research. In a study conducted by Gary Gereffi and Vivek Wadhwa at Duke University [69], the numbers of engineering bachelor's degrees and associate degrees awarded annually by India were reported to be 112,000 and 103,000, respectively. For China the numbers were 351,537 and 292,569, respectively, about 2.5 to 3.5 times higher than in the United States. In addition, in China the number of doctorates in domestic science and engineering has increased rapidly. From 1975 to 2005, China's global share of science and engineering (S&E) doctorates increased from near zero to 11 percent; at the same time, U.S. global share fell from half to roughly 22 percent. [70]

Another component of the talent pools in China and India is doctorates earned overseas. In 2001, the number of Chinese S&E doctorates earned in Japan, United Kingdom, and United States equaled 72 percent of the total S&E doctorates earned by American citizens and permanent residents. [70] From 1986 to 1998, of all S&E doctorates earned in U.S. universities, Chinese students accounted for 8.4 percent in biological and agricultural science and 9.1 percent in engineering. [71] From 1993 to 2000, the total number of engineering doctorates awarded in U.S. universities fell slightly, from 2,228 to 2,206; however, doctoral awards in engineering to Chinese citizens increased 30 percent from 543 to 711 in the same period. [72]

In recent years, with the booming economies in China and India, more and more Chinese and Indian scientists and engineers, especially in high-tech fields such as biomedical studies, have chosen to leave the United States for home and have taken their technical skills with them. According to China's Bureau of Education, since 1978, about 700,000 Chinese college graduates and scholars have gone abroad for advanced degrees, and about 170,000, or 24 percent, have returned. A high portion of the graduates earned degrees in chemistry and life sciences. Currently, 40 to 60 percent of postdoctoral students in the United States are from China and Taiwan. Within 10 years, there may well be a reverse brain drain in U.S. biotechnology. [9]

The scientific disciplines most relevant to the pharmaceutical industry are chemistry and biology. Graduates in chemistry in both China and India outnumber their U.S. counterparts by more than fivefold at the bachelor's level and more than threefold at the master's level. [18] Even correcting for variations in quality, these large numbers provide an impetus for moving higher value work offshore.

Enablers

Resources

Another important factor driving MNPCs to offshore their R&D work to China is that valuable resources might be discovered from traditional Chinese medicine (TCMs), 12,807 medicinal materials derived from natural sources, about 5,000 of which may have some proven clinical efficacy. TCMs' share of the global market in herbal medicines ($60 billion in 2002) is expected to rise to $5 trillion by 2050. [18] The expected advantages of TCMs for MNPCs is that they may provide drug-discovery leads and diversify an MNPC's pipeline.

In addition, both China and India offer access to the broad human gene pool and patient population. Data on different populations is becoming increasingly important as the industry shifts from developing blockbuster drugs to drugs targeted at patient populations with specific genetic polymorphisms. [26] The large patient pool (and large number of treatment-naïve patients) makes it easier and faster to enroll patients in clinical trials.

Infrastructure

China has 185 bio-related institutes and research laboratories, 1.4 million doctors, more than 1 million nurses, and 20 facilities with GLP (Good Laboratory Practices) certification. As many as 300 contract research organizations (CROs) now offer support for clinical trials, which also provides an infrastructure to support the offshoring of R&D work. In India, half a million doctors, 171 medical colleges, and 16,000 hospitals provide a broad infrastructure for offshoring clinical R&D. [27] Six laboratories in India have secured GLP certification, and a dozen more are about to. In addition, more than 20 CROs in India now handle Phase II through Phase IV trials.

Sustainers

Government Support

According to the director of the Pharmaceutical Department, which is overseen by the State Economic and Trade Commission of China, the Chinese government encourages foreign pharmaceutical companies to expand their businesses from just manufacturing to include R&D. They promise that foreign-funded research centers will be exempt from import tariffs and custom taxes. In addition, companies that transfer technology to China will be exempt from business taxes. [28]

The list of key focus areas in the current Five-Year Plan includes biotechnology and innovative drug discovery. Funding in some areas of biomedicine and biotechnology—most notably genetics—has increased rapidly in the past few years. [29] From 2000 to 2005, an average of $600 million in public funds went to China's biotechnology sector. India's Department of Biotechnology has funded more than 1,800 R&D projects, helped to develop 12 vaccines, and transferred 54 technologies to the biotechnology industry, 17 of which have been commercialized. [18]

Many life-science parks, such as Shanghai Zhangjiang Life Science Park, have been established to encourage foreign investment in the pharmaceutical and biological sectors. These parks, which are focal points for the clustering of similar companies, offer MNPCs basic amenities and fiscal and regulatory incentives. A good example is the Beijing Economic and Technological Development Zone, in which both domestic and foreign companies are exempt from taxes for two years after they start making profits. For the next three years, they are taxed at half the normal rate. [30] By the end of 2005, there were 60 such parks in China and five fully operational parks in India and 17 more at various stages of planning or construction. [18] These special economic zones attract foreign direct investment (FDI) in knowledge- and manufacturing-based businesses, and thus attract offshoring by foreign firms.

Improvements in Patent Protection

Sustainable development of offshore pharmaceutical R&D requires a well regulated business environment and a well established legal system to protect MNPCs from misappropriations and infringements of patents and from counterfeit drug makers. A new law in China, the New Medicine Examining Statute, encourages innovation by controlling prices and protecting intellectual property. First, it extends the protection period for new medicines, in some cases from eight years to 12 years. During the protected period, only licensed companies can produce the drug in question. Second, profit margins for new medicines can be higher than for other products, so manufacturers can recoup the costs of R&D more quickly. Third, the government is reducing bureaucratic red tape by contracting out the licensing of new medicines and production plants. All of these measures will stimulate investment, improve R&D, and cut the time-to-market for new medicines. [31]

In September 2003, the Chinese government also passed a regulation for implementing the Law on Drug Administration. The regulation defines new drugs as "drugs that have not appeared in the domestic market," a stricter standard than the old rules that defined new drugs as "drugs produced in China for the first time." The new standard has unnerved many domestic pharmaceutical research institutions, whose main products are imitations of sophisticated foreign drugs. According to the old rules, these drugs could be patented as new drugs only if foreign drug makers had not manufactured the originals in China. [32] Although the trend appears to be toward greater protection of IP, it will take some years for sufficient case law to establish how the government will actively protect the IP rights of foreign firms.

Barriers

Regulatory Barriers

Even though the business environments of China and India have improved in recent years, some regulatory barriers still impede MNPCs' offshore R&D activities. For example, in India, new chemical entities discovered outside the country must undergo initial Phase I trials outside the country; only then can a Phase I trial be conducted in India. [18] This delays the time-to-market for new drugs.

In China, slow approval time (usually 9 to 12 months) is a serious problem. The process of registering a drug and obtaining production and sales permits involves numerous central, provincial, and local authorities and can take several years. [26] In India 3 to 4 months is the norm.

Supply of High-Quality Talent

The large pool of scientists and engineers in China and India is one of the attractions for offshore R&D. However, with the rapid growth of offshore activities and competition from the growing number of domestic companies, the demand for qualified engineers is increasing. For example, in India, the share of global clinical trials is expected to rise from the current level of 1.5 percent to 15 percent by 2011. In addition, the number of global trials is increasing by 10 percent per year. At current training levels, India will turn out only one-tenth the required numbers of clinical research assistants. [18] Thus early movers in offshoring of clinical trials will have an advantage; later entrants will have to work harder to find trained staff. In addition, this competition for skills will accelerate wage inflation and erode some of the cost advantages of offshoring.

Another problem for MNPCs is that, although the potential supply of talent in low-wage countries is large and growing rapidly, only a fraction of potential job candidates are qualified to work for foreign companies. The reasons for the lack of suitability are inadequate language skills, poor quality of education, and limited practical experience. Another problem is cultural differences, which are especially apparent in interpersonal skills and attitudes toward teamwork and flexible working hours. [1]

According to Wadhwa and Gereffi's survey results, multinational and local technology companies in China felt comfortable hiring graduates from only 10 to 15 elite universities across the country and complained that the supply of these graduates was limited. [73] Interviews with 83 human-resource managers in multinational companies reveal that, on average, only 17 percent of engineers and 14 percent of researchers in the life sciences were suitable for hiring by foreign companies. Among all candidates, only 10 percent in China and 25 percent in India would be suitable for offshore R&D by MNPCs. As McKinsey reports, only 2.8 to 3.9 million—or 8 to 12 percent—of young professionals in low-wage countries are suitable for hire by export-oriented services companies, compared to 8.8 million in the sample of high-wage countries. [1] The scarcity of experienced and skilled middle-management-level workers for offshoring companies is even more serious.

In China, for cultural and historical reasons, students are not encouraged to think innovatively. However, innovative thinking is the quality that pharma R&D thrives on. The ratio of graduate students to professors in China can be as high as 20 to 1, compared with a 3-to-1 ratio in the United States. Physicists, chemists, and engineers dominate the talent pool in China. Although the output in applied biology has increased rapidly over the past decade, the percentage of biotechnology- and biology-related fields in China is still modest. Furthermore, it is estimated that just half the potential talent pool in China is geographically accessible to multinational companies.

Protection of Intellectual Property

As discussed earlier, R&D work conducted in developing countries is fragmented and concentrated mostly in relatively lower value-added areas of chemical synthesis and routine analysis. MNPCs tend not to offshore their most proprietary R&D activities because of uncertainties about the protection of intellectual property. [67] These same uncertainties may encourage MNPCs to pursue only fragmented work offshore and not to work across the entire value chain.

Venture-Capital Funding

Besides establishing fully owned subsidiaries in China and India, MNPCs can offshore R&D work to CROs and through partnerships with local firms. Establishing a high-technology company, such as a pharmaceutical research firm, is capital intensive, and there may not be a short path to profitability. For such firms to be established, funding sources and legal and business infrastructure must be available. Therefore, the current lack of established venture capital (VC) firms and funds in China and India represents a barrier to offshoring activity.

The VC industry is at an early stage in both China and India, where most funding has traditionally come from government, financial institutions, and individuals. Currently, most VC funding is from foreign firms, although domestic VC companies are emerging. As wealth accumulates in China and India, private-equity funding may play a larger role. Government policy toward the regulation of finance and investment will certainly influence the extent of domestic and foreign investment.

In India, returning expatriates, particularly from Silicon Valley, have encouraged the establishment of a regulated VC industry. [33] The Securities and Exchange Board of India (SEBI), which regulates the stock market, is now also responsible for regulating VC funds. The first regulations, issued in 1996, offer tax benefits similar to those of U.S. limited partnerships. There are currently 84 VC funds and 54 foreign VC funds registered with SEBI, and many other funds are still unregistered. [34] Although the goal is to promote an exit strategy, the mechanism by which venture capitalists recoup their investments through an initial public offering (IPO), most investment exits are currently realized through mergers and acquisitions. The barriers to VC funding in India include the reluctance of businesses to give up their majority stake to an investor, the lack of fund-management experience, and the lack of infrastructure to provide legal and business support. However, the presence of domestic stock exchanges, a history of domestically managed mutual funds, and a growing entrepreneurial spirit are contributing to confidence in VC investments. [33, 35]

In China, VC funding has been growing rapidly, from just $418 million in 2002 to $1.27 billion in 2004 [36], and the Chinese Venture Capital Association (CVCA) has become an umbrella organization to promote the industry. [37] Exits from venture investing are predominantly in the form of IPOs on foreign exchanges; some are realized through mergers and acquisitions. [36, 37] The main concerns about VC funding in China are the lack of a domestic exchange for IPOs, the lack of experienced fund management and legal capability, and, given the weak IP regime, the inability of companies to retain value from technology.

OFFSHORING IN PHARMACEUTICAL MANUFACTURING

Pharmaceutical manufacturing encompasses a variety of process technologies on different scales. Primary manufacturing involves synthesis of the drug substance, also called the active pharmaceutical ingredient (API) or bulk drug substance. This is followed by secondary manufacturing, which involves drug-product formulation; in this stage the drug is produced in its final dosage form. The last stage involves the filling, finishing, and packaging of drug products for distribution to patients. These stages are often performed at different sites and may be broken down into further steps. For example, in API manufacturing, it is common for chemical intermediates to be supplied by one company to another. The technical and regulatory requirements for the manufacturing facility depend on whether the drug is a chemical or a biological product. High-potency drugs and biologics typically require more containment, hence more infrastructure and stricter maintenance procedures. The volume of the drug depends on its potency and the frequency of dosing. Low-potency drugs that require frequent dosing are produced in large volumes. High-potency drugs that are used sparingly are produced in low volumes. Thus there is a continuum in the size and scale of manufacturing facilities.

Manufacturing of Active Pharmaceutical Ingredients

Manufacturing of the API is frequently offshored by outsourcing to a third party. The primary motivation is cost efficiency. Because FDA approval is required for facilities and processes in the United States or abroad that supply product to the United States, the quality of the API is guaranteed. In China, the world's largest producer of APIs, sales are expected to increase by 17.6 percent in the next few years, from $4.4 billion in 2005 to $9.9 billion in 2010. In India, the third largest global manufacturer (after Italy), sales are expected to increase by 19.3 percent per year, from $2 billion in 2005 to $4.8 billion by 2010, according to a study conducted by Italy's Chemical Pharmaceutical Generic Association. [38] APIs accounted for 60 percent of pharmaceutical exports from India in 2001. [39]

The APIs manufactured in offshore facilities are almost all generic, and thus off-patent products, the point at which cost savings on manufacturing provides a competitive advantage. Patent protection for these products has expired and non-infringing processes can be developed and used for manufacturing them, thus lowering IP concerns. The expansion of the generics market is expected to continue, both in absolute terms ($2 billion growth between 2000 and 2002) and as a percentage of contract manufacturing in India (expected to increase from 20 percent in 2000 to 62 percent in 2010). [5]

An interesting change may be in the wind, however. Dishman Pharmaceuticals and Chemicals (Ahmedabad, Gujarat, India) recently announced that it is the first Indian firm selected by an MNPC as primary manufacturer of an API for a brand new drug. [40] With this business model, the innovator firm can leverage low-cost production before the drug has generic status. Success will depend on protection of IP for the product and process.

Indian companies are becoming more sophisticated. Companies that started as contract manufacturers for intermediates, and then APIs, are becoming vertically integrated and moving into drug-product formulation. This is possible because of improvements in R&D skills, which have enabled them to challenge patents and adopt an aggressive acquisition, IP-based approach to expansion into regulated markets. For example, Ranbaxy USA has submitted more than 20 abbreviated new-drug applications (ANDAs) to the FDA for review of generic products. The Ranbaxy group acquired OHM Laboratories (USA) manufacturing facilities in 1995 and European generics, including Bayer AG, RPG (Aventis), Terapia SA, and Ethimed NV. [41] Nicholas Piramal acquired Avecia and Pfizer's manufacturing site in Morpeth, U.K. [42]

The rise of the Indian pharmaceutical industry, with expertise in reverse engineering and patent challenging, could have a significant impact on the global generics market. In effect, these firms are practicing reverse offshoring by reaching back to U.S. and Western European firms for skills to fill out the value chain.

Indian and Chinese companies are increasingly interacting with each other to leverage their unique strengths. For example, India has emerged as a preferred trading partner with China; India's imports of pharmaceutical products from China increased by 172 percent in 2004 to $303 million in 2005. China is also the leading pharma export market for India. In 2005, imports from India were valued at $58 million. By contrast, U.S. drug product exports to China were valued at $29.5 million. [43]

Manufacturing of Final-Dosage Products

The growth in offshore secondary manufacturing appears to be driven by a combination of both low-cost manufacturing structures and the growth of domestic pharmaceuticals markets. Low-cost manufacturing in China and India enable companies to sell pharmaceuticals at prices affordable to the local population. Low-cost manufacturing also enables penetration into other developing markets where the costs of pharmaceuticals are prohibitive, such as in Southeast Asia and Africa.

Final-dosage manufacturing in India and China is done by a mix of third-party outsourcing and foreign direct investment in manufacturing facilities run by Indian subsidiaries of MNPCs. The company websites of GSK, Pfizer, Wyeth, Aventis, and Abbott (five of the top six MNPCs by domestic sales in India) [36] indicate that they have established manufacturing sites in India to cater to the Indian market and for exports, mainly to Middle Eastern and Asian markets. They also provide some external manufacturing services, including API manufacture, but the focus of these operations is on secondary manufacturing. It stands to reason that, if they can meet the tough cost demands for local sales, they can also leverage higher margins on sales in the regulated markets.

Manufacturing of Biologics

Although both India and China have substantial and growing biopharmaceutical industries, offshoring of biopharmaceuticals manufacturing is still small by global standards because of the nature of these products and processes, the lack of regulatory clarification, and the relative immaturity of the industry. Patents on the first generation of biopharmaceuticals are beginning to expire, but the FDA has not yet issued clear guidelines for how bio-similar or follow-on biologic products should be assessed for safety and efficacy as generic-like substitutes. Such products can enter the marketplace but only after clinical trials have been completed.

Thus the concept of generics does not apply to biological products as it does to chemical drugs. Biological therapeutics cannot be as easily characterized by physico-chemical methods or bioassays; hence their safety and efficacy depend more strongly on the manufacturing process. Thus it can be difficult to transfer a product to a different manufacturing site, which may require clinical evaluation.

However, the technology for characterizing biologicals is evolving rapidly. There is a continuum of molecular complexity in biologicals reflected in the molecular weight and extent of post-translational modification of the molecule during synthesis. Some smaller molecules, such as insulin, which have been manufactured for a long time, are sufficiently well characterized that injectable insulin can be manufactured by numerous companies. [44] The European Medicine Evaluation Agency has published guidelines, including comparison guidelines, for products manufactured at multiple sites. [45] However, the lack of clarification by the FDA poses a barrier for companies interested in producing bio-similar or outsourced products for the U.S. market.

DRIVERS, ENABLERS, BARRIERS, AND SUSTAINERS FOR OFFSHORE MANUFACTURING

Drivers

The primary drivers for offshore manufacturing are low-cost operations and access to rapidly growing pharmaceutical markets in India and China. If a company can manufacture and produce at a cost low enough to be competitive in emerging markets and still be in compliance with FDA requirements, then that company can expect to be cost competitive in regulated markets. In the future, as the efficiencies of manufacturing processes by emerging Chinese and Indian companies improve, MNPCs will have more opportunities to offshore their non-core manufacturing activities. In addition, as Indian and Chinese companies become more innovative, competition to supply the global market will increase, driving improvements in both cost and technology.

Low Cost Structure

Both India and China have lower capital, labor, and raw-material costs than manufacturers in Western Europe or the United States. The largest savings (approximately 60 percent) for these companies is in labor costs. Total cost savings are estimated to be $10,000 per million tablets. [46] Arthur D. Little Benelux estimates annual per-person labor costs at $3,000 in India and $4,000 to $6,000 in China. The cost in Western Europe is well over $50,000. Outlays per installed cubic meter of reactor capacity are at least 40 percent lower than in the West and can be as much as 90 percent lower. [42]

Growing Markets

The value of Chinese and Indian pharmaceutical markets is considerably less than the value of the market in the United States. However, with an expanding, increasingly affluent middle class willing to pay out of pocket for treatment, the markets in India and China are growing. Increased sales of existing drugs at low prices and a wider range of new products in the market are reflective of the growing number of people who can afford more therapies and are demanding world-class treatment. Thus opportunities abound for pharmaceutical companies to expand their operations.

India's pharmaceutical market, which was estimated to be worth $4.5 billion to $4.9 billion in 2004, has grown steadily for the past 15 years. It is estimated that value will rise from $5.3 billion in 2005 to $16 billion in 2015. [18]

In China, the pharmaceutical industry is one of the fastest developing sectors, driven by the medical needs of the country's 1.6 billion people. During the 9th Five-Year Plan (1996–2000), the average annual growth rate of the pharmaceutical industry was 17 percent. For comparison, the rate worldwide is 13 percent. Biotech-based pharmaceuticals in China were worth about 20 billion RMB in 2002, or about 6 percent of the total value of the pharmaceutical industry. This share is predicted to rise to 12 percent in 2006. [47] Estimates of the Chinese market vary widely. IMS estimates that the value was $11.7 billion in 2005 and will be the seventh largest in the world by 2009. [5] In a BCG report, it is estimated that China will become the fifth largest drug market, with a value of $37 billion, by 2015.

Enablers

Experience and Existing Manufacturing Infrastructure

Domestic chemical and pharmaceutical industries grew rapidly in India following the passage of the 1970 patent law recognizing process patents but not composition-of-matter patents. Similarly, in China companies have developed expertise in the reverse engineering of drugs available in Western markets. As a result, there is now a large, experienced workforce with considerable knowledge about the process science and engineering of pharmaceuticals. This talent pool for MNPCs makes it possible for domestic companies to be innovative in designing non-patent-infringing processes. There is also considerable manufacturing infrastructure already in place, such as manufacturing plants and equipment vendors to supply the industry.

Consolidation and Standardization

Medium-sized industries in both China and India are consolidating, and many smaller manufacturing units are closing down. The top 20 companies in India increased their market share from 29 percent to 56 percent in 2004, reflecting this trend. [5] As a result of these consolidations, the remaining facilities are increasingly able to meet international operating standards, which is likely to increase confidence in India as a global supplier. The Drug and Cosmetics Act of 1940 was modified to encourage the standardization of drug manufacturing. [43] Many plants in India are also approved by regulatory bodies, such as FDA, EMEA, MCA-UK, and TGA-Australia. In fact, India has the largest number of FDA-approved facilities outside the United States. Ernst and Young predict that in the future Indian companies will fall into one of three categories:

- global companies that offer both generic and brand-name drugs and co-promotion deals
- medium-sized and large companies resulting from the consolidation of equally sized small to medium companies
- companies that have reduced their scope of operations and specialize in a niche activity

In China, the number of pharmaceutical manufacturers is decreasing, but the productivity and scale of manufacturing is increasing. It is estimated that there are 3,000 GMP-certified manufacturing facilities in China today. [43]

Skilled Workforce

China and India have large and growing numbers of suitably trained graduates in engineering, life sciences, and pharmaceutical science. However, only a fraction of this population is suited to working in international companies. Because both countries are large, part of the talent pool may be inaccessible at the desired locations. A McKinsey report that provided data on the supply of engineers and life-science researchers in China, India, and the United States for 2003 projected the compared annual growth rate for 2003–2009 (Table 2). [1]

The pharmaceutical-science talent pool in India can be estimated based on the number of academic institutions. The All India Council for Technical Education has approved 445 institutes with a combined annual intake of 24,670 students for the diploma or bachelor's degree in pharmacy. In addi-

TABLE 2 Supply of Engineers and Life-Science Researchers in China, India, and the United States, 2003

Theoretical Maximum Talent Supply (in thousands)	Engineers	CAGR	Life-Science Researchers	CAGR
China	1,589 (159)	6%	543 (54)	6%
India	528 (132)	6%	674 (101)	4%
United States	667 (538)	2%	852 (692)	–2%

Note: Numbers in parentheses are workers suitable for recruitment.
Source: Adapted from Das, 2006. [1]

tion, 132 institutes have been approved for students pursuing master's degrees in pharmacy; these institutions take in 2,680 students annually. [48]

Health Insurance

The health-care systems in China and India are largely market based. In China, employer insurance is mandatory in urban areas, although the value is capped and the law is not always implemented. Domestic private insurers have also emerged. Government primary health-care insurance exists in rural areas, but the coverage is inadequate to meet most people's needs. Overall, only 29 percent of people in China have some form of health insurance, and out-of-pocket expenses accounted for 58 percent of health-care spending in 2002. [49]

In India, almost all expenditures for health care are out of pocket. The easing of regulatory restrictions has allowed the entry of some multinational insurers into the market. Although life insurance has been available for some time, private health insurance schemes are just appearing. One example is a Prudential-ICICI product that covers serious procedures, such as heart-bypass surgery, organ transplants, and cancer treatment. [50] It is anticipated that increases in private insurance will expand the market, particularly at the high-value end.

Sustainers

Supportive policies in host countries are necessary to sustain and develop offshoring manufacturing activities. These policies include (1) a commitment to education to ensure the supply of high-quality workers and (2) lowering of barriers to international trade to encourage companies to offshore and to make long-term offshore operations profitable.

Educational Infrastructure

The Indian government is supporting the development of a growing number of international-class academic institutions to support growing industries. The Indian and National Institutes of Technology are already recognized for producing high-quality engineering students. In addition, the foundation of the National Institute for Pharmaceutical Education and Research (NIPER) was established in 1998 to produce graduates and research similar in quality to the standard in the pharmaceutical sciences. There is a demand for at least 10 more NIPER-like institutes. [51]

Another accelerating field is biotechnology. The Department of Biotechnology (DBT), established in 1986, is responsible for developing a scientific and technical workforce. [52] The focus of NIPER and DBT is (1) to produce more graduates and improve standards and (2) to develop post-graduate education (see Table 3). High-quality workers will not only provide a workforce for MNPCs operating in India, but will also enable the development of Indian companies that can compete on a global level.

Government Trade Policies

The government of India is taking several steps to encourage the contract manufacturing of pharmaceuticals. Grants and incentives are offered in the following categories:

- domestic manufacturing for sale in a domestic tariff area (DTA)
- domestic manufacturing/service unit for export of goods and services less than 100 percent (export oriented unit [EOU] or software technology parks of India [STPI] schemes)
- manufacturing/service activity from a special duty-free enclave (SEZ)
- investment in R&D

The concept of a SEZ is modeled on earlier, highly successful initiatives by the Chinese government to increase FDI. FDI restrictions have been eased so that FDI of up to 100 percent is now permitted for bulk drugs and their intermediates and formulations (including bulk drugs produced using recombinant DNA). [54] In addition, biotechnology

TABLE 3 Programs in India Supported by the Department of Biotechnology [53]

	Number of Universities	Annual Intake of Students
General biotechnology	41	530
Agricultural biotechnology	9	110
Medical biotechnology	1	10
Marine biotechnology	2	30
Neuroscience	3	25
Industrial biotechnology	1	10
Masters In Technology Biotechnology	9	140
Masters in Veterinary Science	2	15
Post MD (Doctor of Medicine)/Master of Science Certificate	2	9
Post-graduate diploma	4	56
Totals	74	935

parks are being set up across the country and the Small Business Innovation Research Initiative (SBIRI) has been set up to encourage public-private partnerships in the biotechnology sector. [55]

The government has reduced the costs associated with international trade to make offshoring more attractive to MNPCs. Exporters are allowed to import inputs on a duty-free basis for products that will be exported. In addition, excise duties on pharmaceutical products are being lowered. Currently, the excise duty is 16 percent, but since January 7, 2005, the excise duty has been levied on only 60 percent of the maximum retail price of the drug. There are plans to reduce the excise duty from 16 percent to 8 percent. [54]

Barriers

Insufficient protections of intellectual property and price controls have deterred MNPCs from manufacturing and distributing their products in China and India. In addition, complicated and opaque bureaucracies can also be challenging, particularly to new entrants. The quality of infrastructure for utilities, transportation, and communications is also poor in some places, particularly away from major cities. The poor quality of infrastructure can pose risks to supply chains in a partnership-type offshoring model and may require significant investment by an MNPC setting up in-house facilities. However, these barriers are not specific to pharmaceutical manufacturing and, therefore, are not addressed further here.

Intellectual Property

In the United States, it commonly costs $800 million and takes 10 years or more to launch a new drug. It is impossible for most Chinese drug makers to develop new pharmaceutical compounds, which cost hundreds of millions of dollars. According to the president of Beijing Kevin King Management Consulting Company Ltd., "In a rather long period of time, copying foreign drugs after their patent protection is over, or, for some drug makers, seeking legal loopholes in the patents of foreign drugs to legally produce generic medicines will be a major development strategy of Chinese drug makers. This may lead to frequent legal disputes." [56]

Counterfeiting remains a problem for foreign firms in China. According to Chinese law, domestic firms can produce imitations of foreign drugs awaiting administrative protection from the State Drug Administration (SDA). While SDA reviews the application for protection, it makes information on the drug available to domestic companies to ensure that the foreign drug is not similar to drugs already being produced in China. In 2000, the *China Daily* newspaper reported 50,000 cases of counterfeit or inferior pharmaceutical products in China, which led to the closing down of 1,345 factories. [57]

Membership in the World Trade Organization (WTO) requires compliance with international intellectual-property regimes. As soon as China joined the WTO in 2001, the Trade Related Intellectual Property Agreement (TRIPS) went into force (India joined the WTO in 1995 but did not implement TRIPS until January 2005). [58] Under China's New Pharmaceutical Administration Law, which went into effect in December 2001, stronger measures are being taken against counterfeiters. In 2003, 994 manufacturers and distributors of counterfeit drugs were ordered to cease operations, and counterfeit drugs and facilities with an estimated market value of $60 million were seized. [59]

However, both in China and India compliance with these laws is a concern. Past enforcement efforts have often been impeded by municipal and provincial authorities that profit from counterfeiting activities. Examples of high-profile failed patent disputes in China are Prozac (Eli Lilly, 1999); Viagra (Pfizer, 2004); and Avandia (GSK, 2004). However, it is noteworthy that these patents were disputed by the manufacturers in a court of law, rather than simply copied, as had been done in the past.

Price Controls

The price of drugs is controlled by the Chinese and Indian governments for the purpose of making them affordable to the broad population. However, price controls have been found to delay the introduction of new products because they limit profits and create large price disparities between markets, which increase the likelihood of arbitrage [68]. Although the number of drugs under price controls has been reduced (Table 4), it is unlikely that price controls will be abolished. In fact, recommendations by the Indian prime minister's task force on drug affordability may further reduce profits. For example, the task force recommended the "de-branding" of drugs, so that only the manufacturer's identification and the generic drug name are displayed on packaging. [54]

Wage Inflation

Labor costs are a major source of cost advantage in pharmaceutical manufacturing. Low costs are also the basis for competitive advantage for Indian and Chinese firms competing in the global market. Currently, there are large differences in labor costs in the United States and Western

TABLE 4 Number of Drugs under Price Controls in India since 1970 [54]

Year	Number of Drugs
1970	Almost all bulk drugs and their formulations
1979	347 bulk drugs
1987	142 bulk drugs
1995	74 bulk drugs

Europe on the one hand and India and China on the other. However, with increasing offshore activity, it is likely that there will be wage inflation in the pharmaceutical manufacturing industry as there has been in other industries, such as business process offshoring and information technology, which will reduce the cost differential. According to the Culpepper Pay Trend Survey, the base salary increase for technical employees is about 3 to 4 percent in the United States and 6.3 percent in China. India and the Philippines project salary increases of 9.2 percent and 11.2 percent, respectively, which are much higher than in most other countries. [60] These rates may increase further and thus diminish the labor cost leverage.

Distribution

The success of new entrants in the offshoring market depends not only on their product range and marketing, but also on their ability to access customers. Both China and India cover vast geographical areas and have large rural populations, which can pose challenges. More than 17,000 distributors were operating in China in 1997, channeling medicines to hospitals, retail pharmacies, and stores. In India, almost all pharmaceutical sales take place through a complicated network of more than 6,000 wholesalers and more than 500,000 independent retailers. [5]

Foreign firms must use domestic distributors, but, because they are not exclusive agents, the distributors simply take orders for hospitals and retailers but do not promote their products. The large number of intermediaries makes launching products difficult and increases cost pressures. It also introduces multiple points for the entry of counterfeit drugs.

Some uncertainties remain as to how China's WTO obligations will apply to drug distribution. Furthermore, because the Chinese government has been slow to reform its health care system, it may be difficult for foreign drugs to get on the all-important reimbursement lists; thus they may not be able to supply the largest Chinese buyer—the state hospital system. [25] In China, about 85 percent of drugs are sold in hospitals (mostly private); [49] the rest are sold through retail outlets. Because of consolidation in the retail distribution chain, the top 100 drugstores owned 36,420 outlets in 2005. [61] This will certainly facilitate penetration into the domestic market.

IMPACT OF OFFSHORING

U.S. Employment

The slow, evolutionary changes in labor markets in developed economies will continue in response to continued offshoring. [1] It is estimated that in 2008, 160 million jobs, or about 11 percent of the projected 1.46 billion service jobs in all sectors worldwide, could, in theory, be carried out remotely. Some occupations are more amenable to remote employment than others. In the United States today, about 80 percent of workers are employed in services, about 19 percent in manufacturing, and only 1 percent in farming. [62] The Bureau of Labor Statistics reports that employment in U.S. manufacturing has decreased by two million jobs in the past 20 years. Over the same period, manufacturing output has increased, meaning that factories have higher productivity than before, leading to higher national income and a higher standard of living. Net employment increased by 43 million jobs in other areas, such as educational and health services, professional and business services, trade and transportation, government, leisure and hospitality, and financial services (see Figures 8 and 9). [63]

Does offshoring of R&D create the risk of a rapid loss of high-wage jobs and wage suppression? According to the McKinsey report, offshoring will have little effect on wage levels in developed countries, but local wage inflation will probably continue in some offshoring locations as long as companies concentrate demand on a few cities.

Over the past 30 years, the United States has experienced an 11 percent decline in manufacturing jobs, but wages have remained stable. By comparison, it is estimated that a total of 9 percent of jobs in services in the United States could theoretically be performed offshore. Assuming that half of these service jobs are actually relocated offshore in the next 30 years, the resulting job turnover would be around 225,000 jobs per year, or 1 to 2 percent of the 16 million jobs created every year in the U.S. economy. The theoretical maximum global resourcing of full-time employees in the pharmaceutical industry in 2003 was approximately 200,000, about 13 percent of total employment in the industry. The actual offshore employment in 2003 in low-wage countries was about 10,000. The number is projected to double by 2008, to 21,000 (see Figure 10). Thus offshoring in the pharmaceutical industry will have a small impact on overall employment. [1]

In research innovation and development, the United States remains the unchallenged leader. Today, almost one-third of science and engineering researchers in the world are employed by U.S. firms. Thirty-five percent of the science and engineering research papers are published in the United States, and the United States accounts for 40 percent of global expenditures for R&D. [74] In addition, in a survey by Duke University of 58 U.S. companies that outsource engineering jobs, 61 percent of the respondents said that U.S. engineering employees are equivalent or more productive than offshore engineering employees, and 78 percent said U.S. engineering employees produced equivalent or higher quality work. [73]

Although there may not be an imminent threat to American leadership in technology, the number of young professionals in emerging markets is growing by 5.5 percent annually, while growth in developed countries is only 1 percent. By 2008, the supply of suitable young engineers is expected to be nearly the same in developing and developed coun-

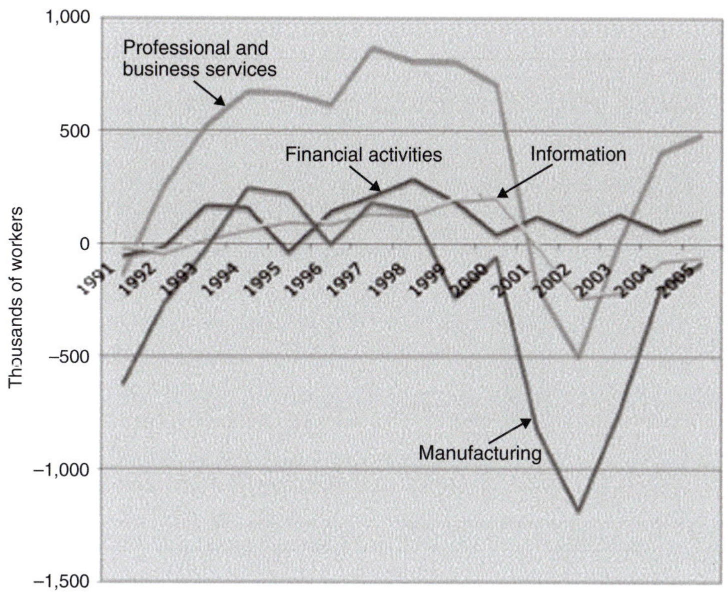

FIGURE 8 Net annual change in employment for selected sectors in the United States, 1991–2005. [64]

FIGURE 9 U.S. employment levels in professional and business services, 1996–2005. [64]

tries. [1] The United States must ensure that its workforce is trained to meet that demand.

U.S. Industry

The study by the McKinsey Global Institute shows that, far from being a zero-sum game, offshoring is a game of mutual economic gain. [63] The study found that every dollar of corporate spending outsourced to a low-wage nation had the following benefits for the United States:

1. U.S. companies captured more than three-quarters of the benefits and gained as much as $1.14 in return. The rest of the benefits ($0.33) was captured by the receiving economy (e.g., India) in the form of wages paid to local workers, profits earned by local outsourcing providers and their suppliers, and taxes collected from second- and third-tier suppliers to the outsourcing firms.
2. U.S. companies saved $0.58 because of cost advantages in offshore countries.

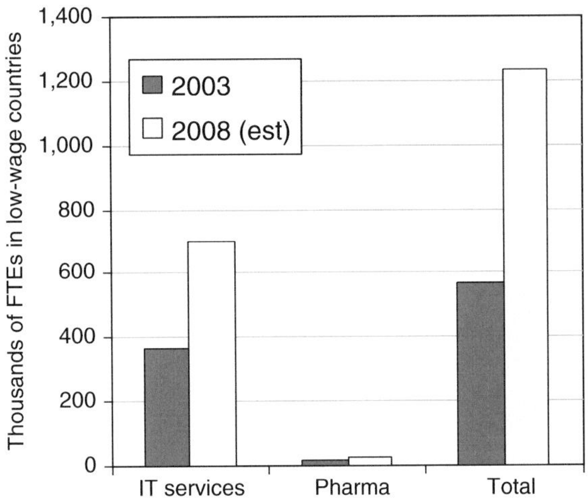

FIGURE 10 Offshore employment in IT, pharmaceuticals, and overall analyzed for 2003 and 2008. Adapted from [1].

3. Corporate savings invested in new business opportunities boosted productivity and created new jobs. Direct benefits to the United States from corporate savings, new exports, and repatriated profits totaled $0.67.
4. U.S. consumers benefited from goods and services at lower prices.

In 2004, U.S. imports of services amounted to $296 billion, and exports of services amounted to $343 billion, giving the United States a balance-of-trade surplus of $47 billion in services. In manufacturing in 2005, the United States had a deficit; the U.S. exported $807 billion and imported $1.47 trillion. With more than 100 million U.S. workers now working in the services sector, outsourcing is expected to increase at an exponential rate in the next decade, constituting a larger share of the U.S. trade balance and giving the United States a comparative advantage in services. [62] This trend may be mirrored in the pharmaceutical industry.

Reverse Offshoring

The offshoring of manufacturing has greatly enhanced the capabilities of the pharmaceutical industries in India and China. Indian companies, in particular, are becoming increasingly sophisticated and expanding globally. Companies, such as Ranbaxy and Dr Reddy's, that started in contract manufacturing of intermediates, and then APIs, are becoming integrated by moving into final-dosage formulations and becoming highly skilled in R&D. This has enabled them to challenge patents in the United States and Europe and to follow an aggressive path of acquisition.

Thus many major Indian companies are pursuing acquisitions of companies that manufacture generic products for regulated markets in the United States and Europe, a strategy of "reverse offshoring" (see Table 5). As the Indian pharmaceutical industry with expertise in reverse engineering and patent challenging grows, it could have a significant impact on the global generics market.

Indian investor companies use revenue generated by generics manufacturing to build up their R&D capacity with the goal of becoming innovator firms themselves. Because of the high level of expertise required to develop a new drug and the associated high costs and risks, alliances with Western companies have become an effective tactic for developing this capability. This strategy can also be advantageous for MNPCs, because collaborative R&D is one way for companies to diversify the risks in their product pipelines. As Table 6 shows, these alliances cover all stages of the pharmaceutical value chain. One striking example of reverse offshoring is Ranbaxy's recent decision to license a product developed in house to the level of Phase I to an American contract research organization, PDD, for preclinical and clinical development and commercialization.

There are no fully integrated Indian or Chinese innovator pharmaceutical companies today, so MNPCs do not face direct competition in this area. However, this situation could change, as offshoring of non-core activities in manufacturing and R&D continues, enabling MNPCs to focus on product development, marketing, and distribution, which may have the effect of shrinking the workforce based in the United States. As long as American and European markets are among the largest and most lucrative in the world, smaller pharmaceutical firms or subsidiaries of MNPCs will be attractive investment targets for growing Indian firms seeking a foothold in regulated markets. Hence it is not clear if the net effect will be a decrease in U.S.-based activities, but the costs of some drugs may fall.

FUTURE TRENDS

Offshoring is an increasingly hot topic that generates controversy about its impact on U.S. employment and the U.S. economy. Growing interest in offshoring is reflected in the increase in both research papers and articles in the press. Data, however, are often sparse and not well documented and must be supplemented by anecdotal evidence. The quality of these data is further compromised by the absence of standard definitions. This was a particular problem in this analysis of the offshoring of pharmaceutical research and manufacturing because much of the data we found was highly aggregated or anecdotal. More specific statistical data on employment and the demand for engineering positions broken down by academic majors, degrees, and functions in the industry in both the United States and abroad will be necessary to produce a clearer picture of how offshoring will impact the science and engineering workforce in the U.S. pharmaceutical industry.

Major European pharmaceutical companies, such as Novartis and GSK, have shifted their R&D centers and

TABLE 5 Indian Acquisitions of U.S. and European Pharmaceutical Companies

Acquirer	Acquirer's Expertise	Target	Date	Target Activities	Reference
Reliance Life Sciences	Biopharmaceuticals	GeneMedix Ltd. (UK)	2007	Manufacture of biosimilars	*The Economic Times*, February 8, 2007
Matrix Laboratories	API and dosage-form manufacturing	DocPharma (Belgium)	2005	Manufacture of API and dosage-form g	Indian Chapter for Democratic Convergence, *www.icfdc.com*, accessed March 2007
		Explora (Switzerland)	2005	API R&D	The Hindu Business Line, September 20, 2005
Jubilant Organysys Ltd.	Products and services for global life-science industry	Target research (USA)	2005	Contract R&D	*www.jubl.net*, October 2005
Malladi	Generic API manufacturer	Novus Fine Chemicals (USA)	2005	Generic API manufacturer	PR Newswire, October 5, 2005
Dr Reddy's Laboratories	Development and manufacture of generic and branded pharmaceuticals and bulk pharmaceutical ingredients	BMS Laboratories Ltd. (UK)	2002	Manufacture and marketing of generics	Pharmabiz.com, April 20, 2006, *www.drreddys.com*, accessed March 2007
		Meridian Healthcare Ltd (UK)	2002	Marketing and distribution	Pharmabiz.com, April 20, 2006, *www.drreddys.com*, accessed March 2007
		Betapharma (Germany)	2006	Generic drug manufacturer	*The Guardian*, February 6, 2006
Ranbaxy Laboratories	Research and international generic manufacturing	Terapia (Romania)	2006	Manufacture of generics	*www.terapia.ro*, June 8, 2006
		Allen Generics (GSK, Italy)	2006	Manufacture of generics	*www.ranbaxy.com*, accessed March 2007
		RPG (Aventis, France)	2003	Manufacture of generics	
		Basics (Bayer, Germany)	2000	Manufacture of generics	
		Ohm Laboratories (USA)	1995	Manufacture of generics	
Sun Pharmaceutical Industries Ltd.	API and dosage-form manufacture	Caraco Pharmaceutical Laboratories (USA)	1996	Generic dosage manufacturer	*www.sunpharma.com*, accessed March 2007
Wockhardt Ltd.		Pinewood Laboratories Ltd. (Ireland)	2006	Manufacture of generics	*www.wockhardt.com*, accessed March 2007
		Wallis (UK)	1998	Manufacture of generics	
		CP pharmaceuticals (UK)	2003	Manufacture and marketing of generics	
		Esparma (Germany)	2004	Manufacture and marketing of generics	
Dishman Pharmaceuticals	Contract and custom manufacture of APIs and intermediates	Synprotec (UK)	2005	Specialty chemicals	*www.pharmaceutical-technology.com*, accessed March 2007
Nicholas Piramal India Ltd.	Research and generic manufacturing	Pfizer, Morpeth (UK)	2006	Finished-dosage packaging, supply chain	*www.nicholaspiramal.com*, 2006
Aurobindo Pharmaceutical	API and dosage form manufacture	Milpharm (UK)	2006	Manufacture of generics	*www.aurobindo.com*, accessed March 2007
		Pharmacin (Netherlands)	2006	Manufacture of generics	*The Times of India*, December 30, 2006

TABLE 6 R&D Alliances between Indian and Western Pharmaceutical Companies

U.S./European Party 1	Expertise of Party 1	Indian Party 2	Expertise of Party 2	Announcement Date	Activities	Reference
Merck (USA)	MNPC	Advinus Therapeutics Ltd (Tata group)	Drug discovery and contract services	2006	Drug discovery and clinical development	*www.merck.com* accessed March 2007, R.T. Badam, Associated Press Newswire, November 16, 2006
Bristol Myers-Squibb (USA)	MNPC	Syngene International Private Ltd (subsidiary of Biocon)	Research	2007	Research	R. Guha, *Market Watch* by Dow Jones, March 14, 2007
PPD Inc (USA)	CRO	Ranbaxy Laboratories Ltd.	Research and international generic manufacturing	2007	License to PPD for development and commercialization, including preclinical and clinical studies	*PR Newswire Europe*, February 27, 2007
GlaxoSmithKline (UK)	MNPC	Ranbaxy Laboratories Ltd.	Research and international generic manufacturing	2003 (extended in 2007)	R&D and commercialization	*PR Newswire U.S.*, February 6, 2007
Eli Lilly (USA)	MNPC	Nicholas Piramal India Ltd. (NPIL)	Research and generic manufacturing	2007	Clinical development, marketing	*The Times of India*, January 14, 2007
Biovitrum (Sweden)	Biopharmaceuticals	Orchid Chemicals	Custom manufacturing	2006	Medicinal chemistry	*A. Krishnan, Global Insight Daily Analysis*, October 30, 2006
ClinTec (UK)	Clinical research	Dr Reddy's Laboratories	Research and generic manufacturing	2006	Clinical development and commercialization	*Business Standard*, January 13, 2006
Wyeth (USA)	MNPC	GVK Biosciences	Contract research	2006	Synthetic chemistry	*Express Pharma Pulse*, March 17, 2005
AstraZeneca (Sweden, UK)	MNPC	Torrent Pharmaceuticals Ltd.	Manufacturing	2005	Drug discovery	*Reuters News*, February 22, 2005

manufacturing facilities to the United States, which is an interesting trend that is not addressed in this report. If this trend continues, the basis on which offshoring estimates are made will be altered. Further work is also necessary to clarify the reasons for, and the impact of, the reverse offshoring phenomenon, that is, firms in India and China looking to acquire operations in the United States and Western Europe.

Special care must be taken in future studies when data from different sources are compared. For instance, China and India have different definitions of "engineering" that may not be consistent with the definition used in the United States. [69] In addition, when assessing the competitive advantages of an engineering workforce, it is important to consider the quality, as well as the number of engineers. Standards and criteria in different countries for qualified engineers may vary with the specific job requirements. Thus data on the engineering workforce must be specified by skills and functions.

This report addresses location-specific factors related to offshoring that make them attractive destinations (the pull) for offshoring for specific parts of the overall value chain. The factors that drive a particular company (the push) to consider outsourcing should be examined in detail. These factors might include the high cost of operations in the United States/Europe, the pressures of operating in regulated markets, trends in R&D productivity, etc. Identifying and understanding these factors may be particularly important in assessing the impact of offshoring on the U.S. pharmaceutical industry.

CONCLUSIONS

The major leverage points for offshoring pharmaceutical R&D are cost, time, and access to scientific and engineering talent. An additional advantage in moving clinical trials offshore may be access to treatment-naïve patients. Not all parts of the pharmaceutical value chain are being moved offshore at the same rate. Thus offshoring activities differ across the R&D value chain. We could find no examples of an end-to-end offshore R&D model. Operating offshore provides MNPCs with access to innovative human resources, although the competition for skilled labor is increasing, and wages are rising above inflation.

Offshoring of pharmaceutical manufacturing provides MNPCs with cost advantages because of the reduced cost of goods sold, and tax leverage, especially if the offshoring location includes a science and engineering zone. Offshoring also allows flexibility in capacity management. The focus of offshoring so far has been on generic products, which minimizes intellectual property risk. Operating offshore provides access to new and developing markets, which gives companies a strategic advantage. We have noted a domino effect in the pharmaceutical supply chain, as suppliers in India offshore to China to reduce costs even further. An increasingly sophisticated local industry is evolving from companies that develop core competency in contract manufacturing.

An emerging trend is reverse offshoring, that is, Indian companies with strong manufacturing bases and positive cash flows investing in U.S. and European acquisitions to improve their access to technological innovation and markets. These companies are also developing in-house R&D capabilities with the intent of becoming major global players in the industry.

Overall, offshoring in the pharmaceutical industry is taking place further afield as companies seek access to the lowest cost resources in the supply chain. MNPCs that are being pressured by domestic health care systems to lower their costs are attracted by growing international markets. China and India are of particular interest because of their rapid economic growth. Government policies on intellectual property, education, health care, and FDI incentives add to a location's attraction for companies considering offshoring.

Although offshoring in the pharmaceutical industry is expected to have a minimal effect on U.S. employment, particularly in R&D, the offshoring of both manufacturing and R&D is likely to increase as the global industry grows. This growth is expected to lead to corporate savings, new exports, and repatriated profits. The corporate savings can be reinvested in new business opportunities to boost productivity and create new jobs. Hence there is likely to be a shift toward higher value-added services from the United States. For U.S. customers, offshoring represents the benefits of lower prices for goods and services.

REFERENCES

[1] Das, S. 2006. Pharma's Prescription: Offshore R&D. Voice & Data, May 01, 2006. Available online at *http://www.voicendata.com/content/bporbit/106050102.asp*.

[2] Farrell, D., M. Laboissiére, R. Pascal, J. Rosenfeld, C. de Segundo, S. Sturze, and F. Umezawa. 2005. Emerging Global Labor Markets. Washington, D.C.: McKinsey & Company.

[3] Kearney, A.T. 2004. Making Offshore Decisions: A.T. Kearney's 2004 Offshore Location Attractiveness Index. Available online at *http://www.atkearney.com/shared_res/pdf/Making_Offshore_S.pdf*.

[4] Kearney, A.T. 2005. FDI Confidence Index, Global Business Policy Council. Vol. 8. Available online at *http://www.atkearney.com/main.taf?p=5,3,1,140,1*.

[5] Ernst and Young. 2004. Progressions: Ernst & Young's Global Pharmaceutical Report, Industry Defining Facts and Events. New York: Ernst and Young International Publishing.

[6] Datamonitor. 2005. Pharmaceuticals in China, industry profile. December 2005.

[7] Merriam-Webster. 2002. Webster's Third New International Dictionary, Unabridged. Springfield, Mass.: Merriam-Webster, Inc.

[8] Oxford English Dictionary. 2006. Draft Entry, September 2006. Available online at *http://dictionary.oed.com*.

[9] Wong, J., S. Goodall, C. Liao, K. Wagner, and R. Lee. 2005. A Game Plan for China: Rising to the Productivity Challenge in Biopharma R&D. BCG Focus. Boston, Mass.: Boston Consulting Group. Available online at *http://www.bcg.com/publications/files/A_Game_Plan_for_China_Rising_to_Productivity_Challenge_in_Biopharma_RD_Dec05.pdf*.

[10] Pocha, J.S. 2006. Growing a Lab Culture. Boston Globe, August 14, 2006.

[11] Outsourcing in Drug Discovery, 2nd ed., Kalorama Information, January 1, 2006.

[12] Pharmaceutical Giants Target Chinese Market. People's Daily, March 20, 2001.

[13] Introduction to Foreign Investment in Medicine. China Daily, April 21, 2006.

[14] Online Extra: Lilly's Lab Go Global, Special Report-Outsourcing. BusinessWeek Online, January 30, 2006.

[15] PricewaterhouseCoopers. 2006. Investing in China's Pharmaceutical Industry. March 2006.

[16] ABPI Annual Review: Working for a healthier world, 2003. Available online at *http://www.abpi.org.uk/publications/publication_details/annualReview2003/ar2003_32-healthier-world.asp*.

[17] Pfizer China takes giant step in local market. China Daily, November 1, 2004.

[18] Bhalla, V., S. Goodall, B. Janssens, R. Lee, C. Liao, K. Wagner, and J. Wong. 2006. Looking Eastward: Tapping China and India to Reinvigorate the Global Biopharmaceutical Industry, August 2006. BCG Focus. Boston, Mass.: Boston Consulting Group. Available online at *http://www.bcg.com/publications/files/Looking_Eastward_Aug06.pdf*.

[19] Executive Agenda. 2004. Pharma Explores Uncharted Territory. Executive Agenda VII(4): 45.

[20] Adecco. 2006. Offshoring and the Global Labour Market. Briefing paper available online at *www.adecco.com*.

[21] Wong, J., V. Bhalla, S. Goodall, P. Vaish, K. Wagner, and B. Janssens. 2006. Harnessing the Power of India: Rising to the Productivity Challenge in Biopharma R&D. May 2006. BCG Focus. Boston, Mass.: Boston Consulting Group. Available online at *http://www.bcg.com/publications/files/Rising_to_the_Productivity_Challenge_in_Biopharma_RD_May06.pdf*.

[22] Goodall, S., B. Janssens, K. Wagner, J. Wong, W. Woods, and M. Yeh. 2006. The promise of the East: India and China as R&D options. Bioentrepreneur, July 25, 2006. Available online at *http://www.nature.com/bioent/building/planning/072006/full/bioent910.html*.

[23] Management Consultant Blog. August 9, 2004. Offshore Outsourcing of Pharmaceutical R&D. Available online at *http://www.iproceed.com/blog/2004/08/offshore-outsourcing-of-pharmaceutical.html*.

[24] Sinja, G. 2004. Outsourcing drug work: pharmaceuticals ship R&D and clinical trials to India. August 16, 2004. Available online at *http://www.sciam.com/article.cfm?articleID=00033282-DBF5-10F9-975883414B7F0000*.

[25] Pharmaceutical Market in China. STAT-USA Market Research Reports, July 31, 2001.

[26] PricewaterhouseCoopers. 2004. China: Prescription for Growth. New York: PricewaterhouseCoopers.

[27] Das, N. 2004. Global pharma eyes India as BPO hub. April 26, 2004. Available online at *http://www.domain-b.com/industry/pharma/200404apr/20040426_bpo.html*.

[28] China Encourages Foreign Investment in Pharmaceutical R&D. People's Daily, May 12, 2000. Available online at *http://english.people.com.cn/english/200005/12/eng20000512_40657.html*.

[29] ATIP (Asian Technology Information Program). 2000. Developing Biotech in China. Asian Technology Information Program Report, February 28, 2000. Available online at *http://www.atip.org/atip-research/atip00.010r.pdf*.

[30] Wang & Wang. 2004. Greater China Intellectual Property Protection. Available online at *http://www.wangandwang.com/rcntnews.htm*.

[31] DECHEMA (Society for Chemical Engineering and Biotechnology). 2004. Rapid growth in pharmaceuticals. Available online at *http://www.dechema.de/2_Pharmaceuticals-lang-en.html*.
[32] News Guangdong. 2003. New drug rule for pharmaceutical labs. Available online at *http://www.newsgd.com/business/laws/200305140334.htm*, accessed January 10, 2003.
[33] Dossani, R. 1999. Accessing Venture Capital in India: Report of a Conference. Asia/Pacific Research Center, Stanford University, Stanford, Calif.
[34] Advisory Committee on Venture Capital. 2003. Report of Advisory Committee on Venture Capital. Available online at *http://www.sebi.gov.in/Index.jsp?contentDisp=SiteMap*.
[35] IVCA (Indian Venture Capital Association). 2006. Welcome to Indian Venture Capital Association. Available online at *www.indiavca.org*, accessed December 3, 2006.
[36] Ernst and Young. 2005. Creating a Technology Hotbed in China: Lessons Learned From Silicon Valley and Israel. Summary Report and Suggestions to the Chinese Venture Capital Association. Available online at *https://eyo-iis-pd.ey.com/drivinggrowth/unprotected/downloads/Creating_Technology_Hotbed_In_China.pdf*.
[37] CVCA (Chinese Venture Capital Association). 2005. China Venture Capital Performance Report, 1994–2005. Available online at *http://www.cvca.com.hk/template/reaearchtemplate.asp?ArticleID=604*.
[38] The Hindu Business Line, May 25, 2006.
[39] Ravindranath, P.K. 2001. The Indian Pharmaceutical Industry since Independence. India: English Edition Publishers and Distributors Pvt Ltd.
[40] Scott, A. 2006. Indian firm secures major API contract for new drug. Chemical Week 168(18): 27.
[41] Ranbaxyusa.com, accessed September 2006.
[42] Mullin, R. 2006. Indian Firms Go Global, Chemical & Engineering News 84(39): 62–65.
[43] Jayakumar, P.B. 2006. Interview with Shen Xianji, vice director of Advisory Committee, Chinese Pharmaceutical Industry Association. PHARMABIZ.com, June 22, 2006. Available online at *http://www.pharmabiz.com/article/detnews.asp?articleid=33903&sectionid=50*.
[44] Coan, T., and R. Ellis. ABN AMRO, Generic Biologicals: The Next Frontier, June 2001.
[45] Schellekens, H. 2004. When biotech proteins go off-patent. Trends in Biotechnology 22(8): 406–410.
[46] Mishra, N. Credit Suisse First Boston. Equity Research, Sector Review: Indian Pharmaceuticals. October 12, 2006.
[47] ACHEMA Worldwide News. 2004. Trend Report: China's Pharmaceutical Industry, 2004.
[48] The All India Council for Technical Education. List of accredited programs and approved institutes. Available online at *http://www.aicte.ernet* in 2006.
[49] Blumenthal, D., and W. Hsiao. 2005. Privatization and its discontents—the evolving Chinese health care system. New England Journal of Medicine 353 (11): 1165–1170.
[50] ICICI Lombard Insurance Direct, *www.icicilombard.com*, 2006.
[51] National Institute for Pharmaceutical Education and Research. *www.niper.ac.in*, 2006.
[52] Department of Biotechnology. 2006b. Biotechnology: A Vision (Ten Year Perspective). Department of Biotechnology, Ministry of Science & Technology, Government of India. Available online at *http://dbtindia.nic.in/publication/vision/morevision.html*.
[53] Department of Biotechnology. 2006c. Fact Sheet. Department of Biotechnology, Ministry of Science & Technology, Government of India. Available online at *http://dbtindia.nic.in/factsheet/factsheetmain.htm*.
[54] Department of Chemicals and Petrochemicals, Government of India, National Pharmaceuticals Policy, 2006, Dec 28th, 2005.
[55] Department of Biotechnology. 2006a. Annual Report 2005–2006. Department of Biotechnology, Ministry of Science & Technology, Government of India. Available online at *http://dbtindia.nic.in/publication/publicmain.html*.
[56] Hepeng, J. 2004. Drugmakers wrestle over medicine patents, China Business Weekly, August 17, 2004. Available online at *http://www.chinadaily.com.cn/english/doc/2004-08/17/content_366249.htm*.
[57] Kermani, F. 2004. Unlocking China's Pharma Potential. Detail-Direct.com, September 24, 2004. Available online at *http://www.detail-direct.com/Features/feature.asp?ROW_ID=494*.
[58] World Trade Organization. *www.wto.org*, 2006.
[59] CCPIT Patent and Trademark Law Office. 2004. Protecting IP Rights of Pharmaceutical Manufacturers. Available online at *http://www.ccpit-patent.com.cn/News/2004060301.htm*.
[60] Culpepper Compensation & Benefits Surveys. 2005. Global Base Salary Increase Projections for 2006. September 13, 2005. Available online at *http://www.culpepper.com/eBulletin/2005/SeptemberPayTrendsArticle.asp*.
[61] Research in China. 2006. China Pharmaceutical Chain Industry Report, 2006. August 2006. Dublin, Ireland: Research and Markets.
[62] BusinessWeek Online. 2006. Readers Report: The Upside And Downside Of Outsourcing. BusinessWeek Online, February 20, 2006.
[63] Bailey, M.N., and D. Farrell. 2004. Exploding the Myths about Offshoring. April 2004. New York: McKinsey&Company. Available online at *http://www.mckinsey.com/mgi/publications/myths.asp*.
[64] U.S. Department of Labor. Data from Bureau of Labor Statistics. Available online at *www.bls.gov*.
[65] Economist Intelligence Unit. 2004. Scattering the Seeds of Invention: The Globalization of Research and Development. New York: Economist Intelligent Unit.
[66] Thursby, J., and M. Thursby. 2006. Research and development: where is the new science in corporate R&D? Science 314(5805): 1547–1548.
[67] Grace, C. 2004. The Effect of Changing Intellectual Property on Pharmaceutical Industry Prospects in India and China: Considerations for Access to Medicines. June 2004. London, U.K.: DFID Health Systems Resource Center.
[68] Lanjouw, J.O. 2005. Patents, Price Controls, and Access to New Drugs: How Policy Affects Global Market Entry. Working Paper 11321. Cambridge, Mass.: National Bureau for Economic Research.
[69] Gereffi, G., and V. Wadhwa. 2005. Framing the Engineering Outsourcing Debate: Placing the United States on a Level Playing Field with China and India. Master of Engineering Management Program, Duke University, December 2005.
[70] Fuller, D. 2006. The Fact Remains—U.S. Tech Leadership Must Be Reinforced. The Walter H. Shorenstein Asia-Pacific Research Center, Stanford University, April 7, 2006.
[71] NSF (National Science Foundation). 1998. Statistical Profiles of Foreign Doctoral Recipients in Science and Engineering: Plans to Stay in the United States. NSF 99-304. Arlington, Va.: NSF; and NSF. 1997. Science and Engineering Doctorate Awards: 1996. NSF 97-329. Arlington, Va.: NSF.
[72] NSF. 2001. Science and Engineering Doctorate Awards: 2000. Detailed Statistical Tables. NSF 02-305. Arlington, Va.: NSF.
[73] Wadhwa, V., G. Gereffi, B. Rissing, and R. Ong, 2007. Where the engineers are. Issues in Science and Technology (Spring).
[74] Gereffi, G., and V. Wadhwa. 2006. More Engineering Grads in China and India? Not so fast. Report developed by graduate students of Duke University's Master of Engineering Management Program under the guidance of Dr. Gary Gereffi, and Vivek Wadhwa with consulting assistance from Katzenbach Partners LLC. Innovation: America's Journal of Technology Commercialization. February/March 2006. Available online at *http://www.innovation-america.org/archive.php?articleID=40*.

Impact of Globalization and Offshoring on Engineering Employment in the Personal Computing Industry[1]

Jason Dedrick and Kenneth L. Kraemer[2]
University of California, Irvine

EXECUTIVE SUMMARY

Globalization has changed the nature, organization, and location of engineering work in the personal computing industry. As a consequence, lower skilled and lower paid engineering jobs that might have been created in the United States are instead being created overseas, while higher skilled and higher paid jobs remain in the United States. The engineering work that remains in the United States requires skills in traditional engineering disciplines, as well as in the intersection of engineering and computer science and new specialties, such as small form-factor design, communications and networking, software engineering, and the interfaces between these. Software engineering in particular is becoming more important in engineering for innovative new products, such as smart phones and handheld devices that add functionality through tightly integrated hardware and software. For personal computers (PCs) and components, embedded software enables large-scale, low-cost production of standard products that can be provided with different features, tailored to particular markets, and continually updated to extend product life.

Work done by branded PC makers has changed from physical engineering concerned with building, testing, and mass production, to conceptual design, planning, and product management. Physical engineering is now done largely outside the branded firms. PC firms initially performed all phases of new-product development in house, but they subsequently outsourced the manufacturing of desktops to contract manufacturers (CMs) in various regions of the world and outsourced the development *and* manufacturing of notebooks to original-design manufacturers (ODMs), mainly in Taiwan. Today, much desktop development is also being handed off to ODMs.

As production and development were outsourced, the location of engineering jobs also shifted. For instance, notebook development and manufacturing were originally done mostly in Japan, and in some cases in the United States, but these activities have moved steadily to Taiwan, which developed the required skills and had lower costs. Recently, Taiwanese ODMs have begun moving engineering work to mainland China where costs are even lower and manufacturing facilities are nearby.

Interviews with executives in charge of new-product development in branded PC firms indicate that relatively few jobs remain in the United States, and those jobs require highly skilled, innovative people with considerable experience. Thus salaries for U.S. engineers have increased steadily to be commensurate with their skill, experience, and productivity.

Historical data and national statistics on the entire com-

[1]This report is based on research conducted by the authors over a 15-year period on the PC industry. They have interviewed more than 200 individuals from 25 companies in the Americas, Europe, and the Asia-Pacific region, including PC makers, contract manufacturers, original-design manufacturers, suppliers, and distributors. For this report specifically, they conducted eight interviews and a small survey of five U.S. companies in the summer of 2006 to collect primary data and gain insight into globalization and its impacts on the engineering workforce in the PC industry. In addition, secondary data were collected on the industry and on engineering employment from government statistics, private research companies, and articles in business and professional trade publications.

[2]The authors gratefully acknowledge the assistance of the National Academy of Engineering in arranging for interviews with senior executives and the insights provided by those executives.

puter industry show no significant change in the number of engineers since 2002. There are no comparable data for the PC industry, per se. However, although the PC industry continues to grow in scale and PCs increase in complexity, thus increasing the need for engineering work, there appears to be little or no increase in engineering jobs in the United States. This can be explained partly by the increasing productivity of engineers, but mostly by a large increase in engineering jobs in CMs and ODMs, especially in Taiwan and China.

Engineering work that remains in the United States is being tailored to the needs of newer, smaller personal computing products, such as wireless notebooks, tablet notebooks, PDAs, MP3 players, and smart phones. This work requires not only knowledge of engineering design for small form factor, but also new engineering specialties related to communications, networking, embedded software, and particularly the interfaces between these and hardware engineering.

Interviewees in PC companies said that generally there was a good balance between the supply and demand for engineers in the United States, but noted shortages in experienced managers (product managers, engineering-discipline managers, project managers, high-level design mangers) and, particularly, in the engineering subdisciplines mentioned in the body of this report. A few firms carefully develop engineers by hiring graduates of elite engineering schools, but most PC firms prefer to hire experienced engineers from other firms.

All of the firms we interviewed hire at least some engineers outside of the United States, some primarily to reduce cost, others for their specialized knowledge. In some cases, companies hire engineers working in offshore facilities, but more often they hire foreign-born engineers to work in the United States, often from U.S. universities. All of the executives considered U.S. immigration policies flawed for failing to consider industry needs, treating all engineering jobs/levels alike, and making it difficult for graduates to stay in the United States. They also faulted limits on the number of visas. At the same time, most executives believe that the offshoring of lower skilled engineering jobs was inevitable and that the United States should concentrate on maximizing its strengths in the dynamic and analytical skills necessary to retain its leadership in the development and commercialization of innovation.

INTRODUCTION

The personal computing (PC) industry includes desktop and notebook PCs, PC-based servers, and various handheld computing devices, such as PDAs, personal music players, and smart phones. Worldwide revenues for the industry totaled $235 billion in 2005, including $191 billion in desktop and portable PCs, $28 billion in PC servers, and $16 billion in smart handheld devices. In addition, PC software accounts for a large share of the packaged-software industry, which had sales of $225 billion, and PC use drives sales of information technology (IT) services and other hardware, such as storage, peripherals, and networking equipment (IDC, 2006a).

In 2005, more than 200 million PCs were shipped worldwide, including 135 million desktops and 65 million notebooks (IDC, 2006b). The United States has the largest PC market (61 million units shipped), followed by Western Europe (47 million units), Asia-Pacific (40 million units), Japan (14 million), and the rest of the world (38 million). The United States is not only the leading market but is also home to the top two PC vendors, HP and Dell, as well as Microsoft and Intel, which continue to set the key technology standards for the global industry. However, competition is becoming increasingly global, with non-U.S. firms holding the next five spots (Table 1) since IBM's PC division was acquired by China's Lenovo in 2004.

As the cost of displays and other key technologies has fallen and as customer demand for mobile products has increased, notebooks and various handheld devices have become the fastest growing product categories. These products are less standardized than desktop PCs and require more engineering in the new-product development phase. In addition, PC models and form factors have proliferated as vendors try to provide customers with more choices, which also increases the engineering requirements of the industry. Finally, PC-based servers account for the largest and fastest growing share of the server market, also requiring more engineering effort to develop cheaper hardware that can handle the work formerly done by expensive proprietary systems.

Unlike the mainframe computer industry, which consisted of vertically integrated firms, the structure of the PC industry is based on specialization, with most firms concentrating on one segment, such as components, systems, software, distribution, or services. Most PC makers today have focused their efforts even further by outsourcing manufacturing, logistics, and other functions and concentrating their own efforts on high-level design, marketing, and branding. Subassembly and final assembly have been outsourced to CMs since the

TABLE 1 Worldwide PC Market Share, 2005

Company	Market Share (%)
Dell[a]	18.2
HP[a]	15.7
Lenovo	6.3
Acer	4.7
Fujitsu/Fujitsu Siemens	4.1
Toshiba	3.5
NEC	2.9
Apple[a]	2.3
Gateway[a]	2.2
Sony	1.6

[a]U.S. companies.
Source: Adapted from IDC, 2006b.

early 1990s. Some parts of the product-development process for notebook PCs were outsourced to Taiwanese ODMs.

PC makers that produce industry standard, or "Wintel" PCs, based on the Windows operating systems and Intel-compatible microprocessors, do not require much innovation. These products are based on hardware and software interface standards set by Microsoft and Intel, and all of the necessary components are available from outside suppliers. Thus most of the R&D in the industry is done by makers of software and components, such as semiconductors, displays, hard drives, and storage.

Nevertheless, although PC makers do not generally create new technologies, they play a critical role in their integration and adoption. PC makers decide which technologies are brought to market, in which combinations, and at what price. Although they have little choice in operating systems (Microsoft dominates here), PC makers make critical choices about which innovations to integrate and which standard to support (when multiple standards are being promoted, as is often the case). To make these choices and to develop and produce successful products, PC companies must have a combination of technical and market knowledge.

ENGINEERING WORK IN THE PC INDUSTRY

Most engineers in the PC industry are involved in new-product development rather than R&D. Spending on R&D by Dell is just 0.9 percent of revenues. HP spends more for R&D as a company, but much of it is concentrated on HP's printing business. Even companies such as Apple or Palm, which spend proportionately more on R&D, are engaged more in product development and the integration of new technologies than in research. Most core innovations in the industry are made at the component level for semiconductors, displays, and hard drives. R&D in the PC industry is focused more on systems engineering, power management, heat dissipation, software tools, and security and data protection (e.g., locking the hard drive if a notebook PC is dropped).

The emphasis has shifted over the past decade as outside suppliers have provided standardized chip sets, integrated more functionality into microprocessors, and developed standard motherboard designs. In the past, some PC companies were involved in the design of application-specific integrated circuits (ASICs), but today these firms either use standard chip sets or work with chip-design companies to customize ASICs for their products. PC companies also used to do their own board layouts, but now they mostly use standard motherboards for desktops and outsource board layout for notebooks. Most engineering work in the industry today involves new-product development for desktop and notebook PCs; work on new products, such as tablet PCs, blade servers, and smart handheld devices is also increasing.

Product development in the industry has become quite standardized. As outlined by Wheelwright and Clark (1992), most product development consists of three phases: design, development, and production. Each of these phases is further divided into specific activities, with outputs and gates that must be passed before the next phase can begin. Design refers to envisioning and defining a new product based on outside innovations and on customer needs. Development is the making and testing of a working product based on the design. Production is the building and shipping of the product, which involves knowledge of process engineering, cost-reduction measures, logistics, and so on.

Product Development for Desktop PCs

Although product development processes have been standardized in the industry, the nature of the engineering varies significantly by product category. Developing a desktop product is primarily a problem of system integration (i.e., incorporating new technologies into products and ensuring that they work together). In terms of physical design, most desktop models are still based on industry standard form factors, such as the bulky but flexible midtower chassis. Standard motherboard designs are available from Intel and various third-party manufacturers. Other components, such as drives and add-on cards are built to fit into standard enclosures.

For desktop PCs, the emphasis is on the development of a new chassis as a basis for multiple models, or stock-keeping units (SKUs), which can be designed for different markets and with different configurations. A PC company executive explained that the design of a new chassis takes about nine months, but a new model based on an existing chassis can be built and tested in as little as two weeks. One vendor introduces as many as 1,000 different consumer desktop SKUs in one year.

Development Processes for Notebook PCs

Notebook PCs have different characteristics that add complexity to the design and development process. Notebooks must be able to run on batteries; the display must be incorporated into the unit; the product must be lightweight yet very sturdy; and the product must be appealing visually. Components must be packaged very tightly into a product that is small, thin, light, portable, durable, and energy efficient, and that does not become too hot to handle from the heat generated by its operation. Notebook developers must make choices and trade-offs to optimize a number of factors (a bigger battery will run longer but add weight; more memory will improve performance but increase cost; a faster processor will increase speed but produce more heat).

New-product development involves solving problems as new technologies are added or new form factors are introduced. Figure 1 illustrates the product development process for notebook PCs.

Manufacturability is a major issue for notebooks because they must be produced in high volume and at low cost. There-

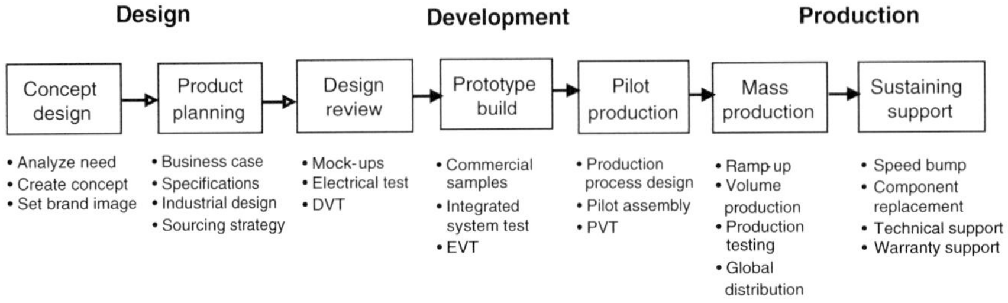

FIGURE 1 Product-development cycle for notebook PCs.

fore, the final assembly must be a relatively simple process in which packing components and subassemblies can be put into a very tight space quickly and with a high level of reliability.

One of the most significant costs for notebooks can be warranteed repairs. Industry sources estimate that as many as 25 percent of notebooks require a warranteed repair during the first year after purchase. A dramatic example was the recall of millions of notebooks in 2006 because of faulty Sony batteries. Both Sony and the notebook vendors who had to deal with the recalls and resulting consumer concerns incurred significant costs.

Product Development for Newer Products

No dominant technology architecture is available for smart phones, iPods, PDAs, and other newer products, most of which are unique to particular companies. Therefore, product development requires more fundamental design choices, such as the selection of core components and operating systems and knowledge is more tacit. In addition, collaboration across engineering disciplines is more important, especially for convergence products, such as smart phones and other mobile devices. Product development for a new device can take as long as 12 to 18 months.

Skill Requirements

Different skills are required for each stage of product development (Figure 2). The design stage requires knowledge of markets and customer demand, as well as an understanding of technology trends. Engineers, usually those who have moved into product management from other engineering jobs, must be able to talk to marketing people and understand how customer demand and technology trends converge. These individuals generally have both experience and advanced degrees.

The teams that develop new-product concepts and manage them through to fruition often include a software engineer, a cost engineer, and a technical product manager, as well as a general project manager and people with business skills, such as finance and marketing. Another key skill at the design stage is industrial design, which is taught in universities but requires a strong sense of the aesthetic tastes of customers in a particular market.

A variety of engineering skills are required at the development stage, primarily in mechanical, electronic and electrical engineering, PCB layout, and software engineering. For notebook PCs, specialized skills are required in thermal dissipation, EMI, acoustics, shock and vibration, power management, materials, and radio frequency. For communications products, such as smart phones, critical skills include radio frequency and software control of telephonic components. These skills require a combination of formal training and experience working in a particular specialty.

At the production stage, the necessary skills are mainly industrial engineering, quality assurance, manufacturing management, and logistics. In addition, this phase requires sustaining engineering, that is, support for products after they are in high-volume production to handle midlife upgrades, such as the addition of a faster processor, end-of-life components, or problems that show up in the field.

In addition to technical skills, firms want engineers who can work in teams that may include people from different

FIGURE 2 Engineering skills for new-product development.

engineering disciplines, as well as marketing people, product managers, and other non-engineering professionals. Non-engineers are particularly important during the design stage, but also throughout development for new product categories for which there are no road maps. Development of these products requires a mix of art and science, what one company refers to as the "Zen" of design, an intuitive understanding gained by working closely in teams led by "Zen masters" who have a sense of the features that should be included and the ones that should be left out.

Experience Requirements

Some firms look primarily for experienced engineers as a way of (1) avoiding the cost of training and (2) immediately increasing productivity. One executive said, "Over the last 15 years, the industry has become so competitive that we have to hire mostly experienced people; we can't wait for junior engineers to learn. We still recruit at colleges but not as much as in the past. It used to be 10 to 15 new hires a year. Now it is more like two per year. Nowadays, engineers get into the field and keep moving around in order to learn."

Not everyone agrees, however. An executive from a nearby competitor said he liked to hire engineers right out of college and had set up an internship program with six universities so students could get experience during summer breaks. Interns in the program become part of core design teams right away, and after a few years are "very self-assured." Most of these students spend two or three summers working with the company. More than half of the interns are offered jobs after graduation. Nearly all students accept, unless they are going on to graduate school.

A similar opinion of the value of new graduates was expressed by an executive at a component-making firm who runs an R&D organization. Most of his new hires, he said, are new Ph.D.s in their first jobs. He prefers to hire people without experience in manufacturing or development because they "don't know that some things can't be done." If they go into manufacturing or development first, they often "learn" that some things can't be done. His company wants people who are not "burdened by experience."

At the other end of the spectrum, there is a shortage in the United States of experienced engineering managers to run projects and departments. Interviewees reported that the shortage is even more acute outside the United States. They defined two types of engineering managers—(1) engineering supervisors who manage engineering teams and (2) technical program managers responsible for getting products to market. The latter do not necessarily have deep technical knowledge, but they are good planners and organizers. The very best of them have a deep understanding of the technology or of how a product will perform in a market. Engineering managers must see that various internal organizations (e.g., engineering, manufacturing, product managers) work together on a product and work with outside firms (e.g., ODMs and component suppliers). According to one executive, "They have to be able to whip people into order."

Changing Requirements

The firms we interviewed reported that the share of jobs in software engineering is increasing. This trend is not obvious in government employment data for the computer industry (Table 2) but is evident in survey data of PC firms (Table 3). More software engineers are needed because functionality in many products is being added through software rather than hardware. This is true for smart phones, music players, and even hard-disk drives that can be customized for specific clients.

Interviewees described a need for people with both software and hardware skills, especially for emerging products that involve close integration of software and hardware functions, such as smart phones and other handheld devices with communications capabilities. A smart phone, for example, may support multiple radio frequencies (e.g., GSM, CDMA, WiFi) and a number of applications, such as e-mail, instant messaging, and Web browsing. The formatting of the bit structure from the applications is different for each radio protocol. Thus software for many products must be written to fit and run on specific integrated circuits, unlike PCs, in which software applications can run on any Intel-compatible hardware running Windows via Windows application programming interfaces. For PCs, software development is largely independent of specific hardware configurations.

Examples of requirements include software engineers who understand telephony and how communication networks function or electrical engineers who know how software controls telephony functions on a smart phone or en-

TABLE 2 Employment Levels for Selected Engineering Occupations in the Computer Industry, 2002–2005[a]

	2002	2003	2004	2005
Computer software engineers-applications (15-1031)	10,250	9,890	12,110	12,800
Computer software engineers-systems software (15-1032)	18,809	18,148	19,430	18,240
Computer hardware engineers (17-2061)	11,140	12,030	11,880	12,940
Electrical engineers (17-2071)	4,580	4,020	3,200	2,900
Electronics engineers, excluding computers (17-2072)	4,360	4,030	3,490	3,710
Industrial engineers (17-2112)	3,520	3,640	3,570	3,430
Mechanical engineers (17-2140)	2,100	2,470	2,160	2,280
Engineering managers (11-9041)	5,270	5,460	5,690	5,630
Industrial designers (27-1021)	260	290	190	180
Totals	60,289	59,978	61,720	62,110

[a]The computer industry is defined as NAICS 334100 (Computer and Peripheral Equipment Manufacturing). Data for years prior to 2002 are based on SIC code 357 (Computer and Office Equipment).
Source: Bureau of Labor Statistics, 2005.

TABLE 3 Survey Results by Job Category (for 5 companies interviewed)

Engineering Job Category	Major Activity	Demand for Engineers	Availability in the United States	Availability in Other Locations[a]	Cost and Quality (relative to U.S.)[a]
Engineering managers	R&D, design, development	Stable or growing	Tight	Tight or enough	Lower cost, lower quality
Engineering product managers	Design, development	Stable	Tight or enough	Tight or enough	Lower cost, same quality
Hardware engineers	Design, development	Stable	Tight or enough	Enough	Lower cost, same or lower quality
Electrical engineers	R&D, design, development	Falling or growing	Tight or enough	Enough	Lower cost, same or lower quality
Electronic engineers	Development	Falling	Tight or enough	Enough	Lower cost, same or lower quality
Mechanical engineers	R&D, design, development	Stable or growing	Tight or enough	Enough	Lower cost, same or lower quality
Software engineers	R&D, design, development	Growing	Tight	Tight or enough	Lower cost, same or lower quality
Industrial engineers	Manufacturing	n/a[b]	n/a[b]	Enough	Lower cost, same quality
Industrial designers	Design	Stable	Enough	Enough	Lower cost, lower quality

Note: Names of firms are confidential. Four were personal computing companies, and one was a component supplier.

[a]Responses regarding availability, cost, and quality for some skills in other locations vary by firm, depending on where these activities are located. We report one response when there was general consensus, more than one if there were different responses. Other locations included Singapore, Taiwan, Malaysia, and Ireland.

[b]Firms interviewed had no manufacturing in the United States, so demand and availability of industrial engineers were not relevant.

gineers who can program a microprocessor to communicate with a network. These skills are currently being taught on the job, because few universities have programs that combine training in computer science and electrical engineering.

Productivity and Demand for Engineers

The productivity of engineers has increased steadily, so fewer engineering resources are required per model/SKU (number of engineers/SKU is used as a productivity measure by some PC makers). However, because of the growth of the industry and the proliferation of SKUs, the overall demand for engineers has grown. For instance, 10 years ago one PC company reported having 50 engineers shipping 50 to 75 SKUs per year in consumer desktops. Today, the company has 165 engineers shipping 1,000 to 1,200 SKUs per year. The increase in productivity is partly due to the use of CAD tools, but it also reflects the outsourcing of development to ODMs.

GLOBALIZATION OF THE INDUSTRY

The PC industry is highly globalized. Final assembly is being done in dozens of countries, but manufacturing is increasingly concentrated in the Asia-Pacific region (Figure 3). The globalization of the PC industry was present almost from its inception in the late 1970s, as early PC makers imported a number of components from Asian suppliers. In the 1980s, leading PC makers, such as IBM, Compaq, Apple, and Dell, set up assembly operations for desktops and notebooks offshore, with production in all major world regions (Ireland, Scotland, and France in Europe; Malaysia and Singapore in the Asia-Pacific region; and Mexico in the Americas).

Subassemblies, such as motherboards and base units, were produced by Asian suppliers or U.S. CMs who located production near major vendors. Final assembly also has been increasingly outsourced to CMs and ODMs. Time-critical, build-to-order production is located in regional markets, and less time-sensitive, build-to-forecast production is located mostly in China.

U.S. PC makers began moving notebook production offshore in the early 1990s. Taiwan developed a homegrown industry focused on notebook PC production, led by a group of ODMs, such as Quanta and Compal, that developed specialized technical knowledge in issues critical to notebook performance, such as battery life, heat dispersion, rugged mechanicals, and electromagnetic interference. Notebooks were produced in Taiwan or Southeast Asia, but as pricing pressure on ODMs increased, the Taiwanese government removed restrictions on manufacturing notebooks in China, and the Taiwanese notebook industry moved en masse to the Shanghai/Suzhou area of eastern China. By 2005, more than 80 percent of the notebook computers in the world were produced by Taiwanese firms, almost entirely in China (DigiTimes, 2006).

Offshoring and Outsourcing of New-Product Development

Branded U.S. PC makers kept product development in house and onshore in the 1980s, but in the notebook market they fell behind Japanese competitors who had superior skills in miniaturizing components and developing small, light, thin products. IBM reacted to Japanese competition by moving notebook development to its subsidiary in Japan, which came up with the very successful Thinkpad design. Compaq worked with Citizen Watch Company in Japan to engineer its notebooks and produce key subassemblies. Apple contracted with Sony for one of the original Powerbook models (Business Week, 1991).

In time, however, most PC makers turned to Taiwanese ODMs for manufacturing, not only to reduce costs, but also to avoid becoming dependent on Japanese partners who

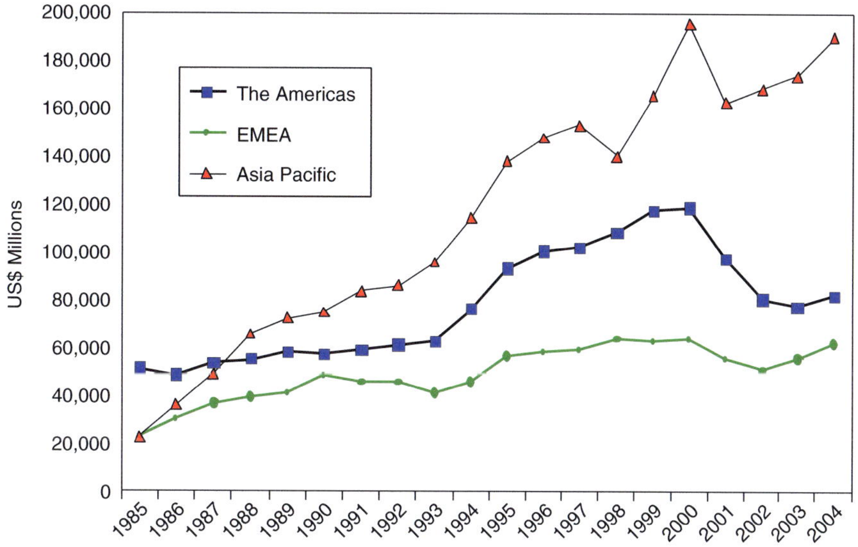

FIGURE 3 Computer hardware production by region. Source: Reed Electronics Research, 2005.

could become competitors. Gradually, Taiwanese ODMs developed specialized engineering skills and began to take over product development as well. Companies such as Dell and Gateway were able to enter the notebook market by working with ODMs on design and development, taking advantage of capabilities nurtured by their competitors.

A major factor influencing the outsourcing of product development was a "pull" from ODMs. Taiwanese ODMs often did not charge explicitly for product development, which they did to win production contracts (according to interviews in Taiwan and China). In addition, once an ODM had a contract, the PC maker had incentives to work with the same ODM for future upgrades and enhancements to its products. A great deal of tacit knowledge, known only by the ODM, was created in the development process. Also, the close linkage of development activities and manufacturing and the feedback to design from manufacturing and sustaining support, created linkages that favored a continuing relationship with that ODM to reduce costs and improve quality.

In addition to the pull from ODMs, there was a "push" by PC vendors. In recent years, some PC makers (notably Dell and HP) have set up their own design centers in Taiwan, thus offshoring some detailed system design, while keeping concept design and system architecture in house. The companies had several motivations—lower cost engineers and programmers, faster development because test facilities were nearby, availability of experienced engineers, government tax incentives, and proximity to emerging markets in Asia. Also, proximity to ODMs made it possible for a design center to send personnel to its ODM for problem solving and to use the ODM's testing facilities. Taiwan also has a pool of skilled, experienced engineers who are less expensive than their U.S. counterparts. In addition, the Taiwanese government provides incentives to attract design centers and strengthen ties to U.S. high-tech companies. For instance, the Industrial Technology Research Institute set up by the Taiwanese government established an incubator in San Jose, California, to link Taiwanese venture capitalists and tech suppliers with entrepreneurs in Silicon Valley (Boudreau, 2006).

At the same time, Taiwanese ODMs have been moving engineering work, as well as manufacturing, to China. ODM design teams in Taiwan are still responsible for the development of advanced technologies and new products that provide competitive advantage. As products mature, however, the development of product variations, incremental improvements, and life-cycle support has moved to China, where they are close to manufacturing and can take advantage of lower costs.

As Figure 4 shows, notebook PC makers and ODMs have also shifted new-product development activities from Taiwan to China, a trend driven by the lower cost of engineers in China and the proximity to manufacturing facilities. Lu and Liu (2004) found that, after access to engineers, the second major factor for locating development activities is proximity to the manufacturing site. For notebooks and other products for which design-for-manufacturability is very important, it is valuable for a company to be able to build and test prototypes on the actual final assembly line. Also, the time frame for ramping up to mass production has been cut dramatically, as have overall product cycles as firms try to introduce new technologies quickly and avoid product obsolescence. If critical manufacturing processes and equipment (particularly tooling equipment) are in place at the manufacturing site, high-volume production can begin almost immediately after a design is finalized. ODMs save time and money by having both pilot and mass production

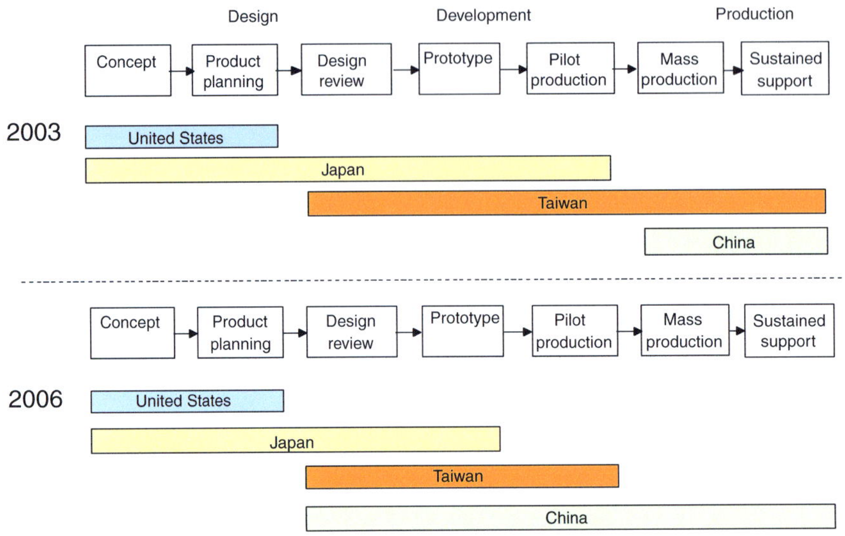

FIGURE 4 Shifting location of product development for notebook PCs. Source: Market Intelligence Center, Institute for Information Industry, Taiwan. Based on figure provided to authors.

in China. Once the crucial decision to move expensive testing equipment to China has been made, it is cost effective to move more development there as well, even if this means bringing in experienced engineers from Taiwan for a year or more to lead development teams.

The shift of product development to Taiwan and China depends not only on the stage of the activity but also on the maturity of the product. The Taiwan design centers of U.S. PC makers are mostly involved in developing new models based on existing product platforms. The development of new form factors or the incorporation of new technologies is still led by teams in the United States. Taiwanese ODMs tend to keep the development of the newest product generations in Taiwan, where they have close working relationships with key component suppliers such as Intel. They are more likely to move the development of more mature products to China.

The activities that are still being done in the United States, which do not appear likely to be moved in the near future, include R&D, concept design, and product planning. All companies, whether American, Japanese, Korean, or other, tend to concentrate R&D in their home countries. Product design benefits from proximity to leading markets where new innovations are first adopted. As long as the United States remains the leading market for innovations in the PC industry and U.S. companies remain leaders in the industry, it is likely that these functions will remain mostly in the United States.

Although R&D activity in the PC industry is limited, design and product planning continue to expand as the market grows and rapid innovation in upstream technologies continues. Some of this work is moving to Taiwan, especially for notebooks, but most is still concentrated in the United States. Foreign PC makers, such as Lenovo, Acer, Fujitsu, and Toshiba design products in their home countries (e.g., China, Taiwan, Japan). However, Lenovo, which acquired IBM's PC business, has left concept design and product planning for the global Thinkpad line in North Carolina, and most development in Japan.

U.S. ENGINEERING WORKFORCE IN THE PC INDUSTRY

Based on data from the U.S. government, the engineering workforce for the entire computer industry remained at about 60,000 from 2002 to 2005 (Table 2). Before 2002, employment numbers were based on Bureau of Labor Statistics data for the broader category, Computers and Office Equipment. Thus the numbers are not comparable in absolute terms. However, employment levels remained stable from 1999 to 2001. About half of the engineers in the industry are employed in the two categories of computer software engineers (applications and system software). The most growth took place in applications engineering, from 10,250 to 12,800. The biggest losses have been in electrical and electronics engineering, where a combined 2,500 jobs were lost. These changes may reflect a shift in focus from hardware to software reported by interviewees.

In the United States, salaries in the computer industry have risen in every engineering occupation (Table 4) since 2002, a pattern also seen in the broader industry category for 1999 to 2001. In the PC industry, our interviews suggest that engineering salaries increased rapidly during the dot-com boom of the late-1990s, then stagnated, and now are rising

TABLE 4 Mean Annual Wages for Selected Engineering Occupations in the Computer Industry, 1999–2005[a]

	1999	2000	2001	2002	2003	2004	2005
Computer software engineers-applications	$70,630	$74,350	$78,240	$81,270	$85,570	$95,180	$94,760
Computer software engineers-systems software	$70,150	$76,130	$81,180			$91,430	$92,030
Computer hardware engineers	$74,880	$78,760	$83,940	$82,820	$96,540	$96,980	$94,690
Electrical engineers	$67,030	$71,870	$73,210	$75,490	$80,180	$82,810	$84,820
Electronics engineers, excluding computers	$68,920	$70,940	$75,580	$76,930	$81,320	$85,270	$86,330
Industrial engineers	$61,660	$64,070	$68,910	$73,330	$76,210	$77,480	$77,710
Mechanical engineers	$59,830	$64,810	$67,310	$68,460	$73,620	$77,250	$78,740
Engineering managers	$97,380	$104,550	$107,290	$125,080	$128,470	$129,450	$130,020
Industrial designers	$59,570	$63,480	$65,180	$66,070	$80,280	$91,850	$94,800

[a]Computer industry is defined as SIC 357 (Computer and Office Equipment) for 1999–2001; NAICS 334100 (Computer and Peripheral Equipment Manufacturing) for 2002, November 2003, and November 2004. Although industry definitions differ, occupational definitions do not. Therefore we include data from the entire 1999–2005 period to show trends in salaries. Source: Bureau of Labor Statistics, 2005.

again. Overall, engineering salaries in the computer industry rose from $61,030 in 1999 to $78,210 in 2005, an increase of 28.1 percent, which compares to a 20.1 percent increase in the consumer price index for the same period (*http://data.bls.gov/cgi-bin/cpicalc.pl*). These data suggest that foreign competition is not driving down salaries in the United States, as had been feared. They may also show that U.S. engineering resources are being shifted to higher value activities and that engineers are in fact becoming more productive, both of which would support higher salaries.

Compared to salaries in other major computer-producing countries, salaries for U.S. engineers are very high. For all engineering categories, including technicians, the average salary is $78,210 (Bureau of Labor Statistics, 2005). Salaries for engineering professions that require four-year degrees average more than $90,000 (Table 4).

The average salary for electronics engineers in all industries in the United States is about $80,000, compared to $60,000 in Japan, $20,000 in Taiwan, and less than $10,000 in China (Tables 5 and 6). However, engineering salaries are reportedly rising fast in China, especially in industry clusters, such as the Shanghai/Suzhou area, as MNCs and Taiwanese firms compete with domestic companies for talent. The willingness of MNCs to pay higher salaries gives them access to more experienced engineers and graduates of top universities, but turnover rates are high.

SKILL AVAILABILITY IN THE UNITED STATES AND OTHER COUNTRIES

Limited national data have been collected on production and the availability of engineers in different countries. A Duke University study of engineering graduates in the United States, China, and India showed that even for these data, definitions are often incompatible (Gereffi and Wadhwa, 2005). We could find no international data at all on the availability of engineers with skills in specific specialties, such as electrical, mechanical, industrial, or software engineers, so we must rely on interviews, our small survey of companies, and other qualitative information.

Gereffi and Wadhwa distinguish between dynamic and transactional engineers, a classification we found useful for characterizing engineering workforces in different countries based on our interviews. Dynamic engineers are

TABLE 5 Comparative Salaries for Electronics Engineers by Location

	Average Base Salary
United States	$78,000
Japan	$63,000
Taiwan	$20,000
China	$10,000

Sources: For U.S., Bureau of Labor Statistics Occupational Employment Statistics. For Japan, Quan (2002). For Taiwan, *EE Times* (2003) and interviews with ODMs in Taiwan. For China, PR Newswire (2004) and interviews with PC makers and ODMs in Taiwan and China.

TABLE 6 Engineering Salaries in China, by Home Base of Notebook PC Companies

Company Home Base	Base Salaries Paid in China
United States	$15,000 (6–7 years experience) $7500 (new graduates)
Japan or Europe	Similar to U.S. companies
Taiwan	$5,000 (new graduates)
China	$5,000 (new graduates)

Source: Interviews with PC makers and ODMs in China, Taiwan, and Japan.

capable of abstract thinking and high-level problem solving using scientific knowledge, are able to work in teams, and are able to work with people from other countries and cultures. Dynamic engineers have at least four-year degrees in engineering and are leaders in innovation. Transactional engineers have learned engineering fundamentals but can not apply this knowledge to solving large problems. Most transactional engineers, who do not have four-year degrees, are responsible for rote engineering tasks.

United States

In our interviews, engineering managers and executives of U.S. companies described engineers in the United States and elsewhere in words very much like those of Gereffi and Wadhwa, with some additional country-level distinctions. In general, U.S. engineers are more dynamic and analytical than their international counterparts, and they have the ability to lead the innovation process.

The team culture in most firms means that most U.S. engineers understand working in cross-functional teams and project management. Even new U.S. graduates have been trained to work in teams as part of their university education. Also, many U.S. engineers have gained some international experience as members of engineering teems sent to Asia to work with local development teams, sometimes for weeks or months at a time.

In addition, a large number of immigrants have earned degrees in the United States and then remained in the country to work for U.S. firms. Because these individuals have knowledge of their home countries, they are often chosen to work with engineering teams in those countries. As part of the entrepreneurial culture in the United States, many U.S. engineers have gained business experience by working on product-development teams or by being involved in start-up companies. Entrepreneurial skills are critical in the early design process when technology road maps must be matched with market demand to develop new products. These skills cannot be easily learned in less entrepreneurial environments farther from leading markets.

Taiwan

Taiwan has a mix of dynamic and transactional engineers, including many mechanical and electrical engineers with strong hands-on experience. Taiwan has the deepest pool of notebook PC developers in the world, as well as engineers with extensive experience developing other products, such as PC motherboards, optical drives, low-end network devices, and add-on cards. In addition, some Taiwanese ODMs are moving into the mobile phone business.

Taiwanese engineers learn mostly on the job and develop great depth in specific disciplines such as EMI, board layout, and thermal and power management. Engineering graduates of Taiwanese universities are said to lack the analytical skills of their U.S. counterparts—skills that are important for working with key component suppliers to define new product architectures. They also have a poor understanding of international markets and generally lack the ability to design successful products on their own. Nevertheless, some Taiwanese engineers are strong managers and team leaders who can manage their own parts of a project and work effectively with PC makers.

China

Most Chinese engineers, even those with four-year degrees, fit the definition of transactional engineers. According to one interviewee, Chinese engineers "work perfectly at doing what they have been told but cannot think about what needs to be done; they lack both creativity and motivation. They are good at legacy systems, but not new things; they can't handle 'what if' situations."

Chinese mechanical and electronic design engineers are well trained but lack the hands-on skills that come with experience. However, they are gaining this experience and receiving significant training on the job from both multinational and Taiwanese employers. One major ODM offers free training courses to engineers and brings in Taiwanese engineers to teach them. ODMs also work with local universities to develop courses in the skills they need. In the words of an ODM manager, "China is a gold mine of human resources, but if you don't train them, you won't be able to take advantage of it." An American executive was equally enthusiastic, "The average might not be high, but there are so many that the cream of the crop must be very good. Chinese engineers feel ownership of the product, pride in it. American engineers will work their tails off on a project if they believe in it passionately, then will want to take off to go skiing or something. The Chinese will just move on to the next project."

Chinese engineers do not have strong design skills or marketing knowledge, especially for foreign markets, but domestic Chinese companies are trying to develop those skills to create products for the fast-growing Chinese market. One interviewee noted that Taiwanese companies are making long-term investments in training Chinese engineers and other professionals, and he expected that his U.S. company would move some of its product development to China as those skills were developed.

Japan

Industrial designers in Japan are not only very good at designing for the Japanese market, but can also create products for the U.S. market if they work with U.S. design and marketing people. Good examples are the IBM Thinkpad line and the successful Toshiba and Sony notebook products.

Notebooks account for more than 50 percent of Japan's PC market, and many products are developed specifically for that demanding market. As a result, Japanese design and development teams have great depth of skills in all design and development areas. They also are very strong in design-for-manufacturability, because most Japanese firms do their own design, development, and manufacturing (although lower value PCs and other products are increasingly being outsourced to Taiwanese companies).

IMPACTS OF OFFSHORING ON U.S. ENGINEERING EMPLOYMENT

Engineering employment in the U.S. PC industry has remained stable in recent years in spite of some offshoring of new-product development. One interpretation is that offshoring may have been well established by the late 1990s and has not greatly affected U.S. engineering employment since then. By 2000, U.S. PC makers had either outsourced development and manufacturing to ODMs or, in the case of IBM, had assigned development to teams in Japan and had offshored manufacturing. As a result, much of the hardware, mechanical, electrical, and electronics engineering required for product development was already offshore, as was the industrial engineering associated with manufacturing. Software engineering, engineering management, and a relatively small numbers of jobs in the various hardware, mechanical, and electrical disciplines necessary to support product design and management were left in the United States.

One result of the offshoring of notebook PC development is that capabilities have been created in Taiwan, such as design-for-manufacturability and designing for small form factors, that can be applied to new product categories, such as handheld devices, smart phones, and digital music players. The fact that U.S. engineering employment in the PC industry is not growing during a time of rapid growth in demand and a proliferation of products and models probably indicates that more engineering is being done outside the United States. ODMs that have gained capabilities in the PC industry are now becoming major suppliers of mobile phones and are likely to become involved in other mobile consumer devices.

The Offshore Scene

Reports and data from our interviews show that Taiwanese CMs and ODMs are rapidly expanding their engineering capabilities. Quanta, the largest notebook ODM, employed about 3,500 engineers in 2003. Since then, Quanta has opened a large new R&D facility outside Taipei that is expected to eventually house 6,000 engineers. The company is also adding engineers in China. Other ODMs have also increased their engineering resources as they take over most of the development and production of the global notebook industry. One interviewee at a U.S. PC maker estimated that the ratio of in-house engineers to ODM engineers on its development projects is about 1:3 for consumer desktops, but closer to 1:1 for notebooks and commercial desktops. A smaller PC maker, by contrast, had only 50 engineers overseeing its ODMs, which develop all of its products.

Most of the work that has moved offshore is transactional engineering, including board layout, tooling, electrical and mechanical engineering, and software testing. These jobs require engineering skills and experience in specific areas, such as power management, EMI, and heat dispersion.

Most engineering work related to manufacturing has also been moved offshore, although there are enough high-level industrial and process engineers in the United States to oversee manufacturing in both places and travel to Asia to troubleshoot when necessary. These jobs do not require great analytical skills, but because a large share of the engineering work required for new-product development falls into the transactional category, the number of engineers offshore can be very high.

For instance, the world's largest CM, Foxconn, is said to have 10,000 tooling engineers, including 2,000 designers (Datamonitor, 2005). Many of these may be technicians with less than a four-year degree. Nevertheless, this example shows how a Taiwanese company can employ large numbers of low-cost engineers for more routine work that must be done very quickly to bring high-volume production on line. As one U.S. executive said, "We don't do much PCB layout, tooling, or testing any more. You can't compete with the large numbers of Asian engineers for that kind of work. The U.S. can't compete on numbers of engineers. We have to take what we're great at in the U.S. and leverage the rest of the world's skills."

The U.S. Scene

The more advanced engineering work is, the less vulnerable it is to offshoring. Taiwanese and Chinese engineers and companies are considered weaker in system-level design and in software than U.S. engineers. In addition, they lack the ability to develop entirely new products that are likely to appeal to the U.S. market. All of the notebook vendors we interviewed agreed that they would not turn over concept design, product management, or product architecture to an ODM and that they only buy off-the-shelf designs from ODMs for low-end products or when they need to fill out a product line very quickly.

One PC maker said that a relatively small number of in-house engineers is necessary for performing the advanced tasks that remain in the United States. Even though these are critical activities, they are not where the bulk of the engineering work is. The same point was made by two top engineering executives at U.S. PC companies. As one of them told us, "The jobs that are really important and are in the U.S. involve product architecture where you need senior engineers, hardware and software engineers generally, and mechanical engineers and industrial design people." The other said, "The core of the design process is in the United States. We define the product—how it looks, how it will be assembled, materials used, features and technologies to incorporate. We determine the mechanical and electrical architecture."

R&D, which depends on high-level researchers with advanced degrees, often Ph.D.s, is also less vulnerable to offshoring. Other reasons for keeping R&D in this country are the strategic importance of some R&D projects and the need to protect intellectual property. Unlike product development, R&D and manufacturing are not necessarily interdependent. Thus R&D jobs have not been "pulled" offshore by manufacturing.

R&D requires highly specific skills, and the key to success is finding people with those skills. If they happen to be offshore, firms are more likely to bring them to the United States, or to hire foreign graduates of U.S. universities, than to move the R&D offshore. One component maker, for instance, has 150 researchers at its R&D lab in the United States, about half of whom are from outside the United States. Unlike companies in other industry segments, such as Intel and IBM, which have R&D labs outside the United States, the U.S. PC industry has kept its R&D in this country.

REFERENCES

Boudreau, J. 2006. Taiwan Towers as Tech Innovator. San Jose Mercury News, January 23. Available online at *http://www.siliconvalley.com/mld/siliconvalley/13692201.htm*.

Bureau of Labor Statistics. 2005. Occupational Employment Statistics. Available online at *http://www.bls.gov/oes/home.htm*.

Business Week. 1991. Laptops Take Off. March 18, pp. 118–124.

Datamonitor. 2005. Hon Hai Precision Industry Company Limited: Company Profile.

DigiTimes. 2006. ICT Report—4Q 2005: Taiwan's Notebooks. Taipei: DigiTimes Research.

EE Times. 2003. EE Times-Asia Salary and Opinion Survey Reveals Mixed Fortunes for Asia's Engineers. Available online at *http://corporate.globalsources.com/INFO/PRESS/ARTICLES/OCT2303.HTM*.

Gereffi, G., and V. Wadhwa. 2005. Framing the Engineering Outsourcing Debate: Placing the United States on a Level Playing Field with China and India. Master of Engineering Management Program, Duke University.

IDC (International Data Corporation). 2006a. Worldwide Black Book: Q2 2006. Framingham, Mass.: IDC.

IDC. 2006b. Worldwide PC Market: 4Q05 and 2005 Review. Framingham, Mass.: IDC.

Lu, L.Y.Y., and J.S. Liu. 2004. R&D in China: an empirical study of Taiwanese IT companies. R&D Management 34(4): 453–465.

PR Newswire. 2004. Salaries of mainland China Electronic Engineers Rising by up to 12 Percent—Global Sources Survey. Available online at *http://www.prnewswire.co.uk/cgi/news/release?id=130057*.

Quan, M. 2002. EE compensation, increases vary worldwide. EE Times. Available online at *http://www.eet.com/story/OEG20020926S0024*.

Reed Electronics Research. 2005. Yearbook of World Electronics Data, Volumes 1–4.

Wheelwright, S.C., and K.B. Clark. 1992. Revolutionizing Product Development: Quantum Leaps in Speed, Efficiency, and Quality. New York: Free Press.

Offshoring of Engineering Services in the Construction Industry

John I. Messner, Ph.D.
Director, Computer Integrated Construction Research Program
Associate Professor of Architectural Engineering
Pennsylvania State University

ABSTRACT

The construction industry is a large contributor to the U.S. and world economies. Participants in the industry are responsible for designing and constructing the built environment including infrastructure, housing, offices, and other facilities. This diverse industry has many project types and requires that many engineering disciplines (civil, electrical, mechanical, chemical, and architectural) work together. Employment in the industry is currently strong and is supported by a strong U.S. and global construction market.

Offshoring of engineering services in the construction industry is not new. U.S. companies have had offshore offices in low-income countries for many years to perform design and construction management services. But with the increase in information technology and the drive to reduce engineering costs on projects, offshoring in the industry has increased recently. In particular, many large capital projects being built by U.S. companies are being designed with some level of engineering work in low-cost engineering centers. To date, offshoring of design services for smaller projects is limited to a relatively small amount of CAD drafting, 3D modeling, and engineering detailing performed by offshore technicians, architects, and engineers.

Although offshoring is having an impact on the U.S. construction industry and the structure of jobs in the industry, the impact is limited at this time. The United States remains a net exporter of design services in the construction industry and employment for engineers remains strong. But the industry is prone to economic cycles that could have a significant impact on this situation in the future. Therefore, it would be prudent to consider taking steps to minimize potential negative impacts of offshoring. U.S. companies will certainly continue to use lower cost labor in other countries to remain competitive globally and to make the construction of more facilities by U.S. companies economically viable.

Measures that should be considered to address the impacts of offshoring include supporting the education and development of globally focused engineers; supporting the export of engineering services from the United States; ensuring that national security and intellectual property are appropriately protected when design services are offshored; and encouraging young people to pursue productive careers in engineering in the construction industry.

The construction industry is a large, diversified industry that focuses on the design, delivery, and renovation of a wide range of facilities, from large petrochemical plants, bridges, buildings, tunnels, roads, and ports to residential units. These facilities play a significant role in housing the population and providing core infrastructure. Engineers from many disciplines perform many different tasks in this diversified industry, including facility programming, design of engineered systems, construction engineering and management, and facility management.

The revenues for the global construction industry total \$3.9 trillion per year (Tulacz, 2005). The United States has the largest construction market of any country with a current annual value of approximately \$1.22 trillion, 9.2 percent of the gross domestic product of the United States (USCB, 2006b). U.S. companies also perform more than \$34 billion per year in international work (ENR, 2006a).

The U.S. construction market has recently grown significantly. Figure 1 shows the annual construction spending from 1993 to 2005. The average annual growth rate during this period was 7.3 percent. Construction spending in the

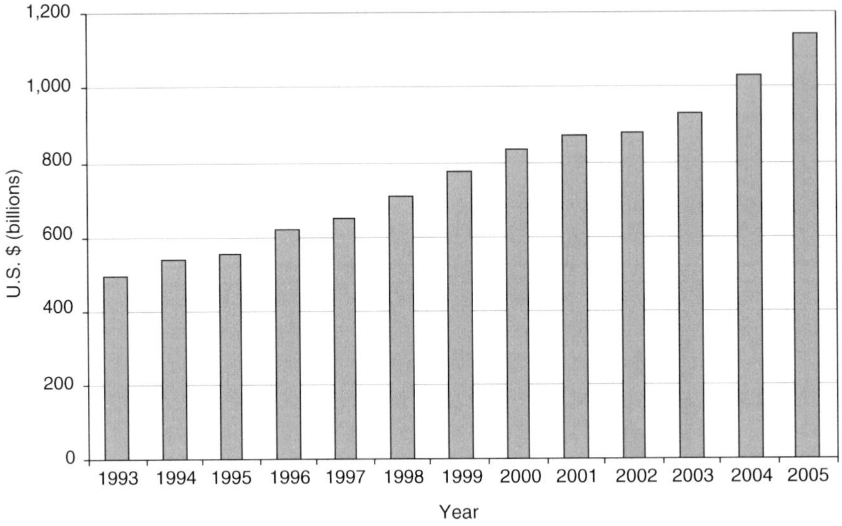

FIGURE 1 Construction spending, 1993–2005. Source: USCB, 2006b.

U.S. market increased by 12 percent from 2003 to 2004 and 11 percent from 2004 to 2005. This rate slowed slightly in 2006 to 8.5 percent (USCB, 2006b).

The global construction market has also been growing in response to the need for infrastructure and housing in developing nations such as China and India, along with continued investments in high-income countries. Data on the overall size of the global industry are limited and not very reliable, but *Engineering News Record* data collected from multiple sources show that the global construction market grew from $3.4 trillion in 1999 (ENR, 2000) to $3.9 trillion in 2004 (Tulacz, 2005), a growth rate of 14.7 percent over a five-year period.

The U.S. share of international work (work performed by a company not headquartered in the country where the construction is done) has been declining. In 2005, U.S. construction companies listed in *Engineering News Record* "Top 225 International Contractors" had revenues of $34.8 billion, or 18.4 percent of the international work done by the largest 225 international contractors (ENR, 2006a). This percentage is down from 36.5 percent in 1985, although it has remained relatively stable for the past 10 years.

One of the most significant challenges facing the U.S. construction industry is the supply of workers, both field employees and professional employees. Fewer people are interested in working in the construction trades, which has raised problems for the consistent delivery of quality facilities. Significant efforts are being made to recruit new design professionals into the industry, but these efforts face many barriers, including a negative perception of the construction industry and low salaries relative to other industries. The limited recruitment of new design professionals, combined with an aging population of experienced engineers who are approaching retirement, is making it difficult for the industry to find employees to design and manage the construction of facilities.

This paper focuses on offshoring of design and construction management services in the construction industry. However, there is no universal definition for offshoring (Trefler, 2005), and the definition is important. The American Society of Civil Engineering (ASCE) has defined offshoring in the construction industry as "the practice of acquiring architectural/engineering services from sources outside of the United States" (ASCE, 2005). But, because some level of design services have historically been performed in other countries for international construction projects, this definition seems incomplete. For example, if a power plant is being constructed by a U.S. contractor in India, some design work has historically been performed in India, and some design work may have also been performed in the country of the large equipment suppliers. Therefore, I propose that we use the following definition:

> Offshoring of design services in the construction industry is the relocation of work that is typically performed in one country to design professionals in the same company in another country, or to a different company in another country, to reduce wage rates.

Sometimes offshoring is performed through offshore outsourcing, that is, when a company hires an external company to perform a service in another country. At other times services are performed by company employees located in a company office in another country. Large international construction companies work in many international locations and have set up offshore offices to perform services on their international projects. Many of these services would not typically be performed in the United States, and thus they are not covered by the definition of offshoring, which considers the potential shift in work from the United States to offshore locations. But recently companies have been either setting up offices, using existing offices, or hiring companies abroad to perform design services that have previously been performed in the U.S. office. These services do fit the proposed definition of offshored services.

ENGINEERING SERVICES IN THE CONSTRUCTION INDUSTRY

The construction industry can be divided into several categories. For the analysis of offshoring, it is helpful to separate the industry into two market sectors: (1) the engineering, procurement, and construction (EPC) sector, and (2) the architectural, engineering, and construction (AEC) sector. Companies that perform work in the EPC sector focus on large industrial or infrastructure facilities. Companies in this sector tend to be large and employ many engineers and engineering technicians to work on the design and construction of large projects, such as power plants, refineries, industrial facilities, offshore platforms, and public works such as water purification plants, wastewater treatment plants, dams and rail projects. Companies in the AEC sector are much more diversified. Engineers in this sector work on the design and construction of buildings and residential facilities. The AEC sector is fragmented and is serviced by a large number of small companies. A number of companies perform work in both the EPC and AEC sectors, but these companies typically have different divisions for each sector.

This paper addresses the offshoring of engineering services in both sectors of the construction industry. The residential construction portion of the AEC sector (approximately 55 percent of U.S. construction) (USCB, 2006a) is not included because, even though there are some large residential developers that construct many units per year, a majority of residential design and construction companies are very small and offshoring remains limited in this sector.

There are almost 2.8 million construction firms employing more than 7 million people in the United States (U.S. Census Bureau, 2002), but the vast majority of these companies are very small; about two-thirds of them have fewer than 5 employees (BLS, 2006b). However, Bechtel, the largest U.S. contractor by revenue in 2005, had total revenues of \$14.6 billion, with \$7.2 billion in international markets (ENR, 2006b). Therefore, approximately 0.6 percent of the U.S. market revenue flows through this one company. The contractor with the largest share of the U.S. domestic market was Centex with \$12.6 billion in U.S. revenue, approximately 1 percent of the U.S. market (ENR, 2006b). The combined revenue of the 400 largest contractors for 2005 totaled \$200 billion (19 percent of the U.S. market) (ENR, 2006b).

As these figures show, the construction industry is very different from many other industries, which are controlled by a small number of large companies. It is also important to recognize that a very large percentage of the revenue for the top 400 contractors is subcontracted to specialty firms. Therefore, the industry is very diverse with many different companies contributing to facility construction.

Design work, which includes architectural and engineering services, is one portion of the overall revenue in the construction industry. According to *Engineering News Record*, which ranks the top 500 design firms in the United States each year, they generated \$59.25 billion in design revenue in 2005, an increase of 11.8 percent over 2004 (ENR, 2006c). Engineering is important to all phases of the construction and delivery of a capital facility. The primary phases for delivering and operating a facility have been defined by Sanvido et al. (1990) as managing, planning, designing, constructing, and operating a facility. The involvement of engineers in each of these phases varies, from the initial facility concept through the operation and renovation of a completed facility.

The projects most likely to involve offshore engineers have certain identifiable characteristics. Engineers typically perform work on large, unique projects. Owners rarely use the same design for multiple buildings or facilities. Even if they do, the design must be modified to accommodate site conditions, and projects must comply with building codes in the location of the project. To design a facility to meet local codes and to take into account local geotechnical, weather, and cultural conditions requires significant local knowledge. Thus local design firms have an advantage. No matter the location of the project, some degree of onsite construction will always be necessary. Thus onsite engineering support is always necessary.

Another important factor is that owners are typically actively involved in the design of their facilities, which requires frequent interaction between owners, or owners' representatives, and architects and engineers. Finally, many owners do not want the detailed design information for their facilities widely distributed to international locations. Thus security of the data is important on many projects. These factors can all make it more difficult to manage engineering teams from various locations, and therefore more difficult to execute a project with offshore engineering labor.

DATA-COLLECTION METHODOLOGY

The data used in this paper to analyze the current status of offshore outsourcing in the construction industry are taken from several sources, including the Bureau of Labor Statistics, the U.S. Census Bureau, the Bureau of Economic Analysis, the National Science Foundation, and *Engineering News Record*. Data are also taken from two surveys performed at Pennsylvania State University. The first survey was completed in 2004 and was sponsored by the Construction Industry Institute (CII). This survey was developed with significant industry input from a research team (CII Project Team 211) with 16 industry and four academic members.

Throughout this paper, this survey is referred to as the "CII survey." Following the survey, more than 20 detailed interviews were conducted with survey participants to gain additional insight into their global sourcing strategies and challenges.

A second survey was distributed in July 2006 to the top U.S. design firms listed in the *Engineering News Record* "Top 225 Global Design Firms." The survey was distributed to the directors of engineering or design of the 82 U.S. firms on the list. However, because only nine responses were received (a response rate of 11 percent), no statistical data will be presented from this survey. The survey did identify current perceptions of several large design firms in the industry, which are incorporated into the recommendations and comments in this paper. The small response to this survey illustrates the challenges of collecting data related to offshoring in the construction industry.

In general, it is difficult to draw accurate conclusions about offshoring based on the available data. No single source of data can be referenced to identify specific information about the current status or trends in offshoring in the construction industry. For example, no reliable data source provides a breakdown of domestic and foreign employees performing engineering and architectural services in construction companies. For future studies, we need to identify and develop methods to improve the collection of accurate data on the offshoring of engineering and architectural services jobs.

ENGINEERING EMPLOYMENT AND EDUCATION

Demand for Engineers

Civil engineering is the primary engineering discipline in the construction industry, but many other engineering disciplines are also important, including electrical, mechanical, industrial, environmental, and architectural engineering (Grigg, 2000). However, the remainder of this analysis focuses on civil engineering, which is the most representative engineering discipline in the industry. The unemployment rate for civil engineers in the U.S. market is only 2.2 percent (Rafferty, 2004). Of the 1.4 million engineers in the marketplace in 2004, 237,000 were civil engineers, the largest percentage (16.4 percent) of any single engineering discipline (BLS, 2006c). (The percentage of electrical engineering and computer science combined is larger.) Although there are many civil engineers in the workforce, the average starting salary for these graduates is one of the lowest for any engineering discipline. As of May 2004, the median salary for graduating civil engineers was $43,679 for a B.S., $48,050 for an M.S., and $59,625 for a Ph.D. (BLS, 2006c). The overall median salary for practicing civil engineers in May 2004 was $64,230, the second lowest of all engineering disciplines. The U.S. Department of Labor projects that an additional 39,000 civil engineers will be needed by 2014, a 16.5 percent increase (Hecker, 2005). This is one of the largest projected increases for an engineering discipline (the percentage increase is larger for environmental and biomedical engineers, but they are much smaller disciplines by quantity).

In addition to the statistics, it is clear from discussions with industry executives that one of the most significant challenges they face is the staffing of projects, which includes the recruitment and retention of engineers. Throughout the construction industry, there is currently a high demand for design and construction professionals in engineering and architecture in the U.S. market. In addition, the engineering workforce is aging, creating a shortage of experienced engineers in many large EPC companies. Because of this, these companies can offshore their engineering work with little impact on the size of the existing workforce in the United States.

In addition to engineers, architects are the other primary professional design participants with a significant impact on offshore outsourcing in the construction industry, particularly in the AEC sector. The Bureau of Labor Statistics (2006a) reported that there were approximately 129,000 architects in the United States in 2004, many of them professionally registered design practitioners. Architects had a reported median salary of approximately $60,300 (May 2004), and average projected growth for 2014 is 22,000 architects (17.3 percent).

It is also important to consider the size of the overall construction workforce. In 2004 5.2 percent of the overall workforce in the U.S. was working in construction supervision or in the construction trades (Hecker, 2005). This does not include manufacturing jobs related to the construction industry through the supply of building materials and equipment.

Supply of Engineers

In the United States, 7,827 B.S degrees were awarded in civil engineering in 2004, a decrease of more than 25 percent from 1981, when 10,678 were awarded (see Figure 2). The decline in civil engineering is similar to the decline in degrees in all engineering disciplines. In 2001, 59,258 B.S. degrees were awarded in all engineering disciplines, 23.6 percent fewer than the high of 77,572 in 1985 (NSF, 2004). But, although the number of engineering graduates in all disciplines has been declining since 1985, with only a slight increase between 1993 and 1995, the number of graduates in civil engineering increased sharply in the mid 1990s. Unfortunately, the number has declined from its peak in 1996.

In a recent study by Duke University, the number of degrees (bachelor's and sub-baccalaureate) awarded for engineering, computer science, and information technology in 2003–2004 was estimated to be 644,106 in China; 222,335 in the United States; and 215,000 in India (Gereffi and Wadhwa, 2005). Obviously, significant numbers of engineers are

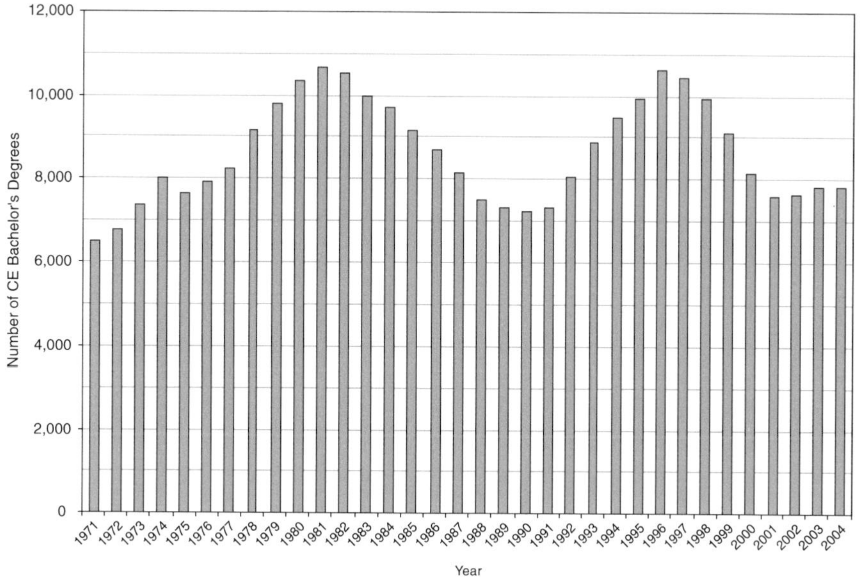

FIGURE 2 Bachelor's degrees in civil engineering, 1971–2004. Source: NCES, 2005.

graduating from universities in lower wage countries, many in civil engineering, although the total number of engineers in these countries is a subject of debate. We do know that several of the largest universities in China are graduating civil engineers, including Tsinghua University with 811; Central South University with 593; and Wuhan University with 219 (Gereffi and Wadhwa, 2005).

U.S. companies can find and employ engineers in other countries (e.g., India, China, and Eastern Europe) for lower wages. Wage rates vary based on region and demand, but figures developed by Hira (2003) show that a typical engineer in the United States receives an annual salary of $70,000, while an engineer in China receives $15,120, and an engineer in India receives $13,580. Thus there is clearly a wage disparity between engineers in different countries. However, based on interviews with engineering directors in several companies, the wages for qualified engineers in Mumbai, India, and some other locations are increasing significantly.

CURRENT OFFSHORING IN THE CONSTRUCTION INDUSTRY

Limited data are available to quantify the current value of work being performed in lower wage, offshore locations. At this time, no single source of data in the public domain documents either the dollar value of offshore engineering work or the amount of engineering time spent by engineers in lower wage locations. The best data sources available at this time are surveys and interviews with industry practitioners. Data collection from these sources has obvious limitations, however, including the potential for inaccurate self-reporting, poor response rates, and reliance on perceptions instead of quantitative data. With these limitations in mind, survey and interview data can provide insights into the current status and future trends in offshoring.

To date, the offshoring of engineering services to lower wage locations has primarily been focused in the EPC sector. Large EPC contractors, and the owners who hire these contractors, were the focus of the CII Survey. Administered in July 2004, the survey had a total response of 46 people representing 33 companies (20 construction companies and 13 large-facility owners) (Messner et al., 2006a).

Some large construction companies have been very active in international markets and have been offshoring engineering work for more than 15 years (Rubin et al., 2004). Compared to several other service industries, the construction industry as a whole has been slow to adopt offshoring, but larger companies, as well as companies in several niche markets in the industry, have started to offshore tasks for some large-scale operations. Some examples of niche markets are the development of 3D models during the design process, the conversion of 2D sketches to CAD models, and the development of engineering shop drawings for trade contractors (e.g., mechanical and steel subcontractors).

The United States is a net exporter of construction, architectural, and engineering services. According to data compiled by the Bureau of Economic Analysis (cited in Nephew et al., 2005), the United States had a trade surplus in construction architecture and engineering services (CAE services) of $2,991 million in 2004[1] (see Figure 3). The

[1]The export value in the Bureau of Economic Analysis data does not include merchandise exports or outlays abroad for wages, services, materials, or other expenses. The import value is a total value.

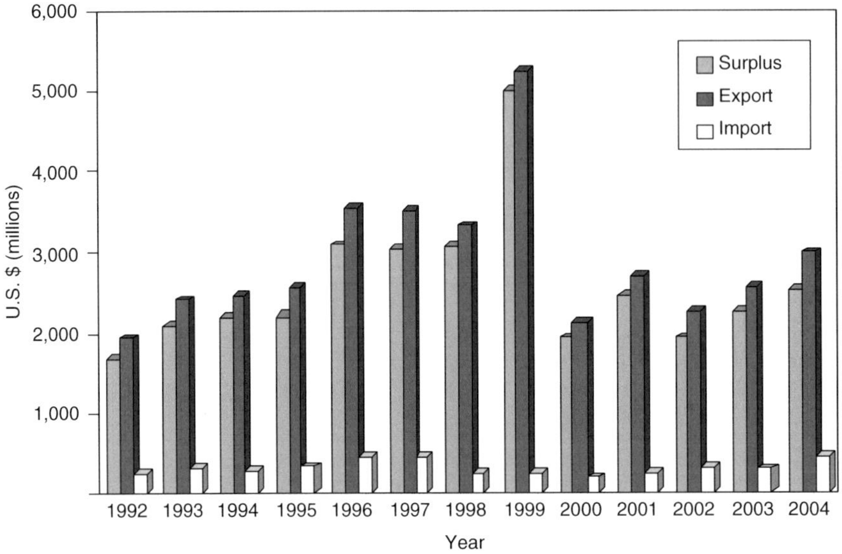

FIGURE 3 Trade surplus for construction, architectural, and engineering services, 1992–2004. Source: Bureau of Economic Analysis, cited in Nephew et al., 2005.

annual values vary widely depending on the number and type of large projects in any given year. It is interesting to note that the United States still has a trade surplus in CAE services with low-income countries that are known for providing low-cost engineering services to the EPC and AEC sectors of the construction industry. In 2004, the United States exported $107 million of CAE services to India (the location of many offshoring centers used by EPC companies) and imported $42 million in services. This trade value does not include company employees located in India, but does include contracted services by Indian companies. Therefore, the value reflects only contracted offshoring, not all architectural and engineering offshoring. For large EPC companies, many of the offshore offices are sole ventures that are not included in the import data. Nevertheless, it is clear that the volume of CAE services performed under contracts with companies in low-income countries is not great, and the United States has maintained a net surplus of services.

Offshoring in the Engineering, Procurement, and Construction Sector

Offshoring in the EPC sector of the construction industry is not new. One survey respondent to the CII study stated that "the use of low cost engineering centers has emerged as a common practice among many large engineering, procurement and construction (EPC) companies. This has primarily been driven by the realization that a large portion of the detailed engineering-design work can be treated as a commodity."

Large capital facility projects in the EPC sector often require many hours of engineering work, much of it related to detailed engineering, including the sizing and routing of piping; the design and location of electrical conduits and wiring; and the detailing of structural elements. This type of repetitive, detailed engineering work makes offshoring more attractive than in some other design practices because it is easier to systematize this type of work and less direct communication is required between the designers.

Of the companies that participated in the CII survey, 74 percent had international offices that were participating in multi-office execution strategies for the delivery of projects. Many had offices in low-cost engineering locations, such as India, China, Czech Republic, Russia, Romania, Poland, Mexico, and Taiwan. Some of these offices were established specifically to provide low-cost engineering services for company projects. Others were developed to perform specific design tasks for domestic construction projects. Large projects in low-wage countries often require that some design work be done locally. In addition, it is sometimes necessary to use engineers in the local environment for code verification and other engineering work that requires a detailed understanding of the local environment. It is important to note that many companies have international engineering offices in high-wage countries, such as England, Finland, and United Arab Emirates, to develop global virtual teams.

One goal of the CII survey was to determine the factors that influence companies to establish global engineering teams for the execution of projects. Table 1, which shows survey results for EPC-sector companies, indicates that the top five factors that drive engineering firms toward the offshoring of engineering work are (1) the need to reduce the costs of engineering services, (2) competition, (3) global customers, (4) the need to locate services close to a project, and (5) the

TABLE 1 Factors Affecting Global Virtual Teaming in the EPC Sector

Drivers of Offshoring	Average Score (1 = low, 5 = high)	Ranking
Need to reduce engineering-services costs	4.3	1
Competition	3.2	2
Global customers or local customers	3.2	3
Need to locate services close to the project location	3.1	4
Need to shorten engineering schedule	2.9	5
Need to expand detailing work for the same cost	2.8	6
Country, client, or funding source requirements	2.8	7
Need to understand/comply with codes and standards	2.7	8
Company policy (e.g., global procurement of services)	2.6	9
Need to balance engineering workload among multiple offices	2.5	10
Developments in technology	2.4	11
Availability of engineers	2.4	12
Need to improve engineering quality	2.3	13
Need to maintain consistency of products/services	2.3	14
Changing education/demographics	2.1	15

Sources: EPC and owner data, Messner, 2006b.

need to shorten the engineering schedule. Of these factors, the need to reduce costs, with an average score of 4.3 out of 5, ranked significantly higher than the other factors.

Companies in the EPC sector face several significant challenges, including an aging engineering workforce. In the EPC sector, there is a growing shortage of engineers with 10 to 25 years of experience. In a study by Gibson et al. (2003), 69 percent of the workforce was 40 years of age or older. The study also concluded that the supply of new engineers would be "insufficient to replace departing engineers and to support the level of growth desired by some owners and nearly all contractor firms."

Offshoring in the Architectural, Engineering, and Construction Sector

Offshoring in the AEC sector primarily affects two professional groups, engineers and architects. Until very recently, very little work on U.S. projects was performed by offshore architects or technicians. This is changing, however, as companies are looking for opportunities to offshore lower skilled technicians' jobs to lower cost markets. An example of this type of service is the transformation of hand-drafted documents into 2D CAD or 3D CAD models. This straightforward task, traditionally performed by CAD technicians or young architects, can be performed without extensive knowledge of a project. Other tasks being outsourced include the development of intelligent building information models and the creation of design details for a completed conceptual design. Little reliable data are available about the extent of these services being performed offshore, but the current perception is that the number of jobs currently performed offshore is relatively small. However, a few companies with larger offices in lower income countries present a different scenario.

The other primary professional services that can be provided offshore are engineering tasks for a building project, such as engineering design for the foundation, structure, mechanical system, electrical system, storm-water management, lighting, and other technical systems. The design of these technical systems requires expertise in both design and analysis. Again, there is no reliable source of data on the size or scale of offshoring in building engineering disciplines. Some companies offshore work, such as steel detailing for fabrication, wood-truss detailing, and mechanical-ductwork detailing; and based on survey results, the number of these companies is growing. Companies that are offshoring are typically not interested or not willing to share detailed information about their initiatives. However, the range of services being offshored is believed to be relatively limited.

Several reasons have been provided for the limited offshoring in the AEC sector:

- Most AEC firms are small, which make the economies of scale for offshoring less attractive when considering an initial investment.
- Some projects involve secure or sensitive information the owner does not want distributed to non-U.S.-based service providers.
- Design professionals must have significant interaction with the owner and other design professionals, which can be challenging when offshoring a project.
- Local knowledge about the project conditions is important (e.g., soil conditions, local codes, standard construction practices, standard materials, and architectural norms in the country).
- Under current market conditions, design professionals can get reasonable fees with their existing labor force, which limits the incentive to reduce costs.

Service providers in low-wage countries are organizing to provide design services with offshore labor to architecture and engineering companies. To date, no large offshore companies, such as Tata Group, Wipro, or Infosys in information technology, have had a significant impact. As more foreign companies and domestic consulting companies provide and manage these services, it will become easier for architectural and engineering companies to become involved in offshoring on a smaller scale (Bryant, 2006). At that point, the primary issue will be how much work companies will be willing to perform with offshore labor. Most architectural and engineering companies are small, and offshoring large parts of their business would be a significant undertaking.

Many architects and engineers are also very aware of their responsibility as service providers to the facility owner, as well as their legal responsibilities for the final design. Therefore, many may hesitate to begin offshoring because of a perceived loss of control over the design process and challenges in communication and oversight. It is much more likely that detailed analysis and modeling work will be performed with offshore labor, because these tasks have traditionally been performed by technicians and lower level engineers who are just starting their careers. Consistent procedures have been developed for these well defined tasks that can ensure quality performance with little oversight. It is interesting to note that these same tasks have increasingly been replaced by software tools that can perform them automatically. For example, the detailing of steel continues to get easier as new computer applications automate the sizing and detailing of steel members and connections and new 3D modeling software makes it easier to develop detailed 3D information models for facilities.

TABLE 2 Perceived Effect of Offshoring on Cost, Time, and Quality by CII Respondents

	Impact on Metric				
Performance Metric	More than 10% increase	0–10% increase	Same	0–10% reduction	More than 10% reduction
Engineering cost	4%	2%	7%	39%	48%
Construction cost	—	4%	75%	17%	4%
Engineering time	2%	18%	48%	24%	8%
Overall project delivery time	—	9%	59%	30%	2%
Engineering quality	6%	11%	65%	18%	—
Construction quality	2%	19%	72%	7%	—

Source: Messner, 2006b.

EFFECTS OF OFFSHORING ON ENGINEERING COST

In the previous section, the primary driver for offshoring was shown to be cost reduction. Therefore, a critical question for the future of offshoring is if or how much offshoring reduces engineering costs. Several indicators suggest that offshoring, when properly executed, can reduce overall engineering costs, at least for large-facility projects and specific, well defined tasks for smaller projects.

The CII survey included questions about respondents' perceptions of how some offshore engineering work affected the cost, time, and quality of projects. For any project, one must consider not only initial engineering-design cost, but also the total delivered-facility cost. Therefore, the survey asked about the effects on engineering and construction costs, as well as on time and schedule. Table 2 shows that most of the contractors who felt that offshoring could reduce costs projected the cost savings to be more than 10 percent. In addition, they believed this reduction could be achieved with no increase (and a potential decrease) in construction cost. Opinions differed markedly about potential savings in time, with the average response being that there was no effect. Most participants felt that engineering quality was the same or slightly lower with offshoring but that construction quality was the same or better.

Cost is one of the main concerns in facility design, and the cost of architectural and engineering-design services varies widely as a percentage of the cost of a project. Typically, these costs are from 7 to 18 percent of the total capital cost of a project, depending on its complexity and size. The cost of design services is impacted by labor rates for design professionals and productivity of the workforce. Even though engineers in lower income countries earn significantly less than U.S. engineers, some costs increase with offshoring, such as added travel, planning time, and information-system costs.

A detailed study of projects by one large owner illustrates the potential savings on large capital facility projects based on the use of low-cost engineering labor. The study analyzed five projects completed between 1992 and 2001. The owner was able to reduce engineering costs on all projects from an average of 16.9 percent for a typical facility to a design cost of only 10.2 percent. This means a total reduction in design-service costs of 40 percent compared to the typical costs (Messner, 2006a). The project team did not notice specific negative impacts for construction costs, although the company had to overcome many challenges in the execution of the projects with engineers from different locations.

This information is not meant to justify offshoring of engineering services or to convince a company to pursue offshoring. It is intended to present findings based on opinions and some quantitative analysis of the impact of offshoring on the cost structure of large capital facility projects. Like many other industries, the construction industry is extremely cost conscious. Therefore, economic factors must be considered when predicting future trends. If design and construction firms can consistently reduce their overall engineering costs through offshoring without negatively impacting quality, then they will certainly continue the current trend of offshoring engineering work to countries that maintain a supply of low-cost engineers.

Different construction industry participants reaped different benefits from offshoring. Facility owners, for example, may attempt to lower the cost for engineering services on a project. During an interview, one executive stated that "some projects become viable due to outsourcing, thereby creating more jobs once the project is complete." Thus offshoring might not only benefit the owner, but might also increase employment in the local economy.

There are also potential costs of offshoring for U.S. citizens, such as a decrease in engineering and architectural jobs in the U.S. market and downward pressure on the salaries of U.S. engineers. Another potential cost is a decrease in tax revenue paid to the U.S. government for services subcontracted to offshore companies. But it is also important to note that if companies get more work because their design costs are lower, the overall tax revenue may increase. One thing is clear—if U.S. companies lose contracts because of their higher cost structure for engineering, they will also lose engineering and architectural jobs and bring in less revenue. In addition, fewer U.S. products will be incorporated into designed projects.

THE FUTURE OF OFFSHORING IN THE CONSTRUCTION INDUSTRY

Predicting what the future holds for the construction industry is difficult, especially because of the limited data on offshoring. There is a clear and consistent perception on the part of executives that the level of offshore outsourcing will increase. When asked about plans for their companies, 92.5 percent of contractors in the CII survey said they plan to increase offshoring. Many interview subjects said they believe an increase in offshoring in the industry is inevitable because of the need to reduce the costs of design services and the limited number of engineers in the U.S. market who can meet the industry's needs.

Some interviewees felt that the increased offshoring would be detrimental to the quality of design services in the industry. As one survey participant put it, "Eventually, all owners will get what they want—low cost designs—high cost problems." Many engineering disasters have been caused by poor coordination, communication, and understanding of design responsibilities. The possibility for these kinds of problems increases when engineering work is done by global virtual teams. Many who had already established operational, low-cost engineering centers abroad believe that they can develop quality engineered solutions and documentation at a lower cost in their design centers, provided the design teams are properly structured and managed. Some even use the lower cost structure to create more detailed designs than they would typically develop in the United States. Because design costs are lower, they argue, they can save in construction costs with added detailing and coordination.

Much of the quality debate associated with offshore outsourcing depends on the industry perspective and industry segment. For example, if you consider the construction industry a service industry, then it is more difficult to provide good service to a client when separated by distance and culture, which cannot be avoided with global engineering teams. But if you view engineering services as well defined tasks (more like a commodity), then you are more likely to consider a low-cost engineering center a viable option for performing cost-effective design services with little impact on quality. Both opinions are predicated on a widely held perception by U.S. practitioners that engineering services performed in low-cost centers is of lower quality. However, a few believe that the quality differential is generated not by lower quality engineering but by poor communications and management.

ADDRESSING THE ISSUES RAISED BY OFFSHORING

Because offshoring in the construction industry will continue to increase, it is important that steps be taken to minimize the negative impacts of offshoring and take advantage of possible benefits. In the following sections, some of these steps are described briefly.

Preparing Engineers for Global Team Responsibility

One very important step that can be taken is to ensure that engineers who enter the construction industry, no matter what their discipline, are prepared to work toward a global design management role. This will require that students learn about global issues along with the managerial skills they will need to manage a global virtual team.

Recent changes in the assessment of education outcomes by the Accreditation Board for Engineering and Technology (ABET) reflect the change in focus from input (or teaching) to outcomes (or learning) (ABET, 2000). Since the implementation of Engineering Criteria 2000 (EC 2000) by ABET, the emphasis on professional skills has increased (Lattuca et al., 2006). International travel by students and participation in study-abroad programs have also increased in the past 10 years (Lattuca et al., 2006). It is critical that these activities continue to be supported and expanded.

Efforts are also under way to add four outcomes for students in civil engineering programs to the 11 EC 2000 criteria. These outcomes, "the knowledge, skills and traits necessary to become a licensed professional engineer," are described in *The Civil Engineering Body of Knowledge for the 21st Century: Preparing the Civil Engineer for the Future* (ASCE, 2004). In this report, outcomes related to business, public policy, the understanding of the role of a leader, and leadership principles are defined and described. These additional criteria would expand the range of knowledge of engineering graduates and help prepare them to participate in global engineering teams. The recommendations in this study are consistent with recommendations developed by the National Academy of Engineering in *The Engineer of 2020* (NAE, 2004).

Leadership and Research by Professional Societies

Professional societies have the opportunity and the resources to analyze offshoring and point the way to changes that will help the U.S. construction industry and other industries address the effects of increased offshoring. Some professional societies have issued policy statements to address the issue. The American Society of Civil Engineering, for example, has approved the following policy statement on offshoring of engineering services (ASCE, 2005):

> The American Society of Civil Engineers (ASCE) believes that the offshoring of engineering services should be accomplished in a manner that protects the public health, safety and welfare. ASCE believes that A/E [architectural and engineering] services must address the following criteria:
>
> - Appropriate homeland security requirements;
> - Licensing laws related to responsible charge;
> - Principles and/or requirements of Qualification-Based Selection using full disclosure of staffing and location; and
> - Fair trade agreement practices which apply.

In January 2004, the National Society of Professional Engineers (NSPE) Board of Directors approved a much more restrictive position statement:

> . . . the outsourcing of engineering should be done only when the talent cannot be found in the US. If outsourcing of engineering work is done, it should be done using the same rules, regulations, and laws that employers and employees are subject to in the US.

In addition, NSPE says that outsourcing should not jeopardize national security and that all parties should be aware of the location of offshore work and the conditions under which it is performed (Boykin, 2004).

These policy statements differ significantly. For professional societies to accurately analyze the impact of offshoring and provide guidance for engineers and companies in the construction industry, they will need additional data to support these policy statements. Professional societies have an opportunity to provide accurate information to their constituents on this topic that could lead to the development of recommendations for public policy.

Government-Imposed Trade Barriers

One thing is clear from surveys and interviews with executives in the construction industry. Whether or not they support increases in offshoring, none of the respondents for this research wants the U.S. government to intervene by establishing trade barriers that would impact the flow of trade in engineering services in the industry. Executives in companies that already use lower cost engineering centers feel that limiting the use of offshore engineers would negatively impact their ability to compete on a global scale. Executives in companies that do not offshore engineering services believe that government restrictions would simply not work over the long term.

Retraining to Meet Changing Demand

As offshoring increases, the demand for engineers, architects, and technicians with particular skills in the industry will change. For example, some technicians are specifically focused on the development of 2D CAD or 3D models from existing paper-based drawings or sketches. This type of work is easy for companies to offshore. Unless these workers are taught some new skills, they risk losing their jobs. Companies and the government should consider providing programs to support the retraining of technical employees in areas that are in higher demand in the U.S. market. As offshore engineers gain expertise (move up the value chain), U.S. engineers will have to continually outpace their lower cost counterparts in productivity or knowledge. It is important that these engineers be provided with guidance and retraining to enable them to remain active participants in the U.S. market.

Government Support for Exporting Engineering Services

For the long-term competitiveness of design professionals in the U.S. market, U.S. firms must remain competitive on a global scale. This will require that the U.S. government facilitate the entry of U.S. engineering and architectural firms into foreign markets. The more work they do in international markets, the more overall work will be managed and executed by U.S. employees, even if some of the design work for these projects is performed by an offshore workforce. The U.S. government already provides some support for the export of architectural and construction services, but not at the same level as some foreign governments (Vonier, 2006). The continued expansion of markets and revenue for companies is critical to maintaining a thriving international and domestic construction industry.

Ensuring Information Security

For national security reasons, data related to U.S. and sensitive facilities abroad must be appropriately managed. This does not necessarily mean that work cannot be performed in international locations, but additional security measures must be implemented when sensitive information is involved. Facility information related to infrastructure systems and building projects in the United States should not be readily available to all people throughout the world.

Recruiting and Retaining Engineers in the Construction Industry

Finally, we must send a realistic message to potential engineering and architectural college students. Many factors, including salary and the image of the industry, impact a student's decision to pursue engineering in the construction industry. Prospective students and their parents receive many negative messages through the media about the potential impacts of offshoring on engineering jobs in the United States. Unfortunately, media stories rarely distinguish between the types of jobs being offshored. Although offshoring will clearly have long-term effects on the structure of the construction industry and the way engineering work is done, there will continue to be a strong domestic demand for well educated, motivated engineers.

We should aggressively encourage young people to enter engineering disciplines that support the construction industry, and universities must continue to work hard to retain students in engineering fields by putting more emphasis on career progression, career coaching, salary comparisons, and even the intangible benefits of seeing the results of a project. Prospective students should be presented with an economic picture of the industry that makes sense, and the industry should find ways to provide more support to students and develop a more robust talent pipeline.

CONCLUSIONS

Offshoring in the construction industry will clearly have an impact, but it may not mirror the trends in other service or manufacturing industries. To date, most offshore outsourcing has been done for large capital facility projects that require many engineering hours. These projects are undertaken only by large engineering companies in the United States and other high-income markets throughout the world. Most U.S. companies are currently aggressively hiring new engineers in the U.S. market, even as they expand their engineering workforces in lower income countries. Therefore, to date offshoring has not had a significant impact on the employment of engineers in the U.S. construction industry, although some lower level engineering technician and engineering work has been relocated offshore.

The fragmented nature of the AEC sector of the construction industry, combined with the sheer complexity and unique qualities of each project and the necessity of understanding owners' requirements, make offshoring on a large scale more difficult in the construction industry than in some other industries. Some companies have focused on the systematization of global virtual teaming processes to benefit from offshore engineers, but many have not yet revised their standard business practices to use lower cost engineers to provide services. A significant number of executives are concerned that the use of lower cost engineers will have a negative impact on the quality of engineering services, thereby decreasing, or even cancelling out, the benefits of reducing engineering labor costs.

So far, employment prospects for new college graduates or experienced engineers in the U.S. engineering workforce in the construction industry have not been much affected by offshore outsourcing. This does not mean that this situation will remain as it is. The construction industry is dependent upon the capital-project spending of other industries (e.g., the oil, housing, and transportation industries, and private-sector companies, etc.) Therefore, construction spending in any particular segment of the industry is constantly changing. If spending declines in market sectors with high rates of offshoring (e.g., the power market), employment by companies in those sectors could easily be impacted.

The United States is a net exporter of design services (architectural, engineering, and construction services) in the construction industry, and the export of these services provides significant benefits to construction companies and suppliers to the construction industry, as well as some other companies. When U.S. design firms perform a project design, they tend to use materials and equipment that are familiar to them, which are likely to be produced by U.S. suppliers, and they tend to support the use of U.S. contractors, who often have working relationships with the U.S. firms. Thus other sectors of the U.S. economy also benefit.

Many large design firms believe that the use of offshore, low-cost engineering centers enables them to remain cost competitive in the low-margin environment typical of engineering projects in the construction industry. Some argue that using lower cost engineers in international locations, such as India, Mexico, and Eastern Europe, gives them an advantage in winning engineering contracts. Without this cost advantage, they argue, those contracts might be awarded to competitors in other countries, thereby impacting the U.S. engineering community, as well as other construction companies and suppliers in the U.S. market.

While offshoring is not currently causing a decline in engineering employment in the construction industry, it is very important that the industry and the country focus now on fundamental changes to address the clear trend toward offshoring. Recommendations for preparing for the future include expanding the range of engineering education to improve teamwork and leadership skills; increasing support for U.S. companies competing for work overseas; ensuring that national security and intellectual property are appropriately protected when companies use offshore design professionals; providing guidance to engineers in the industry; and supporting research to improve our understanding of offshoring and improve the quality of data. Finally, we must encourage young people to pursue careers in engineering in the construction industry.

ACKNOWLEDGMENTS

The author would like to thank the National Academy of Engineering and the Construction Industry Institute (CII) for their generous support of this research. We thank especially Samuel Florman and George Tamaro of the NAE Committee on the Offshoring of Engineering and the CII Project Team 211 members. The author would also like to thank survey and interview participants who contributed to this research. Opinions, findings, conclusions, and recommendations in this paper are those of the author and do not necessarily reflect the views of the sponsors or Pennsylvania State University.

REFERENCES

ABET (Accreditation Board for Engineering and Technology). 2000. Engineering criteria 2000, third edition. Baltimore, Md.: ABET Engineering Accreditation Commission.

ASCE (American Society of Civil Engineering). 2004. Civil Engineering Body of Knowledge for the 21st Century: Preparing the Civil Engineer for the Future. Reston, Va.: American Society of Civil Engineers.

ASCE. 2005. Offshoring of engineering services. Washington, D.C.: ASCE. Available online at *www.asce.org/pressroom/news/policy_details.cfm?hdlid=507*. Accessed August 29, 2006.

BLS (Bureau of Labor Statistics). 2006a. The 2006–07 Career Guide to Industries: Architect, Except Landscape and Naval. Washington, D.C.: U.S. Department of Labor.

BLS. 2006b. The 2006–07 career guide to industries: construction. U.S. Department of Labor, Washington, D.C. Available online at *www.bls.gov/oco/cg/print/cgs003.htm*.

BLS. 2006c. Occupational Outlook Handbook 2006–2007 Edition. Washington, D.C.: U.S. Department of Labor.

Boykin, D. 2004. Offshore outsourcing stirs national debate. Engineering Times 26(3).

Bryant, P. 2006. Decline of the engineering class: effects of global outsourcing of engineering. Leadership and Management in Engineering 6(59): 59–71.

ENR (Engineering News Record). 2000. What the world spent on construction. Engineering News Record, December 4, 38–45.

ENR. 2006a. The top 225 international contractors: 2006 rankings by 2005 revenue. August 21–28, 2006.

ENR. 2006b. The top 400 contractors: 2006 rankings by 2005 revenue. May 22, 2006.

ENR. 2006c. The top 500 design firms: 2006 rankings by 2005 revenue.

Gereffi, G., and V. Wadhwa. 2005. Appendix: Duke Outsourcing Study: Empirical Comparison of Engineering Graduates in the United State, China and India. December. School of Engineering, Duke University.

Gibson, G.E. Jr., A. Davis-Blake, K.E. Dickson, and B. Mentel. 2003. Workforce demographics among project engineering professionals—crisis ahead? Journal of Management in Engineering 19(4): 173–182.

Grigg, N.S. 2000. Demographics and industry employment of civil engineering workforce. Journal of Professional Issues in Engineering Education and Practice 126(3): 116–124.

Hecker, D. 2005. Occupational employment projections to 2014. Monthly Labor Review Online. Available online at *www.bls.gov/opub/mlr/2005/11/art5full.pdf*.

Hira, R. 2003. Global Outsourcing of Engineering Jobs: Recent Trends and Possible Implications. Testimony of Ronil Hira to The Committee on Small Business United States House of Representatives. Available online at *www.ieeeusa.org/policy/policy/2003/061803.html*.

Lattuca, L.R., P.T. Terenzini, and J.F. Volkwein. 2006. Engineering Change: A Study of the Impact of EC2000. Baltimore, Md.: ABET Inc.

Messner, J.I., C. Chen, and G. Joseph. 2006a. Implementing Global Virtual Engineering Teams. Research Report 211-11, November 2006 Austin, Tex.: Construction Industry Institute, The University of Texas at Austin.

Messner, J.I., L. Hankins, and A. Krumins. 2006b. Planning a global virtual engineering team: a tool for success. CII Research Summary 211-1, September 2006. Austin, Tex.: Construction Industry Institute, The University of Texas at Austin.

NAE (National Academy of Engineering). 2004. The Engineer of 2020: Visions of Engineering in the New Century. Washington, D.C.: The National Academies Press.

NCES (National Center for Educational Statistics). 2005. Digest of education statistics: 2005. Washington, D.C.: U.S. Department of Education. Available online at *http://nces.ed.gov/programs/digest/d05/tables/dt05_282.asp*. Accessed August 2006.

Nephew, E., J. Koncz, M. Borga, and M. Mann. 2005. U.S. International Services: Cross-border Trade in 2004 and Sales through Affiliation 2003. Washington, D.C.: Bureau of Economic Analysis.

NSF (National Science Foundation). 2004. Science and Engineering Degrees: 1966–2001. NSF 04-311, March 2004. Arlington, Va: NSF.

Rafferty, H.R. 2004. Civil engineering: looking good this year. Diversity/Careers in Engineering & Information Technology, June/July 2004. Available online at *http://www.diversitycareers.com/articles/pro/04-junjul/chgTech_civil.htm#top*.

Rubin, D.K., P. Reina, M.B. Powers, and T. Illia. 2004. As cost pressures mount, offshoring is making the work go round. Engineering News Record, August 2, 2004.

Sanvido, V.E., S.A. Khayyal, M.M. Guvenis, K.J. Norton, M.S. Hetrick, M.A. Muallem, E.K. Chung, D.J. Medeiros, S.R.T. Kumara, and I. Ham. 1990. An Integrated Building Process Model. Technical Report No. 1, January 1990. University Park, Pa.: Pennsylvania State University, Computer Integrated Construction Research Program.

Trefler, D. 2005. Offshoring: Threats and Opportunities. Brookings Trade Forum 2005: The Offshoring of Services: Issues and Implications, Washington, D.C. Available online at *www.brookings.edu/es/commentary/journals/tradeforum/2005btf_trefler.pdf*.

Tulacz, G.J. 2005. World construction spending nears $4 trillion for 2004. Engineering News Record, January 3, 2005.

USCB (U.S. Census Bureau). 2002. 2002 Survey of Business Owners. Available online at *http://www.census.gov/csd/sbo/index.html*. Accessed August 25, 2006.

USCB. 2006a. Construction spending—February 2006. Available online at *http://www.census.gov/const/www/c30index.html* Accessed April 3, 2006.

USCB. 2006b. Construction spending—June 2006. Available online at *http://www.census.gov/const/www/c30index.html*, accessed August 29, 2006.

Vonier, T. 2006. The export of professional design services by foreign architecture firms. AIA International eNews. Available online at *www.aia.org/nwsltr_intl.cfm?pagename=intl_a_200607_vonier*.

Semiconductor Engineers in a Global Economy

Clair Brown and Greg Linden
University of California, Berkeley

THE CHANGING NATURE OF SEMICONDUCTOR ENGINEERING WORK

The main forces affecting the nature of engineering work in the semiconductor industry are the evolution and globalization of technology. U.S. semiconductor firms are in many cases leading these changes both at home and abroad. But with increased global competition, U.S. chip engineers must continually upgrade their skills, deal with mobility among employers, and rely upon their own resources, rather than their employers, to manage their careers.

At present, global competition does not seem strong enough to undermine the positive employment and wage effects of the industry's continued growth for most workers, although job opportunities for older workers and those at the bottom of the job distribution have deteriorated. Many overseas companies, such as Taiwan's foundries and India's design-services providers, complement U.S. companies and have lowered barriers to entry at a time when the costs of design and manufacturing are skyrocketing. This situation plays to the strengths of U.S. engineering by keeping viable the fabless start-up system for bringing innovation to market. The cost reductions enabled by Asian suppliers of fabrication and design services are also contributing to falling semiconductor prices, and thus supporting the continued expansion of markets, both at home and abroad.

The semiconductor (or integrated circuit [IC] or chip) industry involves three distinct stages of production—design, fabrication, and assembly and packaging. Each stage has been affected differently by globalization and offshoring:

- Design: The design of integrated circuits is carried out primarily by engineers. The offshoring of design activities to low-cost locations has been accelerating since the mid-1990s.
- Fabrication: Wafer fabrication involves a large number of process and equipment engineers, who account for approximately 25 percent of total direct workers at a manufacturing or fabrication facility (called a "fab"). Offshoring and onshoring of IC factories appears to have reached a relatively mature and stable stage.
- Assembly and packaging: The final stage of IC manufacturing is the most labor intensive, but engineers make up only 6 percent of the typical assembly plant workforce. Assembly offshoring began in the 1960s, and assembly and packaging are now performed almost entirely abroad. Assembly and packaging are *not* discussed in this paper because the employment implications for U.S. engineers are insignificant.[1]

The semiconductor industry produces a wide range of products, from relatively simple discrete diodes and transistors all the way to complex "systems on a chip." Most market statistics reported here and elsewhere reflect "merchant" semiconductor sales, that is, sales to unrelated companies. A less visible share of the industry is devoted

This paper was prepared for the National Academy of Engineering Workshop on the Offshoring of Engineering: Facts, Myths, Unknowns, and Implications, October 24–25, 2006, Washington, D.C. The paper is based on research conducted for a forthcoming book by Brown and Linden, *Change Is the Only Constant: How the Chip Industry Deals with Crisis*.

[1]For an analysis of the globalization of assembly, see Brown and Linden (2006).

to "captive" chip design and manufacture internal to a company. This model is most prevalent in Japan but still exists in the United States, primarily at IBM, where nearly 50 percent of chip output in 2000 was for captive use.[2] Other systems companies, such as Apple Computer or Cisco, that don't make or sell chips may nevertheless design them for internal use. These chips may or may not be counted in merchant data depending on whether they are manufactured by a branded ASIC company, such as LSI Logic (which would be counted), or by a manufacturing-services "foundry," such as Taiwan Semiconductor Manufacturing Corporation (which wouldn't be included). All foundry sales are excluded from this analysis to prevent double counting.

The work of engineers who design, manufacture, and market chips has been transformed by the continuous progression of manufacturing technology, which has evolved for more than 30 years along a trajectory known as "Moore's Law," the name given to a prediction made in a 1965 article by Gordon Moore. Moore, who co-founded Intel a few years later, predicted that the cost-minimizing number of transistors that could be manufactured on a chip would double every year (later revised to every two years). The industry has maintained this exponential pace for more than 30 years.[3]

Moore's prediction was based on several factors, such as the ability to control manufacturing defects, but the driving technological force has been a steady reduction in the size of transistors. The number of transistors leading-edge producers can fabricate in a given area of silicon has doubled roughly every three years. From 1995 to 2003, the pace accelerated and the number doubled every *two* years.[4]

This relentless miniaturization is now reaching the molecular level. The smallest "linewidth" (feature on the chip surface) has shrunk from two microns in 1980 to less than one-tenth of a micron (100 nanometers [nm]) a quarter-century later. Viewed in cross-section, the thickness of horizontal layers of material deposited on the silicon surface is currently about 1.2 nm. For an idea of the scale involved, the width of a human hair is about 100 microns, and the width of a molecule is about 1 nm (one-thousandth of a micron).

This progress has involved considerable expense for R&D, and the cost of each generation of factories has steadily increased. By 2003 the price tag for a fab of minimum efficient scale was more than $3 billion.

The Moore's Law trajectory has led to growing complexity of the industry's most important chip designs. The size of a design team depends on the complexity of the project, the speed with which it must be completed, and the resources available. Design teams can be as small as a few engineers, and project duration can vary from months to years. A chip like Intel's Pentium 4, with 42 million transistors fabricated on a 180 nm linewidth process, engaged hundreds of design engineers for the full length of a five-year project.[5]

Functional integration has reached a point at which certain chips encompass most of the individual components that populated the circuit board of earlier systems, giving rise to the name "system on a chip" (SOC). SOC integration offers the benefits of speed, power, reliability, size, and cost relative to the use of separate chips.

Although the manufacturing costs of an SOC are lower than for the separate components it replaces, the fixed costs of a complex design can be significantly higher. A major reason is that system-level integration has drawn chip companies into software development because system software should be generated in parallel with the system-level chip to ensure coherence. Chip companies also offer their customers software-development environments, and even applications, to help differentiate their chips from those of their competitors. In a large chip-development project, software can now account for half the engineering hours.

U.S. chip companies accounted for about half of the industry's revenue in 2005, with Intel alone commanding about 15 percent of the market. The only U.S.-based firms in the 2005 global top 10 were Intel and Texas Instruments, but the United States has a great many mid-size companies that account for about half of the top 50. Some of these are "fabless" companies that design and market chips but leave the manufacturing to other companies, primarily Asian contract manufacturers known as foundries. All new entrants to the chip industry in recent years have adopted the fabless model.

Fabless revenue has grown much faster (compound annual growth rate of 20 percent) than the semiconductor industry as a whole (7 percent) over the last 10 years. In 2005, the largest fabless companies, Qualcomm, Broadcom, and Nvidia, each had revenues of more than $2 billion.

The discussion in this paper of how the labor market for semiconductor engineers, both domestic and worldwide, has been changing in response to changes in skill requirements is based on our ongoing interview-based research on the globalization of the semiconductor industry. Since the early 1990s, the Berkeley Sloan Semiconductor Program has collected data at semiconductor companies globally.[6] In the past seven years the authors have interviewed managers and executives at dozens of semiconductor companies (both integrated and fabless) in the United States, Japan, Taiwan,

[2]IC Insights data reported in Russ Arensman, "Big Blue Silicon," Electronic Business, November 2001.

[3]The revision occurred in 1975 (John Oates, "Moore's Law is 40," The Register, April 13, 2005).

[4]Mark LaPedus, "ITRS chip roadmap returns to three-year cycle," Silicon Strategies, January 21, 2004.

[5]Terry Costlow, "Comms held Pentium 4 team together," EE Times, November 1, 2000. "Linewidth" refers to the size of the features etched on a wafer during the fabrication process. Each semiconductor process generation is named for the smallest feature that can be produced.

[6]The Competitive Semiconductor Manufacturing Program is a multidisciplinary study of the semiconductor industry established in 1991 by a grant from the Alfred P. Sloan Foundation with additional support from the semiconductor industry. Further details are available at *esrc.berkeley.edu/csm/* and *iir.berkeley.edu/worktech/*.

India, China, and Europe. We also use data from the Bureau of Labor Statistics, the Semiconductor Industry Association, and the Institute of Electrical and Electronic Engineers, as well as other published and proprietary sources (e.g., industry consultants).

We begin by looking in detail at data sets on employment and earnings of U.S. semiconductor engineers, H-1B workers, and overseas engineers. We then discuss the factors affecting the U.S. labor market for semiconductor engineers, including technological change, immigration policy, and higher education practices. A discussion of globalization follows in terms of offshoring by U.S. companies, the availability and quality of low-cost engineers in Asia, and the development of the semiconductor industry in Taiwan, China, and India. In the final section we consider the outlook for the U.S. chip-industry workforce.

THE U.S. LABOR MARKET FOR ENGINEERS

Factors that have affected the semiconductor industry in the past six years include a severe recession during 2001, a recovery that stalled in 2004, a large decline in venture funding for start-ups that picked up again in 2006, changes in the number of H-1B visas, and a drop and subsequent recovery in foreign student applications to U.S. graduate engineering schools since 9/11. In light of these changes in government policies and swings in the business cycle, disentangling an underlying, long-term trend in the offshoring of engineering jobs is extremely difficult. Readers should keep this caveat in mind when reading the following analysis of the U.S. labor market for semiconductor engineers, as well as the discussion of engineering jobs in selected countries.

Because of inadequacies and gaps in the available data, we use more than one source for our analysis. To identify trends in the employment levels and earnings of semiconductor engineers, we use two major national data sets that have different strengths and weaknesses. The Bureau of Labor Statistics' Occupational Employment Statistics (OES) (*www.bls.gov/oes/home.htm*) provides a large job sample collected from establishments that report detailed occupational characteristics. However, comparisons of data from different years are not exact because OES is designed for cross-section comparisons rather than comparisons over time.[7] Moreover, OES does not provide educational characteristics.

The American Community Survey (ACS) (*http://www.census.gov/acs/www/*), a relatively new household survey started in 1996 to update the census between decennial surveys, provides not only detailed educational characteristics of workers, but also occupational and industry characteristics of their jobs. Thus ACS is much better suited to our labor market analysis. However, the sample size for ACS for 1996–2002 is too small for detailed analysis. For these reasons, we look at both the OES and ACS data sets in our analysis. Because they yield somewhat different results, however, we caution the reader against drawing strong conclusions based on either data set alone. The inconsistencies and gaps reflect a need for better data collection by government agencies.

We also use the very large Census Longitudinal Employer-Household Dynamics (LEHD) data set that links employees and employers to describe semiconductor career paths and firm job ladders between 1992 and 2002. This enables us to look at how workers form career paths by piecing together jobs offered by semiconductor firms.

Employment and Earnings (OES Data)

We begin by looking at employment levels and annual earnings for selected engineering jobs in 2000 and 2005, based on OES data. For the semiconductor industry, we use the North American Industry Classification System (NAICS) "Semiconductor and Other Electronic Component Manufacturing" (NAICS four-digit level 3344), which includes relatively low-value components such as resistors and connectors. The most relevant subcategory, "Semiconductor and Related Device Manufacturing" (NAICS 334413), accounted for 39 percent of employees (and 45 percent of nonproduction workers) in the 3344 category in 2003, but occupation-specific data are not available at this level of industry detail.[8]

In 2005, 2.4 million people were employed nationally in "engineering and architecture" occupations,[9] with average annual earnings of $63,920 (see Table 1). Another 2.9 million people were employed in "computer and mathematical" occupations, with average annual earnings of $67,100. National employment in engineering and architecture fell 7.5 percent from 2000 to 2005, and average annual earnings of these workers rose 18.2 percent (more than the CPI-urban, which rose 13.4 percent).[10] Computer and mathematical jobs increased slightly (0.7 percent) from 2000 to 2005, and average annual earnings of these workers rose 15.6 percent, slightly more than inflation.

The semiconductor industry (NAICS 3344) employed 450,000 workers in 2005, with 21 percent in engineering and architecture occupations (36 percent of them as technicians or drafters) and 6.4 percent in computer and math occupations (40 percent of them in computer support or administrative positions). These two groups do not include managers,

[7]The OES survey methodology is designed to create detailed cross-sectional employment and wage estimates for the U.S. by industry. It is less useful for comparisons of two or more points in time because of changes in the occupational, industrial, and geographical classification systems, changes in the way data are collected, changes in the survey reference period, and changes in mean wage estimation methodology, as well as permanent features of the methodology. More details can be found at *http://www.bls.gov/oes/oes_ques.htm#Ques27*.

[8]U.S. Census Bureau, "Statistics for Industry Groups and Industries: 2003," Annual Survey of Manufactures, April 2005.

[9]This is the broad occupational category used for engineers in the OES.

[10]*http://data.bls.gov/cgi-bin/surveymost?cu.*

TABLE 1 Employment Levels and Earnings for Engineers in All Industries and in the Semiconductor Industry, 2000 and 2005

	2000		2005			
	Employment	Average Annual Earnings	Employment	Average Annual Earnings	Percentage Change in Employment	Percentage Change in Earnings
Architecture and Engineering Occupations (total)	2,575,620	$54,060	2,382,480	$63,920	–7.50%	18.24%
—in Semiconductors	132,150	$52,100	95,520	$68,720	–27.72%	31.90%
Electrical Engineers (total)	162,400	$66,320	144,920	$76,060	–10.76%	14.69%
—in Semiconductors	10,050	$69,560	10,620	$82,400	5.67%	18.46%
Electronic Engineers (total)	123,690	$66,490	130,050	$79,990	5.14%	20.30%
—in Semiconductors	14,170	$65,400	15,700	$82,430	10.80%	26.04%
Aerospace Engineers (total)	71,550	$69,040	81,100	$85,450	13.35%	23.77%
Chemical Engineers (total)	31,530	$67,160	27,550	$79,230	–2.62%	17.97%
Civil Engineers (total)	207,080	$58,380	229,700	$69,480	10.92%	19.01%
Computer Hardware Engineers (total)	63,680	$70,100	78,580	$87,170	23.40%	24.35%
—in Semiconductors	5,990	$70,780	14,440	$89,870	141.07%	26.97%
Industrial Engineers (total)	171,810	$59,900	191,640	$68,500	11.54%	14.36%
—in Semiconductors	12,580	$64,420	11,030	$74,250	–2.32%	15.26%
Mechanical Engineers (total)	207,300	$60,860	220,750	$70,000	6.49%	15.02%
Computer and Mathematical Occupations (total)	2,932,810	$58,050	2,952,740	$67,100	0.68%	15.59%
—in Semiconductors	27,080	$66,660	28,770	$77,800	6.24%	16.71%
Computer Programmers (total)	530,730	$60,970	389,090	$67,400	–6.69%	10.55%
Software Engineers, Applications (total)	374,640	$70,300	455,980	$79,540	21.71%	13.14%
—in Semiconductors	5,890	$72,680	8,250	$86,860	40.07%	19.51%
Computer Software Engineers, Systems (total)	264,610	$70,890	320,720	$84,310	21.20%	18.93%
—in Semiconductors	8,280	$76,660	7,090	$90,820	–14.37%	18.47%

who represent 8.2 percent of semiconductor employees. Nationally, some 12 percent of electronics engineers, 7.3 percent of electrical engineers, 18 percent of computer-hardware engineers, 5.8 percent of industrial engineers, and approximately 2 percent of computer-software engineers (applications and systems) are employed in the semiconductor industry. Together these six occupations account for 54 percent of engineering jobs in the semiconductor industry (or 85 percent if techs, drafters, and computer-support jobs are excluded).

Engineering jobs ("Architecture and Engineering Occupations") in the semiconductor industry fell a surprising 28 percent between 2000 and 2005 (Table 1, line 2).[11] However, if we look at the major categories for semiconductor engineers, jobs increased for electrical engineers (6 percent), electronics engineers (11 percent), and computer hardware engineers (141 percent). Semiconductor jobs for industrial engineers fell 2 percent, the only specialty in which job growth for semiconductor engineers was lower than for engineers nationally.

Jobs for software engineers ("Computer and Mathematical Occupations") in the semiconductor industry increased by 6 percent between 2000 and 2005, while all jobs in these occupations increased less than 1 percent nationally. The increases were unevenly distributed, however. Semiconductor industry jobs for software-applications engineers increased by 40 percent, while jobs for software-systems engineers fell by 14 percent.

On average, engineers in the semiconductor industry command higher salaries than their counterparts in other industries. In 2005, semiconductor industry engineers earned 7.5 percent more than engineers nationally, and software engineers in the semiconductor industry earned 16 percent more than software engineers nationally. In any given specialty, engineers in the semiconductor industry had average annual earnings of 3 percent (for electronics engineers) to 9 percent (for computer software engineers, applications) higher than engineers in other industries. Engineers in the six main semiconductor engineering specialties all experienced average growth in real earnings (i.e., above the inflation rate of 13.4 percent for the period), ranging from 1.9 percent for industrial engineers to 14 percent for computer-hardware engineers. Note that these comparisons are not adjusted for education or experience, which are taken into consideration in the next section using a different data set.

Of course, employment levels between 2000 and 2005 did not increase continuously. Applications software engineers experienced a dip in employment in 2004 after strong employment growth in 2003, and electrical and electronics engineers experienced a dip in employment in 2003 followed by very strong growth in 2004. This is consistent with the jump in the national unemployment rate for electrical and electronics engineers to 6.2 percent in 2003, as it converged for the first time in 30 years with the general unemployment

[11]Comparison of 2000 and 2005 is not exact because SIC 367 was used in 2000 for the industry code and NAICS 334400 was used in 2005.

rate, before falling back in 2004 to a more typical rate of 2.2 percent.[12]

Overall we can say that the labor market for semiconductor engineers appeared to be relatively strong in the five years after the dot-com bust in 2000, when earnings nationally were mostly stagnant during the economic recovery, with income gains going mainly to the top decile (especially the top 1 percent). Semiconductor engineers even experienced better job and earnings growth than engineers in the same specialties in other industries. Although employment for industrial engineers and software-systems engineers in the semiconductor industry fell, employment for the other four specialties increased. Although earnings growth was relatively high only for computer-hardware engineers and electronics engineers in the semiconductor industry, all six specialties had relatively high average annual earnings in 2005, ranging from $74,250 for industrial engineers to $90,820 for software-systems engineers.

Age-Earnings Profiles by Education and Experience (ACS Data)

To analyze the earnings structures of U.S. semiconductor engineers by education and experience, we use another data set, the ACS (*http://www.census.gov/acs/www/*). We calculated age-earnings profiles for three educational levels, less than a bachelor's degree (< B.S.), a bachelor's degree (B.S.), and a graduate degree (M.S./Ph.D.),[13] using ACS data for 2000, 2002, and 2004 for a sample of workers defined as follows:

- age 21 to 65
- industry code 339 (electronics components and products, comparable to NAICS 3344 and 3346)
- occupation codes (selected electrical and electronics, software, and other engineering occupations and selected managerial occupations)[14]

The age-earnings profiles for the B.S. (Figures 1 and 2) and M.S./Ph.D. groups (Figures 3 and 4) show how the annual earnings of semiconductor engineers increase with knowledge and skill levels (educational level) and experience (age) for 2000 and 2004.

The results are also given in Table 2, which shows earnings profiles for all three educational levels for 2000, 2002, and 2004, with earnings adjusted for inflation (in 2004 dollars using CPI-urban).[15] One cautionary note: because the sample size for 2000 is small, the results for that year are less reliable than for 2002 and 2004. Also some of the age-education groups were too small to show full results.[16]

Returns-to-Experience

Median and average real earnings increased with experience (age) for all educational groups through the prime ages. After that, median (but not necessarily average) earnings declined for older workers (age 51–65). However, average earnings did not decline for older workers in any education group in 2000 or for older M.S./Ph.D.-level workers in 2002, and median earnings did not decline for older < B.S. workers in 2004. The general increase and subsequent decline in median earnings implies that these engineers typically received a positive return-to-experience until they were in their fifties and sixties, when earnings for many of them declined. The decline can be explained, at least in part, by the number of weeks worked (Table 3). Workers older than 50 were much more likely than younger workers to work less than a full year (defined, conservatively, as less than 48 weeks of paid work).

Comparing degrees, engineers with B.S. degrees typically had higher returns-to-experience than engineers with advanced degrees. B.S. holders earned one-half to three-fourths more in their peak years (age 41–50) than in their entry years (age 21–30). Engineers with graduate degrees (M.S./Ph.D.) earned 10 to 20 percent more in their peak years (age 41–50) than they did a decade earlier (age 31–40), shortly after their entry-level years.

The variance in earnings increased with age for prime-aged and older engineers (see 90/10 ratio in Table 2). The increase in variance is typically thought to reflect faster growing pay for higher performers, and pay for top earners would be expected to increase as engineers become managers.

12Data were provided by Ron Hira. BLS redefined occupations beginning with the 2000 survey covering 1999, but there is no evidence that the redefinition has contributed to the post-bubble unemployment rise. See also Kumagai (2003).

13< BS includes workers with a high school degree or GED but no B.S. degree (the proportion of this group that did not have an associate degree was 41 percent in 2000, 27 percent in 2002, and 13 percent in 2004); BS includes college graduates who do not have a higher degree; MS/PhD includes workers with a Masters or Ph.D. degree (the proportion of this group that had only a Masters was 90 percent in 2000, 81 percent in 2002, and 82 percent in 2004). Workers without a high school degree and workers with professional degrees (e.g., MD, DDS, LLB, JD, DVM) are excluded.

14We used several different samples of occupation codes in order to test for sensitivity of age-earning profiles to the definition of semiconductor engineer occupations. In the results presented here, we included SOC 172070, 172061, 151021, 151030, 151081, 172131, 172110, 172041, 119041, 113021, 111021, 112020, 113051, and 113061. When we restricted the sample to fewer occupation codes, the age-earnings profiles remained mostly stable, with the earnings of the top 10 percent increasing for older groups with the inclusion of more managerial occupations.

15Earnings for *n* percent represents the earnings where *n* percent of observations are below this value and (100 – n) percent of observations are above this value. Earnings for the 50th percentile represent the median.

16For education-age-year cells (3 × 4 × 3 = 36) with fewer than 10 observations, no results are shown (two cells). For cells with fewer than 20 observations (and at least 10 observations), only mean and median income and full weeks worked are shown (six cells).

The sample sizes by year and education (not age) are as follows:

	2000	2002	2004
< BS	44	129	127
BS	151	367	363
MS/PhD	78	250	271

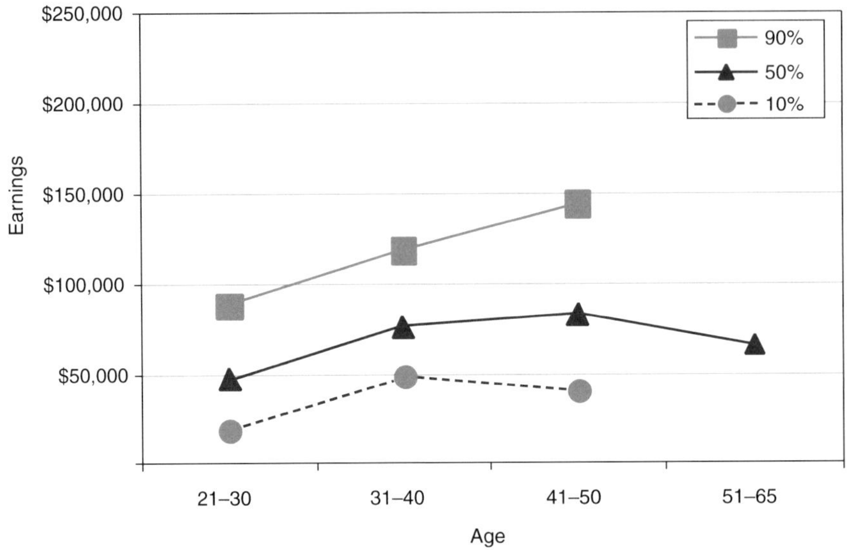

FIGURE 1 Age-earnings profile for B.S. holders in 2000.

However, the increase in variance between prime-age and older engineers reflects a sharp drop in pay at the bottom end of the scale (the 10th percentile group), especially in 2004. These profiles indicate that many older engineers are facing declining and inadequate job opportunities.

Returns-to-Education

As expected, median and average earnings increased with education. Comparing real median earnings for the younger groups, we see that the return for a B.S. degree has been fairly high, with college graduates typically earning 20 percent to 65 percent more (depending on age and year) than those who finished high school but not college. Put another way, in 2002 and 2004, a typical young engineer (age 21–30) with a B.S. degree earned the same pay as a typical engineer without a B.S. but with 10 years more experience (age 31–40).

The graduate-degree premiums over a B.S. (median earnings for M.S./Ph.D. compared to B.S.) were not stable over the short time period shown, so it is difficult to determine the trend for returns for graduate education. The graduate-degree premium for the youngest group, when many were still in school, was 36 percent in 2002, but fell to 8 percent in 2004. The graduate-degree premium for workers in the early stages of their careers (age 31–40) was 7 percent in 2000, then shot up to 25 percent in 2002 and 36 percent in 2004, confirming our interview-based findings that the relative

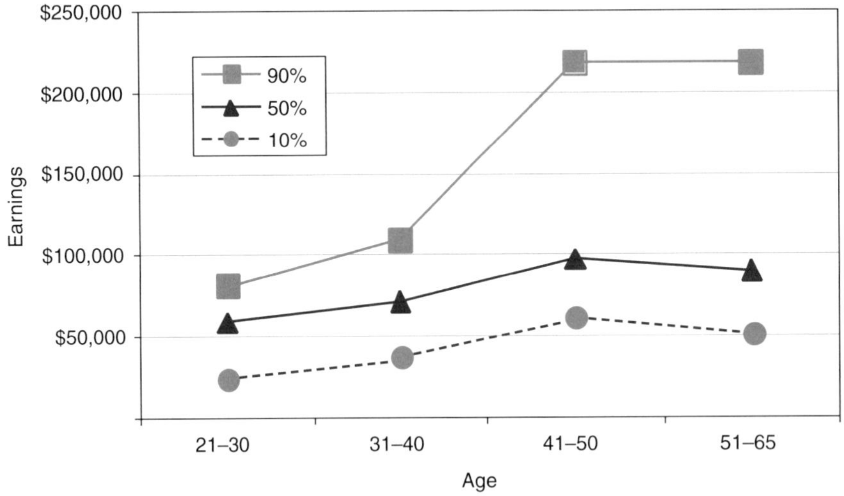

FIGURE 2 Age-earnings profile for B.S. holders in 2004.

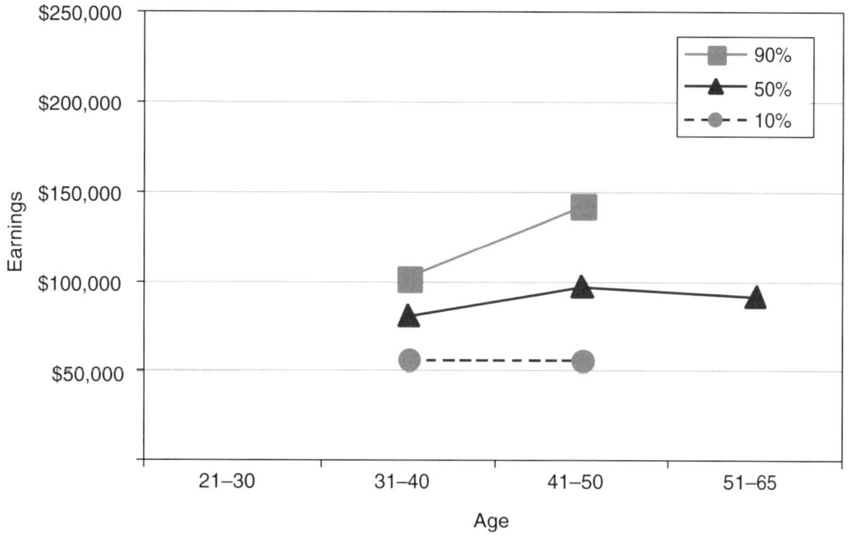

FIGURE 3 Age-earnings profile for M.S./Ph.D. holders in 2000.

demand for younger M.S. and Ph.D. holders is increasing as a result of increasing technical complexity in manufacturing and design. A typical engineer (age 31–40) with an M.S. or Ph.D. earned slightly less than the average engineer with a B.S. but with 10 years more experience (age 41–50).

For workers in their peak years (age 41–50), the graduate-degree premium fell from 16–19 percent in 2000 and 2002 to 9 percent in 2004. For the oldest workers, the graduate-degree premium fell even more dramatically, from 38–49 percent in 2000 and 2002 to 13 percent in 2004. For engineers older than 40 in 2004, the graduate degree premium was only 10 percent, indicating weak incentives for domestic workers to pursue graduate degrees, even though our fieldwork indicates that the industry needs them.

The variance in earnings was higher for engineers with graduate degrees than for engineers with B.S. degrees in 2004. In both 2002 and 2004, the variance in earnings for older engineers with B.S. and graduate degrees was very high, with the 90/10 ratio ranging from 4.3 to 7.6.

Earnings over Time

The ACS earnings profiles showed slower growth of average earnings between 2000 and 2004 than the OES data

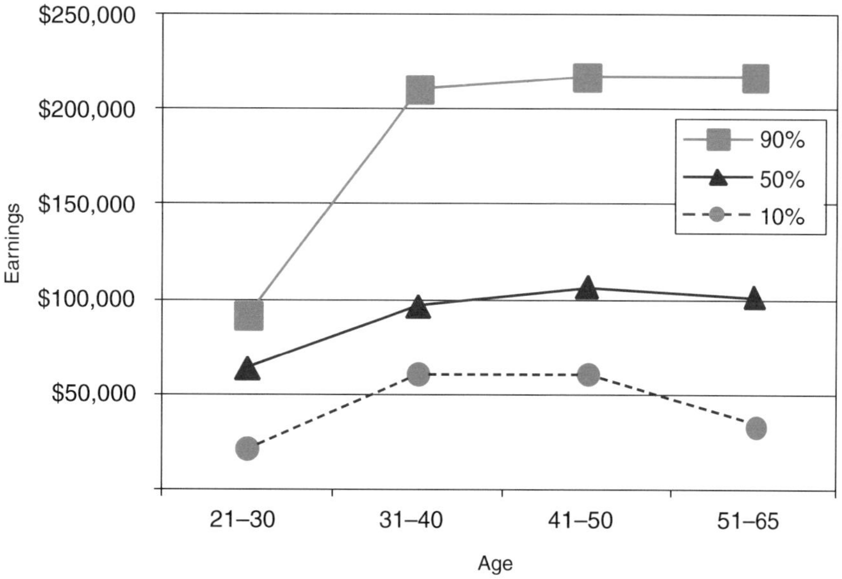

FIGURE 4 Age-earnings profile for M.S./Ph.D. holders in 2004.

TABLE 2 Age-Earnings Profiles (adjusted for inflation) for 2000, 2002, and 2004[a]

		2000				2002				2004			
Age		21–30	31–40	41–50	51–65	21–30	31–40	41–50	51–65	21–30	31–40	41–50	51–65
Less than a Bachelor's Degree	10th percentile					$6,051	$2,9245	$23,194	$32,270		$32,421	$35,461	$34,448
	50th percentile		$34,966	$60,899	$48,973	$48,405	$57,481	$57,481	$49,515	$40,526	$60,790	$68,895	$70,415
	90th percentile					$90,759	$80,675	$85,717	$72,607		$121,579	$193,513	$7,770
	90/10 ratio					15.00	2.76	3.70	2.25		3.75	5.46	2.84
	Mean		$4,606	$53,693	$70,505	$46,649	$57,127	$56,069	$52,402	$41,612	$68,819	$84,736	$64,523
Bachelor's Degree	10th percentile	$20,710	$53,444	$44,536		$30,496	$37,061	$49,026	$32,825	$24,316	$36,575	$60,790	$50,658
	50th percentile	$52,052	$83,505	$91,299	$72,372	$58,239	$72,005	$88,946	$70,945	$58,763	$70,921	$97,263	$89,665
	90th percentile	$96,867	$130,270	$158,104		$95,299	$127,066	$158,832	$158,832	$81,053	$109,421	$217,829	$217,829
	90/10 ratio	4.68	2.44	3.55		3.12	3.43	3.24	4.84	3.33	2.99	3.58	4.30
	Mean	$58,127	$89,949	$107,758	$109,566	$60,867	$79,222	$104,635	$87,555	$57,470	$76,809	$116,220	$109,410
Master's Degree or Ph.D.	10th percentile		$61,238	$61,238		$61,945	$55,062	$63,533	$45,002	$21,276	$60,790	$60,790	$32,320
	50th percentile		$89,073	$106,331	$100,207	$79,417	$90,005	$105,888	$105,888	$63,322	$96,250	$106,382	$101,316
	90th percentile		$111,341	$155,878		$95,299	$137,654	$158,832	$339,901	$91,184	$210,737	$217,829	$217,829
	90/10 ratio		1.82	2.55		1.54	2.50	2.50	7.55	4.29	3.47	3.58	6.74
	Mean		$89,360	$114,175	$121,988	$79,769	$95,060	$120,872	$127,819	$61,167	$112,238	$127,075	$124,065

[a]The repetition of earnings in some cells, especially for the 90th percentile group, appears to be a coincidence and not a mistake. A check of the data indicates that many workers with different levels of education and in different occupations reported the same earnings, which are not top coded.

TABLE 3 Engineers Working Less Than a Full Year (48 Weeks), by Degree Level, for 2000, 2002, and 2004

	Age Ranges			
	21–30	31–40	41–50	51–65
2000				
Less than a Bachelor's Degree	[a]	10%	0	35.71%
Bachelor's Degree	25%	3.28%	2.56%	10.53%
Master's Degree or Ph.D.	[a]	3.23%	4.55%	12.5%
2002				
Less than a Bachelor's Degree	14.81%	0	14.89%	31.82%
Bachelor's Degree	13.7%	11.11%	9.24%	28.57%
Master's Degree or Ph.D.	13.33%	16.13%	3.7%	26.09%
2004				
Less than a Bachelor's Degree	35.71%	7.69%	3.70%	20%
Bachelor's Degree	15.85%	10.62%	9.82%	10.71%
Master's Degree or Ph.D.	25%	7.34%	12.35%	17.78%

Note: The value in each cell is the proportion of engineers in that age group with the indicated degree who worked less than 48 weeks in the indicated year.

[a]<10 observations (not shown)

showed between 2000 and 2005, primarily because the ACS earnings were higher than in OES data in 2000 and comparable in 2004 and 2005. However patterns varied across occupations. In the ACS data, average computer science earnings grew much faster than average electrical and electronics earnings, where growth did not keep up with inflation (not shown in tables). In comparison, the OES data showed comparable positive earnings growth for these occupations between 2000 and 2005.

Although ACS data were developed to be compared over time, while OES data were not, the small sample sizes of the ACS data make them less representative and less reliable than the OES data. For these reasons, we cannot say with confidence how much earnings by semiconductor engineers grew from 2000 to 2005.

Summary

Overall the earnings data indicate potential problems in the high-tech engineering market. Although the graduate-degree premium appears to be adequate for younger workers, the low returns-to-experience for engineers with graduate degrees make returns on investment in a graduate degree inadequate over an engineer's entire career, especially the returns implied by the 2004 ACS data. The returns to a BS degree were adequate for engineers younger than 50. However, older workers at all three educational levels experienced a troubling drop in median real earnings. The data also indicate that the variance in earnings for high-tech engineers is growing, partly because earnings at the bottom of the distribution are rising very slowly, or even falling, as engineers age. Thus, although the high-tech engineering labor market appears to be strong nationally, data by age and education indicate that engineering jobs at the bottom end may be deteriorating and that older engineers may be finding fewer high-quality job opportunities.

Career Paths for Semiconductor Professionals (LEHD Data)

We now look briefly at how the jobs and earnings of semiconductor workers, including engineers, changed from 1992 to 2001 based on a very large linked employer-employee data set, the Census Bureau's LEHD.[17] The data cover all occupations, engineers as well as office workers, technicians, managers, and others. We focus here on prime-age male and female workers (ages 35–54) in two educational groups—medium (some college) and high (college graduate and above).

The career paths are shown for modal groups, that is, the largest groups of workers who had held one, two, or three jobs, with at least one job in a semiconductor establishment during the decade. Other (smaller) groups of workers also changed jobs but had different career paths.

For those who had held two jobs; the first job was outside the semiconductor industry and the second job in it. For those who had held three jobs, the first two were outside the semiconductor industry, and the last one was in the industry.

Career Paths

Semiconductor workers followed two distinct types of career paths—loyalist and job changer (see Table 4). Workers who already worked for semiconductor employers and had good job ladders (high initial earnings and good earnings growth) tended to become loyalists, that is, they did not change jobs during the period studied. The career paths of loyalists were considerably better than the career paths of job changers.

Workers on inferior job ladders outside the semiconductor industry tended to become job changers, and most of them eventually ended up on a relatively good job ladder. Job changers had relatively low initial earnings in jobs outside the semiconductor industry and experienced substantial earnings growth (usually 20 to 30 percent for younger and 10 to 20 percent for older workers) by taking jobs in the semiconductor industry. Among job changers, two-jobbers began with higher pay outside the industry and were able to enter the semiconductor industry sooner than three-jobbers. Although highly educated three-jobbers experienced healthy earnings increases when they changed jobs outside the semiconductor industry, the increase was smaller than when they got jobs in the industry. Because the overall earnings growth

[17]This material is taken from the Sloan-Census project that produced the book *Economic Turbulence* by Brown et al. (2006) and related papers (see *www.economicturbulence.com*). See Chapter 5 for an overview of job ladders and Chapter 6 for an overview of career paths in the semiconductor and four other industries (software, finance, trucking, and retail food).

TABLE 4 Semiconductor Career Paths, Workers Age 35–54

		Males			Females		
		Loyalists	Two Jobs	Three Jobs	Loyalists	Two Jobs	Three Jobs
Medium Education	A	$32,564	$15,046	$12,458	$13,084	$8,148o	$7,314
	B	.054	.056	.058	.039	.030	.041
	C	$55,780	$25,926	$21,998	$19,641	$10,999	$10,999
High Education	A	$36,084	$22,893	$18,197	$14,990	$10,132	$9298
	B	.059	.048	.047	.044	.028	.030
	C	$65,207	$36,925	$29,068	$23,569	$13,356	$12,570

Notes: A = mean initial earnings (2005 dollars, inflated from 2001 dollars using the CPI-urban).
B = net annualized earnings growth rate (in log points) over the 10-year simulated career path.
C = simulated 2001 final average earnings (2005 dollars).
Source: Adapted from *Economic Turbulence* (Brown et al., 2006), Chapter 6, Table 6.1. Original calculations by authors from Census LEHD data. These career paths are for all workers in all occupations in the industry. They include engineers, as well as office workers, technicians, managers, and other occupations.

of two-jobbers and three-jobbers was about the same over the 10-year period, the two-jobbers usually maintained their initial earnings advantage.

Although job changers usually experienced higher earnings growth over the decade than loyalists, the growth did not offset their much lower initial earnings. Thus loyalists ended the period with substantially higher earnings. The legendary job hoppers in Silicon Valley (engineers who left good jobs for even better ones), constituted a smaller group than the job changers shown here, who left relatively low-wage jobs for jobs that paid slightly more.

Job Ladders

Data (not shown here) indicate that large firms provided 85 percent of semiconductor jobs. Firm fortune matters in the job ladders offered by large, low-turnover firms, as we see by comparing firms with growing employment to firms with shrinking employment. *Large growing firms with low turnover* provided 50 percent of the jobs in the industry, and these firms are typically known for providing good jobs. Semiconductor jobs in these firms tended to last a relatively long time—27 percent lasted for at least five years during the decade studied.

Large shrinking firms with low turnover provided an interesting contrast. Even though these firms were reducing employment, new hires still accounted for 30 percent of jobs; however, less than 20 percent of jobs lasted more than five years. Thus these firms appeared to be replacing experienced workers with less-expensive new hires. When we compared ongoing and completed long-term (more than five years) jobs, we found that shrinking large firms tended to shed experienced workers with lower earnings growth, because annualized earnings growth was higher (by half a percentage point) in ongoing jobs than in completed jobs for all groups.

These patterns marked a change in the way big companies deal with difficulties. IBM provides a good example of how downsizing programs evolved from the 1980s to the 1990s. In 1983, IBM offered workers at five locations a voluntary early retirement program in which workers with 25 or more years of experience would receive two years of pay over a four-year period. IBM offered voluntary retirement programs again in 1986 and 1989.[18] Because these programs were voluntary for the general workforce, rather than for targeted job titles or divisions, the change in workforce usually did not turn out as the company might have chosen: better workers often opted to leave, and weaker workers, without good job opportunities elsewhere, often opted to stay.

The deep recession in the early 1990s finally pushed IBM, DEC, and Motorola, once known for providing employment security, to make layoffs.[19] The new approach to downsizing included voluntary programs for *targeted* workers. If these workers did not accept the termination program, they could be subject to layoffs, making the program less than voluntary in reality. In 1991 and 1992, IBM selected workers eligible for termination, which included a bonus of up to a year's salary. In this way, more than 40,000 workers were "transitioned" out of the company. Downsizing continued through 1993, and by 1994 IBM was actually laying off workers.[20]

With the dot-com bust in the early 2000s, semiconductor companies undertook massive layoffs. By the end of 2001, Motorola had laid off more than 48,000 workers from its peak of 150,000 employees in 2000.[21] As swings in demand became more volatile, the idea of lifetime employment in the semiconductor industry became a thing of the past, although selected workers still had excellent job ladders and long careers.

The data in Table 5 show that for large firms with low turnover, growing firms offered higher initial earnings than shrinking firms to both men and women (by 7 to 37 percent),

[18] *http://www.allianceibm.org/news/jobactions.htm.*

[19] Some of the observations about specific firms here most likely reflect divisions of these large, complex firms beyond their production of semiconductors. We think the patterns discussed reflect the impact of globalization on high-tech firms.

[20] *http://www.allianceibm.org/news/jobactions.htm.*

[21] *http://www.bizjournals.com/austin/stories/2001/12/17/daily22.html.*

TABLE 5 Job Ladders for Semiconductor Industry Workers, Age 35–54

		Growing, Large Firms with Low Turnover	Shrinking, Large Firms with Low Turnover	Growing, Large Firms with High Turnover	Growing, Small Firms with Low Turnover	Growing, Small Firms with High Turnover
Males						
Medium Educated	A	$21,462	$18,012	$14,810	$15,517	$17,115
	B	.054	.061	.063	.068	.076
	C	$36,592	$33,266	$27,860	$30,771	$36,592
Highly Educated	A	$23,057	$21,541	$21,388	$21,070	$20,600
	B	.059	.061	.040	.075	.055
	C	$41,582	$39,503	$32,018	$44,493	$35,761
Females						
Medium Educated	A	$13,024	$9519	$10,589	$8,506	$8,879
	B	.039	.036	.021	.048	.085
	C	$19,128	$13,722	$12,890	$13,722	$20,791
Highly Educated	A	$14,080	$10,334	$12,424	$10,692	$9897
	B	.044	.036	–.002	.054	.064
	C	$22,038	$14,970	$12,059	$18,296	$18,712

Notes: A = mean initial earnings (2005 dollars, inflated from 2001 using the CPI-urban).
B = net annualized earnings growth rate (in log points) across the simulated career path.
C = simulated 2001 final average earnings (2005 dollars).

Source: *Economic Turbulence* (Brown et al., 2006), Chapter 5, Table 5.1. Original calculations by authors from Census LEHD data. The career paths are for all workers in all occupations in the semiconductor industry, including engineers, office workers, technicians, managers, and other occupations.

and the growing firms compared to shrinking firms offered lower earnings growth to men and higher earnings growth to women. Overall men's job ladders are more similar in growing and shrinking firms than women's job ladders, and so men seem more protected from economic turbulence than women. A comparison of "stayers" (i.e., ongoing long jobs) and "movers" (i.e., completed 1–3 year jobs) shows that annualized earnings growth for short jobs was only two-thirds that of long jobs in both growing and shrinking large firms. These results indicate that growing firms used high initial earnings to attract talented workers, among whom only a select group was given access to career development with long, steep job ladders.

Compared to growing firms, large shrinking firms paid lower initial earnings but offered higher earnings growth for short jobs; the job ladders for younger men were better relative to those of older men. These results indicate that large firms, both growing and shrinking, used market-driven compensation systems based on salaries in the spot market for engineers. Growing firms appeared to provide long job ladders with career development for a select group, while other workers faced either a plateau or "up or out." Possibly workers not on the fast track left voluntarily for better jobs elsewhere. Shrinking firms appeared to keep selected experienced workers and replaced the others with new hires at market rates. New hires appeared not to have access to long job ladders with career development, even though the older workers still had long job ladders. These findings are consistent with changes we observed in our fieldwork at large U.S. companies in the 1990s.

Small growing firms with low turnover were likely to be early-stage fabless companies that hired mainly technical personnel and offered relatively good job ladders for college-educated workers. Although these firms offered relatively low initial earnings, their earnings growth was high. After 10 years, earnings at these companies surpassed earnings of experienced workers in large shrinking firms and were close to earnings at large growing firms with low turnover. Small, growing firms may be an increasingly important source of good job ladders.

Overall, economic turbulence has had negative effects on job ladders. Over the decade studied, growing large firms with low turnover allowed highly paid new hires to compete for access to long job ladders with career development, while shrinking large firms with low turnover forced experienced workers to compete to keep their jobs, which were either being eliminated or being filled by new hires paid at market rates. In any case, the era of lifetime jobs with career development appears to be over, and many workers must improve their job prospects through mobility.

FACTORS THAT INFLUENCE ENGINEERING WORK AND WAGES

The U.S. labor market for engineers is affected by a variety of long-term forces, including technological change, immigration policy, and educational practices. In this section we consider the effects of each of these.

Technological Change: Wafer Size

Engineering jobs in chip fabs have evolved over the last several technology generations, driven primarily by simultaneous increases in wafer size and automation, both of which have been important for raising productivity and keeping the industry on its Moore's Law trajectory. We look at how engineering work within the fab changed during the transition from 150 mm to 200 mm wafers. Our analysis is based on detailed data gathered in the mid-1990s by the Berkeley Competitive Semiconductor Manufacturing (CSM) Program at a sample of fabs running 150 mm and 200 mm wafers in four countries.[22]

Larger wafer size requires major reengineering of equipment and process technology. In addition, materials handling and information systems must be highly automated to handle the increased weight and value of each wafer safely and to minimize human error. Automation changes the composition of the workforce by increasing the need for engineers and decreasing the need for operators. In the CSM data, the percentage of engineers increased from 15 to 24 percent of the total workforce between 150 mm- and 200 mm-generation plants; at the same time the percentage of operators declined from 73 to 62 percent (see Table 6). The overall employment level of the fab stayed approximately the same at about 750 workers.

The shift in jobs from operators to engineers resulted in an increase in higher paying, high-skilled jobs at the expense of lower paying, low-skilled jobs. However, the earnings structure across occupations also changed (see Table 7). The initial pay of technicians and engineers was more than one-third higher in the 200 mm fabs than in the 150 mm fabs, and their pay premium over operators increased.

In terms of returns-to-experience (i.e., maximum pay compared to initial pay), experienced engineers fared poorly, as their ratio of maximum to initial pay fell from 2.8 (150 mm fabs) to 2.0 (200 mm fabs). The returns-to-experience for technicians and operators remained stable, as the experienced techs and operators had the same pay improvement in the 200 mm fab as the new hires.

Over time, experienced engineers lost out as their average maximum real salary was actually lower in the 200 mm fabs than in the 150 mm fabs. In interviews, we learned that fabs were more interested in having young engineers with knowledge of new technology than they were worried about losing older engineers. Consequently, they were willing to increase wages of new hires without raising the wages of experienced engineers. With rapidly changing technology, an ample supply of new hires, and low turnover, companies were able to flatten engineers' career ladders (see, for example, Figure 4, above) with no adverse consequences.

We do not have comparable data for 300 mm fabs, which have completely automated materials handling and wafer processing. Complete automation is necessary because of the high value of each 300 mm wafer, which has an area 2.25 times that of a 200 mm wafer. The 300 mm wafer is heavier and more awkward to handle, which raises the risk of it being dropped by human handlers—not to mention the ergonomic risk to humans.

TABLE 6 Workforce Composition (mean head count in matched 150 mm and 200 mm fabs)

	150 mm Fabs	200 mm Fabs
Operators	547 (73%)	470 (62%)
Technicians	91 (12%)	107 (14%)
Engineers	114 (15%)	181 (24%)
Total	752	758

Source: Brown and Campbell, 2001.

TABLE 7 Workforce Compensation (mean wage or salary in matched 150 mm and 200 mm fabs)

	150 mm Fabs		200 mm Fabs	
	Initial Pay	Maximum Pay	Initial Pay	Maximum Pay
Operators (hourly)	$5.88	$15.47	$7.12	$18.44
Technicians (hourly)	$6.68	$11.50	$9.12	$15.83
Engineers (monthly)	$1,785	$5,019	$2,381	$4,689

Source: Brown and Campbell, 2001.

Because new 300 mm fabs process advanced circuits, such as circuits with 90 nm or 65 nm processes, the amount of inspection, number of metrology steps, and number of in-line engineering-related activities are significantly higher than for their older 200 mm counterparts for the same wafer throughput. As a result, most of the labor savings achieved through the automation of materials handling, which requires approximately 30 percent less labor input, is reapplied to new engineering tasks, which are much higher value-added and more intellectually challenging and require more troubleshooting.

The overall number of workers is not reduced as a result of advanced factory automation. Instead, there is a shift in task composition. The percentage of workers with higher engineering and technical problem-solving skills is greatly increased, while the percentage of workers required for wafer movement and equipment starting and stopping is greatly decreased. However, the proportion of engineers remains the same.[23]

H-1B Visas

U.S. visa and educational policies directly impact the supply of engineers, especially those with advanced degrees, in the domestic market. In this section, we look at the earnings of H-1B visa holders. The H-1B visa is used by foreigners employed temporarily in positions that require specialized

[22]Twenty-three fabs in four countries were part of the CSM survey. For this table, the 150 mm wafer fabs were matched to the 200 mm wafer fabs by company, so that human resource policies could be compared for the two groups. This reduced our sample to 14.

[23]Personal communication, April 2005.

knowledge and at least a bachelor's degree. H-1B visas are granted to companies (rather than workers). A company must submit an application that includes a job title and the intended wage or earnings, which must reflect the prevailing wage rate. With various application fees and legal expenses, the initial cost to an employer is in the range of $2,500 to $8,000 per application.[24] H-1B employees can work only for the sponsoring U.S. employer[25] and only do the activities described in the application. A foreigner can work for a maximum of six continuous years (including one extension) on an H-1B visa.

The current law limits the number of H-1B visas that may be certified to 65,000 per fiscal year. Many companies think this number is too low, and businesses have lobbied for higher limits. The numerical limitation was temporarily raised to 195,000 in FY2001, FY2002, and FY2003.[26] Note that only initial applications are included in the annual limit; requests for extensions beyond the initial three-years are not included. Applications by universities and nonprofit research institutions are also not counted against the cap. In addition, there are 20,000 special exemptions for foreigners with master's and Ph.D. degrees from U.S. universities. Even in 2003, before U.S. graduates with advanced degrees were exempted, many H-1B visa holders had advanced degrees (M.S. 29 percent; Ph.D. 14 percent; professional degree 6 percent).[27] H-1B visas are granted for a wide range of occupations, including engineering, medicine, law, social sciences, education, business specialties, and the arts.

We collected data from the H-1B applications certified[28] to the top 10 U.S. chip vendors and the top 10 non-U.S. chip companies (referred to here, for convenience, as the top 20 companies) from 2001 through 2005 (U.S. government fiscal years). Companies can provide either a specific proposed pay rate or a minimum and maximum of the proposed pay range, and pay can be annual, monthly, weekly, or hourly.[29] The reasons for choosing either a specific rate or a range are worth exploring in future research. One possibility is that a specific rate may be stated when a company has a specific individual in mind for the visa; a range may be used when an individual has not yet been identified.

During the five-year period, the 20 companies in our sample were granted approval of 15,784 H-1B visa applications, of which 14,035 were granted to U.S. firms. Overall, 49 percent stated a specific salary rate, and 51 percent stated a minimum-maximum salary range (reported separately in Table 8). We analyzed four occupational groups, which represent most of the semiconductor applications: electrical engineering, computer-related jobs, manufacturing-related jobs, and business and administrative jobs. Since most H-1B applications were made by U.S. firms, we focus on these. More of the applications by non-U.S firms were for business and support jobs (15 percent) or for other kinds of engineering jobs (18 percent); 80 percent of these stated earnings rates. Compared to the earnings stated by U.S. companies, the earnings stated by non-U.S. companies for EE and CS applications tended to be slightly higher on average with a larger 90/10 ratio, but lower on average for non-EECS jobs with a larger 90/10 ratio.

U.S. chip companies were most likely to apply for H-1B visas for EE jobs (37 percent with average rate $77,560 or average minimum of $66,944) or CS jobs (52 percent with average rate $78,537 or average minimum of $75,685). The other applications were primarily for other engineering jobs (8 percent with an average rate $79,806, or average minimum of $65,425).

EE applications primarily stated a specific rate; the distribution tended to be approximately 15 percent above the distribution for the minimum when a range was given. In contrast, CS applications primarily stated a range, whose minimum had a distribution close to the distribution of the specific earnings rates, where they were used. A possible interpretation, consistent with the OES data in Table 1, is that the high computer science minimum indicates that software programmers in the chip industry are receiving a premium.

We checked the applications in 2005 by all other companies and industries (called "other firms" here) for EE and CS jobs to see if they used comparable rates and ranges, since H-1B visas might be functioning differently in different industries. The top chip companies accounted for 56 percent of all EE applications but only 5 percent of CS applications.

We can compare H-1B application rates to actual earnings for EE-CS engineers. In the ACS data, EE-CS engineers earned, on average, $69,000 to $96,000 (overall average $86,000) from 2000 to 2004, and in the OES data they earned $66,000 to $84,000 (overall average $74,000) from 2000 to 2005. The average rates on H-1B visa applications granted to the top 20 semiconductor companies fell between these two averages. However, it is difficult to make comparisons of these earnings independent of worker experience and education, because many semiconductor companies hired H-1B visa workers as new EE-CS graduates, often with graduate degrees, from U.S. universities.

Interestingly, the "other firms" mostly specified earnings rates in their H-1B applications for both EE and CS jobs. The rates used on EE applications by "other firms" have a lower mean and 10th percentile compared to the top chip firms; the rates used on CS applications by "other firms" have a con-

[24] GAO (2003).

[25] The U.S. employer may place the H-1B visa worker with another employer if certain rules are followed.

[26] *http://www.uscis.gov/graphics/howdoi/h1b.htm.*

[27] DHS (2004).

[28] During this five year period, 1.6 percent of the applications were denied (including a small number that were put on hold). These applications are not included in our analysis. We also dropped one outlier, which was probably an input error, an application naming $10.6M as the pay for a senior test engineer. The prevailing wage was given as $93,330.

[29] The two methods of applying (rate and range) are reported separately here. Most applications (95 percent) use annual earnings. Monthly, weekly, and hourly rates were converted to annual rates (12 months, 52 weeks, or 2,000 hours).

TABLE 8 H-1B Visa Applications Approved, 2001–2005

		Observations (%)	Mean	Standard Deviation	10%	90%
Top Ten U.S. Chip Firms						
Electrical and Electronics Engineering Job Codes						
Rate given		3436 (24%)	77,560	16255	62,400	96,160
Range given	min		66,944	13991	52,800	85,225
	max	1792 (13%)	102,992	23410	73,375	130,000
Computer Science Job Codes						
Rate given		2106 (15%)	78,537	18275	61,302	98,239
Range given	min		75,685	18318	56,277	100,000
	max	5234 (37%)	96,118	19662	75,000	125,000
Manufacturing Engineering Job Codes						
Rate given		649 (5%)	79,806	16801	58,200	96,000
Range given	min		65,425	14609	48,788	85,000
	max	403 (3%)	104,798	25202	73,006	130,000
Business, Marketing, Administrative Support Job Codes						
Rate given		163 (1%)	87,533	40824	50,400	130,000
Range given	min		73,549	24725	44,200	106,000
	max	252 (2%)	101,535	34193	64,900	140,000
Top Ten Non-U.S. Chip Firms						
Electrical and Electronics Engineering Job Codes						
Rate given		430 (25%)	80,161	18941	59,527	105,694
Range given	min		77,580	18627	55,104	99,808
	max	188 (11%)	106,911	31388	70,900	154,300
Computer Science Job Codes						
Rate given		432 (25%)	79,525	18476	57,500	101,100
Range given	min		68,712	13843	52,361	86,606
	max	124 (7%)	91,773	22201	64,676	120,000
Manufacturing Engineering Job Codes						
Rate given		292 (17%)	73,458	16419	53,600	95,000
Range given	min		69,070	16997	53,100	102,168
	max	19 (1%)	86,217	25232	60,270	132,000
Business, Marketing, Administrative Support Job Codes						
Rate given		230 (13%)	81,882	39447	42,150	134,838
Range given	min		60,406	24271	39,145	88,486
	max	34 (2%)	82,882	36511	50,000	140,000
Other Chip and Non-Chip Firms						
Electrical and Electronics Engineering Job Codes						
Rate given		7701 (6%)	69,302	24175	45,000	100,000
Range given	min		67,737	20807	45,000	95,256
	max	2098 (2%)	84,710	28592	50,000	124,000
Computer Science Job Codes						
Rate given		96720 (71%)	60,698	20371	42,000	87,250
Range given	min		58,523	16860	42,000	81,600
	max	29964 (22%)	77,277	25747	50,000	120,000

Note: companies can submit applications with a specific proposed rate to be paid or can provide a range (min, max). No duplicates were submitted.
Source: U.S. Department of Labor. Available online at *http://www.flcdatacenter.com/CaseH1B.aspx*.

siderably lower distribution compared to the top chip firms. Once again, consistent with the wage data in Table 8, H-1B applications for EE-CS jobs in the chip industry appear to carry a premium compared to other industries.

A GAO study in 2003 of H-1B visa holders compared the annual pay for selected occupations, including electrical/electronics engineers (EEs), to a sample of U.S. workers using the Census Department's Current Population Survey of 2002. As GAO notes, for a variety of reasons the annual salary comparisons are not exact. For one thing, we do not know if the visas were actually used.

The GAO comparison of EEs with H-1B visas and U.S.

citizenship in 2002 showed that the H-1Bs were younger (32 years vs. 41 years; 62 percent under 35 years old vs. 28 percent) and much more likely to have graduate degrees (50 percent vs. 20 percent) (GAO, 2003, pp. 14, 15). When median annual salary of EEs aged 31 to 50 years old were compared, H-1Bs earned less than citizens (H-1Bs with graduate degrees earned $77,000, while citizens earned $88,000; H-1Bs with less than a graduate degree earned $65,000, while citizens earned $70,000) (GAO, 2003, p. 42). For younger EEs (age 18 to 30) without graduate degrees, however, H-1Bs earned more than citizens ($60,000 vs. $52,000) (GAO, 2003, p. 42). These data indicate that H-1B visa holders may be having a downward impact on labor-market opportunities for mature engineers, but probably not for young engineering college graduates.

Applications for Selected Companies

H-1B visa applications for five large U.S. companies, IBM, Intel, Motorola/Freescale, Qualcomm, and Texas Instruments, together accounted for 76 percent of the H-1B granted applications in our sample (Table 9). Since Motorola spun off its chip operations as an independent company, Freescale, in 2004, we combined applications granted to Motorola and Freescale.

Of these five companies, IBM used the most H-1B visas; almost 4,000 were granted during the five-year period. Most IBM applications stated a range of earnings, with an average minimum ($82,072) that was considerably higher than the average minimum stated by the other four companies. Since IBM is now more of a services company then a hardware company, we assume that many of these applications were for jobs that were not chip-related.

Like IBM, Motorola most often stated a range of earnings. Motorola's average minimum ($62,866) was 25 percent lower than IBM's. Even so, Motorola's rates were slightly higher than the national EE-CS salaries in the OES.

The three companies that focus on semiconductors, Intel, Qualcomm, and Texas Instruments, tended to state actual earnings, and their average earnings were within 2 percent of each other. Stated earnings for Intel showed less variance than for QualComm or Texas Instruments. The three companies applied for H-1B visas to fill jobs that required a variety of skills and experience. Overall, their rates seemed to reflect the national EE-CS salaries in the ACS.

By comparing Freescale's applications to Motorola's for 2004 and 2005, we can estimate the extent to which Motorola's applications were for engineers in their semiconductor business. Freescale was granted 11 percent as many H-1B visas as Motorola in 2004 and 18 percent as many in 2005. Freescale's pay rate had a much narrower range than Motorola's, with the ratio of Freescale's sample maximum to minimum rates between 2.5 and 2.7 (compared to Motorola's ratio of 5.0 to 4.6). However Freescale's averages for the minimum and maximum rates were very close to Motorola's averages in 2004 and 6 percent higher than Motorola's in 2005. This indicates that semiconductor engineers had average earnings compared to the broad range of other workers at Motorola.

The proposed wages for the top-20 companies, as well as for these specific companies, indicate that H-1B visas were issued for a wide range of jobs, some of them high-level jobs that paid well over $100,000, and some low-level jobs that paid less than $50,000. To what extent the lower paying jobs are being used to keep semiconductor earnings low for domestic new hires, and to what extent the higher paying jobs are going to foreigners at the expense of qualified experienced U.S. engineers cannot be determined These remain important policy questions.

Inter-year Comparisons. If we compare H-1B visas granted by year, we note that the number granted to each of these five companies jumped, either in 2003 or 2004, and remained high, even as the national limit and fee dropped dramatically. The semiconductor companies seemed to take advantage of the additional 20,000 H-1Bs available for workers with graduate degrees from U.S. universities that went into effect in 2004. Sixty-one percent of the H-1B visas awarded to the top-20 companies were awarded during the last two years (2004 and 2005) of the five-year period.

Intel's applications for H-1B visas increased dramatically during the five-year period. One-quarter of its H-1Bs were

TABLE 9 H-1B Visas Granted to the Top Five U.S. Companies, 2001–2005

Variable		Observations	Mean	Standard Deviation
IBM				
Rate given		395	$88,353.9	33462.38
Range given	min	3599	$82,071.5	18307.34
	max		$96,150.2	19493.39
Motorola/Freescale				
Rate given		264	$66,472.4	28978.98
Range given	min	2256	$62,910.4	12993.04
	max		$92,573.9	18760.1
Intel				
Rate given		1574	$78,065.1	11673.03
Range given	min	1122	$65,921.4	10107.71
	max		$121,519.6	19650.69
Qualcomm				
Rate given		1632	$76,775.5	14152.01
Range given	min	0	0	0
	max		0	0
Texas Instruments				
Rate given		1076	$76,754.3	15717.13
Range given	min	61	$73,352.4	20891.36
	max		$91,727.7	22138.06

Note: Companies can submit applications with a specific proposed rate to be paid or can provide a range (min, max). No duplicates were submitted.
Source: U.S. Department of Labor, H-1B Program Data. Available online at *http://www.flcdatacenter.com/CaseH1B.aspx*.

granted in the first three years and three-quarters in the last two years. The company also shifted from stating minimum-maximum ranges to actual wage rates, although the earnings rates remained comparable.

The H-1B Share of the Workforce. We now look at how H-1B visa applications compare to employment levels at Intel, Motorola, and IBM. In 2005, Intel employed approximately 99,900 people worldwide, with more than 50 percent located in the United States. Motorola employed 69,000 employees, with more than half employed outside the United States, and with 24,000 eligible for stock options. IBM employed 329,000 worldwide, approximately 40 percent of whom were eligible for the U.S. retirement plan (at the end of 2004, when the plan was discontinued).[30]

If we assume that Intel had 50,000 domestic employees and used its 1,280 H-1B visas to hire new workers in 2005, then approximately 2.6 percent of Intel's domestic employees were newly hired H-1B visa holders. If most H-1B visa holders work for Intel for five years, then approximately 5.4 percent of Intel's 2005 domestic employees (and an even larger percentage of engineers) were H-1B visa holders.

If we assume that Motorola used its 728 H-1B visas to hire new workers in 2005 and that these professional specialists held similar jobs to those of Motorola employees eligible for stock options, then almost 3 percent of Motorola's domestic professionals were newly hired H-1B visa holders. If most H-1B visa holders work for Motorola for five years, then 8 percent of Motorola's domestic professionals were H-1B visa holders (or 6 percent of Motorola's domestic workforce, if one-half of the workforce was domestic) in 2005.[31] Comparable calculations for IBM indicate that IBM hired 1,150 H-1B visa holders in 2005, or 0.8 percent of the domestic workforce, and 2.8 percent of its domestic workforce (and a larger percentage of its professional domestic workforce) were H-1B visa holders.

These data indicate that semiconductor companies use H-1B visas strategically in hiring and managing their engineering employees. In large U.S. companies, H-1B visa holders comprise an important part of the domestic professional workforce. One reason for the importance of H-1B visas is that major U.S. universities provide engineering graduate education to many foreign students, and upon graduation, these students are in great demand by U.S. companies.

[30]These employment figures are from the company's 10-K reports to the SEC: Intel at *http://finance.yahoo.com/q/sec?s=INTC*, Motorola at *http://finance.yahoo.com/q/sec?s=MOT*, and IBM at *http://finance.yahoo.com/q/sec?s=IBM*.

[31]This percentage was adjusted downward for the Freescale spinoff. We assumed that 15 percent of the H-1B visa holders hired in 2001, 2002, and 2003 (the proportion of H-1B applications by Freescale compared to Motorola in 2004 and 2005) worked for Freescale (not Motorola) in 2005.

U.S. Education of Foreign Students

Higher education has played an important role in the development of the U.S. semiconductor industry, and M.S. and Ph.D. engineering graduates provide the essential workforce for semiconductor companies. Engineers who graduate at the highest level, the Ph.D., have attained not only state-of-the-art knowledge, but also the ability to conduct research and to keep abreast of the latest technology during their careers.

Many U.S. graduate engineering students are foreign nationals. Figure 5 shows the number of engineering Ph.D.s (not including computer science) awarded at U.S. universities to students from five key Asian countries over a 12-year period. As the figure clearly shows, China has sent a large and growing number of doctoral engineers to the United States. At the other extreme, Japan sent very few students during the same period.

The number of students from Taiwan, which relied on U.S-educated Ph.D.s to develop its semiconductor industry, has declined since 1994. In our fieldwork in Taiwan in February 2005, many semiconductor experts raised concerns about decreasing interest in U.S. graduate study because they still considered Taiwanese doctoral training inferior to U.S. training. The number of advanced engineering graduate students from India and Korea also declined in the late 1990s, although both have been increasing again since 2002.

We also looked at the granting of Ph.D.s in electrical engineering and computer science to U.S.-born and foreign-born students. Figure 6 shows electrical engineering Ph.D.s by citizenship and gender for 1995 to 2004. Noncitizen male students earned significantly more diplomas than their U.S. counterparts throughout the period. Noncitizen female students earned more degrees than U.S. women beginning in 1998.

The same data for computer science students (Figure 7) show that the numbers of degrees awarded to citizens and noncitizens are much closer, although once again noncitizen male students were awarded more Ph.D.s than their U.S. counterparts nearly every year.

These figures clearly show that the United States is training hundreds of foreign advanced engineers every year, thus increasing the ability of foreign chip firms to compete with U.S. companies and making it easier for U.S. firms to find qualified personnel, either for their U.S. operations (when noncitizens can remain in the country) or their offshore subsidiaries.

In our earlier discussion of returns-to-education, we noted that the earnings premium for a domestic BSEE who pursues a graduate degree was relatively low. For foreign BSEEs, however, the financial incentive to pursue a U.S. graduate degree is much greater. A U.S. graduate degree opens the door for these students to high-paid jobs both in the United States and at home. In our fieldwork, we found that advanced degree holders in semiconductor centers like Shanghai and Bangalore, especially if they have some U.S. work experi-

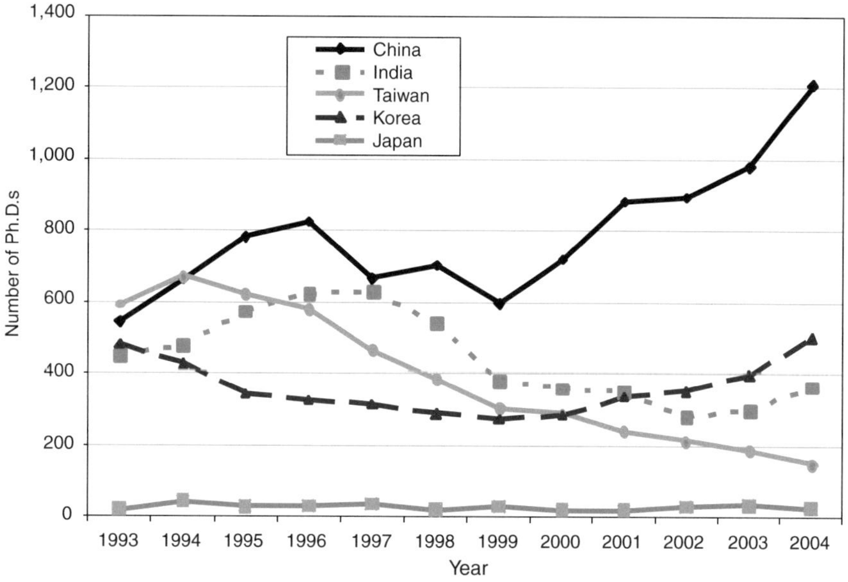

FIGURE 5 Engineering Ph.D.s in the United States by country of origin, 1993–2004. Source: National Science Foundation, Division of Science Resources Statistics, Science and Engineering Doctorate Awards: 2002 (App. Table 5), 2003 (App. Table 11), and 2004 (App. Table 11).

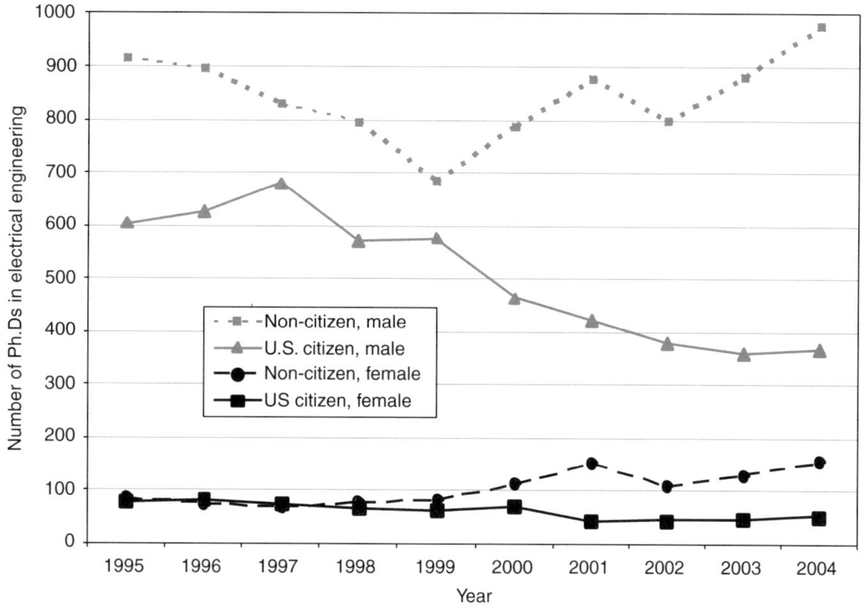

FIGURE 6 Electrical engineering Ph.D.s by gender and citizenship status, 1995–2004. Source: National Science Foundation, Division of Science Resources Statistics, Science and Engineering Doctorate Awards: 2004 (App. Table 3).

ence, earn comparable salaries to their U.S. counterparts. Locally educated EEs earn much less.

GLOBALIZATION

Globalization is one of the primary forces affecting the work and rewards of U.S. semiconductor engineers. In this section we briefly describe offshore investments by U.S. semiconductor companies, provide some data on chip engineers, and then profile the state of the chip industry in Taiwan, China, and India.

Offshoring by U.S. Semiconductor Firms[32]

The three primary reasons for locating value-chain activi-

[32]See Brown and Linden (2006) for a more detailed discussion of offshoring by U.S. semiconductor firms.

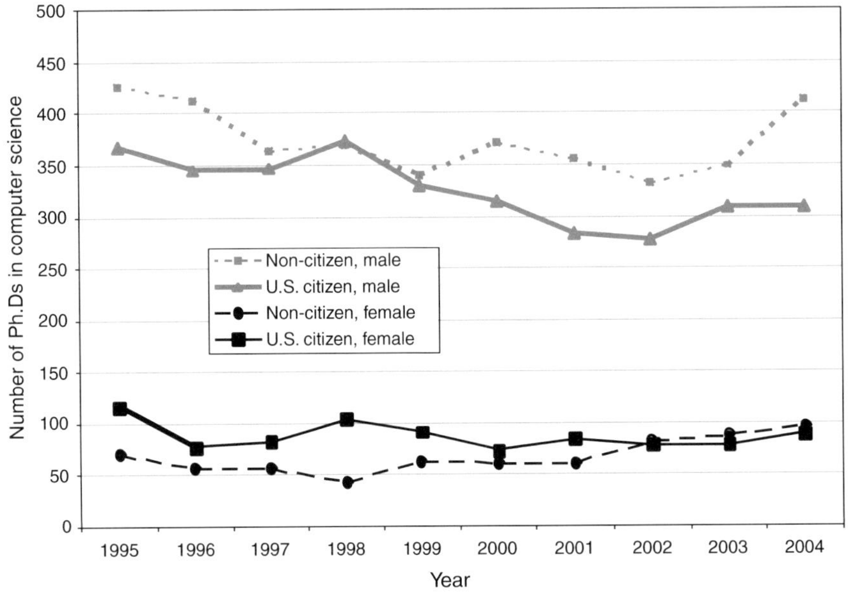

FIGURE 7 Computer science Ph.D.s by gender and citizenship status, 1995–2004. Source: National Science Foundation, Division of Science Resources Statistics, Science and Engineering Doctorate Awards: 2004 (App. Table 3).

ties are (1) access to location-specific resources, especially engineering talent; (2) cost reduction; and (3) access and development of a local market. Often, all three reasons influence the decision of a company to move an activity to a new location via internal investment or outsourcing. For example, a company may move chip design to China to take advantage of low-cost engineering talent with knowledge of customized solutions for regional Chinese telecommunication systems, as well as to gain government approval for market access.

Because chip manufacturing is so capital intensive, offshore investments in chip fabrication have been driven historically by concerns about market access, particularly tariffs, more than by cost reduction. Thus most U.S.-owned offshore fabs are located in developed countries, such as Japan. In 2001, approximately one-third of U.S.-owned capacity was located offshore (Table 10). Conversely, about 22 percent of the fab capacity located in North America was owned by companies based in other regions (not shown). Foreign companies still find the United States an attractive place to invest, as evidenced by Samsung's recent commitment to a new, multibillion-dollar fab in Austin, Texas.[33]

One factor that limits fab investment abroad by U.S. companies is the availability of high-quality fabrication services for hire. In 2005, the outsourced fabrication market was worth $18 billion,[34] most of it accounted for by dedicated contract manufacturers, known as "foundries," in Taiwan. The first foundry, Taiwan Semiconductor Manufacturing Corporation (TSMC), is still the largest. If TSMC sold chips under its own name, it would have been on the chip industry top 10 list in 2005, with $8.2 billion in revenue. Foundries are excluded from our calculations, however, to avoid double-counting of their and their customers' chips. Because the foundry price accounts for about one-third of the final chip value, TSMC actually manufactured nearly $25 billion worth of chips, which would make it number two (after Intel) in the overall chip industry.

On the one hand, the emergence of the foundry model in Asia has meant that less production capacity has been built in the United States. On the other hand, the foundry model has greatly facilitated the growth of the fabless design sector, one of the industry's growth engines, as discussed in Section 1.[35] As a back-of-the-envelope calculation, we estimate that if all foundry production were based in the United States instead of Asia, it might add 11,000 industry jobs, of which about

TABLE 10 Distribution of U.S-Owned Fab Capacity, 2001

North America	65.4%
Europe/Middle East	18.6%
Japan	13.0%
Asia (except Japan)	3.0%

Source: Calculations courtesy of Rob Leachman.

[33]David Lammers, "Analysis: Samsung fab deal ends drought for Austin," EE Times, April 14, 2006.

[34]Gartner Dataquest estimate reported in "Foundry Revenue Drops in 2005, Gartner Reports," Electronic News, March 27, 2006.

[35]See Macher et al. (1998) for a discussion of the factors leading to the U.S. industry's resurgence after its loss of global market share in the mid-1980s.

2,600 would be highly paid engineering jobs.[36] Keep in mind, however, that not all foundry sales are to U.S. customers. In 2003, for example, half of TSMC's gross revenues came from non-U.S. sources.[37]

As a point of comparison, the Fabless Semiconductor Association reported that publicly traded fabless companies in North America employed approximately 45,000 workers as of December 2004.[38] A review of company information suggests that more than half of these employees were software or hardware engineers. The proportion that was located offshore is not known.

On the design side, U.S. chip firms have opened increasing numbers of offshore design subsidiaries in Asia over the last decade. Specialized skills are an important reason U.S. semiconductor companies are investing overseas, particularly in Europe. Britain, for example, has developed expertise in consumer multimedia, and Scandinavian countries are noted for skills in wireless network technology. U.S. firms regularly acquire small European companies to obtain both application know-how and teams of pre-trained engineers.

As is true of fabrication, design offshoring works both ways, and many foreign companies maintain design centers in Silicon Valley or elsewhere in the United States to take advantage of the high skills and productivity available there as well as to be closer to U.S. customers. Philips of the Netherlands, for example, bought VLSI Technology, a major ASIC company with more than 2,000 employees (about one-third of whom were fab workers), in 1999 for nearly $1 billion.[39] Hitachi Semiconductor has a U.S. design group of several hundred strong.[40] Toshiba has a network of seven ASIC design centers around the United States.[41] Even foreign start-up companies may need a U.S. design team to work with U.S. customers or to access leading-edge analog design skills.

However, a reason for design offshoring that has generated a great deal of attention in the industry is cost reduction. For Silicon Valley firms, some cost reduction can be achieved by opening satellite design centers elsewhere in the United States, because some locations have average engineering salaries as much as 20 percent lower than in the Silicon Valley area. But even these salaries are much higher than salaries in India and elsewhere, as discussed below.

Cost-driven, in-house offshoring incurs non-wage costs that partially offset the difference in salaries, especially during the early stages of establishing an offshore design center. One non-wage cost that is often mentioned is the lower quality and productivity of inexperienced engineers, which also adds monitoring costs. Another is the time and inconvenience of communicating across time zones, which can be considerable. Finally, additional control mechanisms may be necessary to protect key intellectual property. According to a venture capitalist, the actual savings from going offshore is more likely to be 25 to 50 percent than the 80 to 90 percent suggested by a simple comparison of salaries.[42]

Nevertheless, U.S. firms are investing regularly in low-cost design centers in Asia, especially in India. The first U.S. company to establish design operations there was Texas Instruments in 1985. Among the top 20 U.S. semiconductor companies, only two (Micron and Atmel) have *not* established design centers in India. Nine of these top companies have opened Indian operations since 2004. The size of these operations varies widely, with Intel employing about 3,000 engineers and smaller companies, like Marvell, employing fewer than 100 engineers.

The net impact on U.S. jobs from the offshoring and outsourcing of fabrication is hard to assess. As we said above, offshore fabs have been at least partially balanced by foreign-owned fabs in the United States. And although Asian foundries have probably contributed to a long-term reduction in U.S. chip manufacturing, the net loss of engineering jobs has probably been offset, at least partly, by the increase in design jobs at fabless companies. The government data in Table 1 suggest that despite the increase in offshoring in recent years, the number of engineers employed in the industry has increased.

Data from the Semiconductor Industry Association (SIA) provide further support for this argument (see Table 11). The SIA data are based on an annual survey of large and medium-sized U.S. semiconductor companies, which together account for approximately 80 percent of the U.S. industry's sales. The results are then extrapolated to represent all U.S. semiconductor firms. Although the data may not be strictly comparable from year to year, they can be used to discuss general trends and confirm other data. The total engineering employment at the top 20 companies has increased significantly over the period in question, with the offshore engineering staff growing slightly faster in most years. The number of engineers located in the United States increased sharply at the end of the 1990s, before the recession caused an employment slump in the early 2000s. Another sharp increase in U.S. employment is shown between 2004 and 2005,

[36]TSMC, which accounts for about half the foundry industry, has one 150mm, one 300mm, and five-and-a-half 200mm fabs outside the United States. These fabs probably have different rated capacities, but we can approximate employment by calculating 750 workers per plant, which works out to 5,625. Doubling that to approximate the entire foundry sector brings us to 11,250.

[37]Note 27c of Form 20-F filed by TSMC with the Securities and Exchange Commission for fiscal year ended December 31, 2003.

[38]FSA (2005).

[39]"Philips to acquire VLSI Technology for $953 million," Semiconductor Business News, May 3, 1999.

[40]"Hitachi Forms North America Semiconductor Systems Solutions Unit," Hitachi Press Release, September 2, 1998.

[41]"Toshiba Expands Soc Design Support Network with Opening of San Diego Design Center," Toshiba Press Release, November 26, 2002.

[42]Interview, May 2004.

TABLE 11 U.S. Semiconductor Engineers by Location, 1997–2005

	1997	1998	1999	2000	2001	2002	2003	2004	2005
U.S.-based Engineers	49,702	46,704	61,856	76,129	72,564	72,860	71,991	66,581	83,167
Offshore Engineers	7,253	19,692	17,446	19,964	27,226	29,813	30,876	34,632	42,193
Total	58,952	68,394	81,301	98,093	101,791	104,675	104,870	103,217	127,365
% in U.S.	87.3%	70.3%	77.9%	79.2%	72.7%	70.9%	69.9%	65.8%	66.3%

Source: David R Ferrell, "SIA Workforce Strategy Overview," ECEDHA Presentation March 2005; 2004 and 2005 data: unpublished SIA survey results provided by Ferrell.

although the OES data for those two years do not confirm this trend.[43]

The number of offshore engineers increased sharply in 1998, and again in 2001, and again in 2005. Even with the ups and downs, the percentage of the workforce in the United States tended to hover between 70 and 80 percent from 1998 to 2003; it then fell to 66 percent in 2004–2005. These data indicate a mild shift in employment of engineers offshore relative to the United States. If it continues, this shift could have a depressive effect on U.S. engineering employment and earnings.

The Semiconductor Industry in Japan, Taiwan, China, and India

Engineers in the U.S. semiconductor industry have long been accustomed to competition from abroad. However, the competition may now be within a single company, for example, between two design groups in different countries. In this section, we look at the availability, quality, and cost of chip engineers outside the United States.

A major problem with comparing semiconductor engineering talent in different countries is that the engineers in China and India, and to a lesser extent in Taiwan, are younger and have less education than engineers in the United States and Japan. In India and China, technicians with two-year degrees are often classified as engineers (this happens much less often in the United States and Japan). Relatively little graduate training is available in semiconductor engineering in India and China, and what is available is not comparable to graduate programs in the United States and Japan. Taiwan is an intermediate case; undergraduate and master's level engineering programs are comparable to those in the United States and Japan, although Ph.D. programs are still catching up.

Taiwan's semiconductor industry was built largely by Ph.D. engineers who returned to Taiwan after receiving degrees and valuable work experience in the United States. A similar process is occurring in China and India. Thus we think Taiwan may provide a model of how semiconductor engineering will develop in India and China as the semiconductor industry in those countries matures, with the important difference that Taiwan is a much smaller country.

The semiconductor industry in India and China is still quite young in terms of design, although both countries are active in this area. In China, domestic companies, often with personnel and funds from Taiwan, are major players in the development of semiconductor design. In China's fabrication sector, both multinational companies (MNCs) and domestic companies (again with input from Taiwan) are very important players. In India, where subsidiaries of MNCs are the major players in the development of the semiconductor industry, fabrication has not yet begun.

Semiconductor Engineering in Asia

With the caveat that comparisons of semiconductor engineers in the United States, Japan, Taiwan, China, and India involve comparing engineers with different education and experiences, Table 12 provides rough estimates (based on a combination of published sources and interviews) of salaries, worldwide fab investment by local companies, and the number of active chip designers (excluding embedded software). We also provide an index of protection of intellectual property (IP), which is an important consideration in deciding which engineering activities might be moved outside the United States. However, the intellectual property protection rating covers all industries; thus low scores in the table may reflect lapses in specific sectors, such as pharmaceuticals, trademark goods, or recorded media, which are not relevant to the semiconductor industry.

The salary figures suggest that engineers in the United States and Japan earn much more than most Asian engineers. These data, however, are imprecise and have high variance; thus they provide only a general guide. The salaries are for engineers with at least five years of experience in the United States and for engineers aged 40 in Japan, the approximate age they leave the union and begin to receive higher salaries. Note that 40 is the age at which the salary trajectory for U.S. engineers begins to level out. Semiconductor engineers in the other countries tend to be younger and less experienced; thus the salaries for engineers in China and India are for individuals with one to three years of experience.

[43]The OES total for all software and other engineer categories was 73,650 in the May 2004 data and 76,300 in the May 2005 data.

TABLE 12 Estimates for Selected Countries

	Annual Salaries for EE/CS Engineers	Value of Fabs Constructed, by Country of Ownership, 1995–2006	Number of Chip Designers	Intellectual Property Protection, 2004 (10 = high)
United States	$82,000	$74 billion	45,000	9.0
Japan	$60,000	$66 billion	—[a]	7.2
Taiwan	$30,000	$72 billion	14,000	6.5
India	$15,000	$0	7,000	5.0
China	$12,000	$26 billion	5,000	3.7

[a]We have been unable to obtain an estimate for the number of chip designers in Japan.

Sources: U.S. salary from 2004 BLS Occupational Employment Statistics web site (average for electronics and software engineers in NAICS 3344); Japan salary (average for circuit designer and embedded software engineers aged 40 years old) from Intelligence Corporation's data on job offers in 2003; Taiwan salary information from March 2005 interview with U.S. executive in Taiwan; China and India salaries are estimated based on a combination of interviews, business literature and online job offerings; value of fabs (when fully equipped) from Strategic Marketing Associates (*www.scfab.com*), reported in "Chipmaking in the United States," Semiconductor International, August 1, 2006; number of chip designers in U.S. from iSuppli as reported in "Another Lure Of Outsourcing: Job Expertise," *WSJ.com*, April 12, 2004; number of chip designers in Taiwan from interview with Taiwan government consultant to industry, March 2005; number of chip designers in India and China are author estimates based on conflicting published sources and discussions with industry analysts in 2005; intellectual property protection data from Gwartney et al., 2006, Chapter 3. All numbers rounded to reflect lack of precision.

As the semiconductor industry quickly expands in China and India, wages are reportedly rising rapidly. For example, the salary range offered by SanDisk in Bangalore (*JobStreet.com*, June 2005) for a design engineer with one to three years of experience was $9,200 to $18,400.[44]

The salary gap is narrower for comparable key employees. One report claimed in 1999 that the salary ratio between the United States and India for experienced design engineers or managers was only 3-to-1.[45] Senior managers with foreign experience are paid a large premium that eliminates any cost advantage; this reflects the critical importance of these managers in implementing new technology and projects.[46] The overall differential between Indian and U.S. salaries has been declining as Indian salaries rise, and the earnings of domestically trained Indian engineers has been doubling in their first five years on the job.

Salaries are also difficult to compare because of different compensation packages. In the United States and Taiwan, profit-sharing bonuses that vary with the business cycle can be an important part of a compensation package. In the United States, benefits, including health insurance, Social Security, and stock options, also make comparisons difficult.

The value of fab construction over the past decade provides a general idea of the presence of this part of the value chain in each country. China, at $26 billion, has made significant inroads since its early public-private joint ventures with Japan's NEC in the mid-1990s. In India, in sharp contrast, not a single commercial-scale fab has been constructed, although several have been proposed.

We also estimate the number of chip designers, a group that is critical to the development of the semiconductor industry. According to some sources, about 400 chip designers are being added each year in India and China.[47] However, that number can be misleading, because there is some confusion about the definition of "chip designer." One industry executive claimed that there were only 500 "qualified IC designers" in China in 2004.[48] A Taiwanese consultant didn't even consider the later (and lower skilled) stage of physical design, called "place and route," to be part of chip design.[49] By those criteria, about 30 percent of the Taiwanese designers shown in the table would be eliminated.

Estimates of Higher Education

As we discussed above, engineering programs in U.S. universities have attracted large numbers of foreign students. The United States leads the world in higher education, especially in graduate training, as the Academic Ranking of World Universities (*http://ed.sjtu.edu.cn/ranking.htm*) by Shanghai Jiao Tong University shows (see Table 13). Fifty-three of the top 100 universities are located in the United States; five are located in Japan. Of the top 500 universities, 168 are in the United States, 34 are in Japan, and only 21 are in China, Taiwan, and India combined.

The numbers for bachelor of science engineering degrees in Table 13 must be treated with caution, because the quality of education varies widely from country to country. The numbers may indicate political and social commitment to advancing technical education rather than actual capability. Also, these numbers are changing as India, and especially China, expand their engineering degree programs. According to a widely cited Duke University study, the annual number of new EE-CS-IT bachelor's degrees in China in 2004 had reached 350,000 (Gereffi and Wadhwa, 2005). But it is an open question how long it will take these new programs to develop quality teaching programs.

Although China and India have large numbers of engineering graduates, according to our interviews graduates from U.S. universities are better trained, especially in

[44]Converted at 43.52 Indian rupees to the dollar.

[45]"Special report: India awakens as potential chip-design giant," EE Times, January 22, 1999.

[46]Interviews at 15 semiconductor design centers in Bangalore in November 2005.

[47]For India: "Designs on the future," Express Computer (India), February 10, 2003; for China: PriceWaterhouseCoopers (2004), p. 7.

[48]PriceWaterhouseCoopers (2004), p. 7.

[49]E-mail exchange, March 2005.

TABLE 13 Estimates of Higher Education for Selected Countries

	Academic Ranking of World Universities, 2005		
	Universities in Top 100	Universities in Top 500	Engineering B.S. Degrees, 2001
U.S.	53	168	110,000
Japan	5	34	110,000
Taiwan	0	5	35,000
China	0	13	220,000
India	0	3	110,000

Source: Academic Ranking of World Universities values tabulated by authors from ARWU 2005 Edition, accessible at *http://ed.sjtu.edu.cn/ranking2005.htm*; engineer B.S. degrees tabulated by authors for "Engineering" and "Math/Computer Science" from Appendix Table 2-33, "Science and Engineering Indicators 2004," National Science Foundation except for India, which is an estimate for 2003–2004 from Appendix "USA-China-India" in Gereffi and Wadhwa, 2005.

teamwork on projects and in using tools and equipment. For example, undergraduate students in India and China usually have no chance to work with automated chip design (EDA) tools, while EE students in the United States do. According to McKinsey, only 10 percent of Chinese and 25 percent of Indian engineering graduates are likely to be suitable for employment by U.S. MNCs (McKinsey Global Institute, 2005).[50]

However, as we have already pointed out, the competition is not only between U.S. students trained in the United States and foreign students trained abroad. A large number of foreign students receive training in the United States.

Country Profiles

Next we look at the evolution of the semiconductor industries in Taiwan, India, and China and compare the technology capabilities of these countries with those of the United States. On the design side, the quality of engineers in Asian countries, both in universities and in companies, has been improving, as is clear from papers submitted to the International Solid-State Circuits Conference (ISSCC), which is IEEE's global forum for presenting advances in chip design (see Figure 8). From 2001 to 2006, submissions from China, India, and especially Taiwan increased noticeably. The number of acceptances for Taiwan also increased dramatically, even as the overall acceptance rate fell from 53 percent to 38 percent, and we expect that acceptances from India and China will increase in the near future as the quality of their university engineering programs improves.

[50]These figures were arrived at by McKinsey based on a survey of HR managers at multinational subsidiaries in these and other countries that asked the question: "Of 100 graduates with the correct degree, how many could you employ if you had demand for all?"

Taiwan

Taiwan has the best-established semiconductor industry of the three Asian countries. According to Taiwan's Ministry of Economic Affairs, the country ranked third (behind the U.S. and Japan) in semiconductor-related U.S. patents.[51] The foundry model originated in Taiwan in 1987, and three of the top five foundries are located there. Taiwan also has rapidly growing production of memory chips and numerous successful fabless chip companies, four of which reported revenues of more than $500 million in 2005.[52]

Table 14 shows the value of Taiwan's semiconductor industry output by stage of production for 2005. Fabrication, at $18.9 billion, accounts for the largest share of the $34.8 billion total, followed by chip design at $8.6 billion. Similar analyses are not possible in most major chip-producing countries where all stages of production are performed by large integrated producers. Taiwanese companies, however, have embraced the disaggregated business model, and only a handful of companies are involved in multiple steps in the value chain.

Since the late 1970s, Taiwan has benefited from focused government programs and the return of U.S.-educated and trained engineers.[53] In 1980, the government created the Hsinchu Science-Based Industrial Park, which is still the island's largest concentration of semiconductor firms. Hsinchu is also home to two of Taiwan's leading engineering universities, and the government's microelectronics lab, ERSO, which played a pioneering role in the development of the industry, including the creation of chip companies such as TSMC and UMC. ERSO conducts some of the most advanced research in the country, and its thousands of alumni are encouraged to commercialize technology via local start-up companies.

The Taiwanese chip-design sector is mostly locally owned, although a few MNCs also operate design subsidiaries there. Taiwanese companies have embraced the fabless model, and some 60 fabless companies were listed on the Taiwan Stock Exchange in December 2004.[54] By comparison, about 70 fabless companies were listed on NASDAQ in 2004. In 2001, the Taiwanese government renewed its efforts (Si-Soft) to improve local chip-design capabilities. As part of this initiative, the faculty teaching chip design more than doubled, from 200 in 2001 to more than 400 by 2005.[55]

One advantage for Taiwan's fabless firms is the availability of an important local market. Many Taiwanese systems companies design, assemble, and procure components for computers, communications equipment, and consumer elec-

[51]Cited in "Taiwan ranks 4th in the world in US patents received," Taipei Times, Oct. 17, 2006.

[52]"Data Snapshot," Semiconductor Insights: Asia (FSA), Issue 1, 2006.

[53]Saxenian (2002).

[54]FSA (2005).

[55]Chikashi Horikiri, "Taiwan Transforms into IC Development Center," Nikkei Electronics Asia, February 2006.

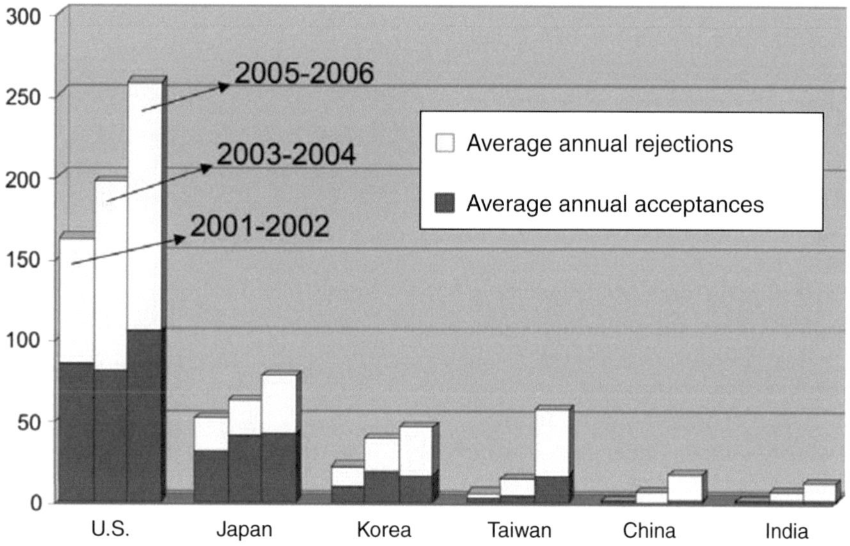

FIGURE 8 ISSCC acceptances and rejections by country, 2001–2006. Source: Tabulated from unpublished ISSCC data.

TABLE 14 Value of Taiwan's Semiconductor Industry, 2005

	Output Value (US$ billions)	Growth Since 2004
IC design	$8.63	5.8%
Foundry services	$18.90	–3.0%
IC packaging	$5.21	6.4%
IC testing	$2.04	13.0%

Source: IEK-IT IS data, reported in "Taiwan IC production value reached US$34.8 billion in 2005, says government agency," *DigiTimes.com*, January 19, 2006.

tronics for world-famous brands, including Hewlett-Packard, Nokia, and Sony. In 1999, 62 percent of Taiwan's chip-design revenue came from local sales.[56] Taiwan is second only to the United States in fabless firms by revenue, with firms specializing in cost-down, fast-follower capabilities. From a U.S. perspective, Taiwanese competition has shortened the market window during which U.S. chip companies can recoup their investments in chips before similar products are produced at a lower price.

Taiwan's design teams were praised in our interviews for their execution, a vital trait in an industry where time-to-market often means the difference between profit and loss. A frequent criticism, however, was that they were not yet truly innovative. Ironically, Taiwanese companies are locked in as technology followers by their reliance on business from local systems firms, which are as much as a generation behind the leading-edge technology.[57]

In the early stage of development of its semiconductor industry, Taiwan depended upon graduate training in the United States. Since the mid-1990s, the number of Taiwanese receiving Ph.D.s in engineering has declined steadily, and today only a few are pursuing graduate training in the United States. Although graduate education has improved in Taiwan, we heard some concerns in our interviews about declining numbers of returnees from the United States. Past returnees brought with them both graduate training and work experience that included management skills as well as practical knowledge.

The Taiwanese government has instituted several programs to improve the local design sector, including a plan to train several thousand new design engineers in Taiwan's universities, the creation of an exchange where local chip-design houses can license reusable functional blocks, and an incubator where early-stage start-ups can share infrastructure and services.[58] Another initiative is intended to attract chip-design subsidiaries of major semiconductor companies; early takers include Sony and Broadcom (a major U.S. fabless company). In 2000, a government research institute created the SoC Technology Center (STC) to design functional blocks that can be licensed to local companies, a model Taiwan has used successfully in other segments of the electronics industry. STC has more than 200 engineers, most of whom have master's degrees or better.[59]

For the Taiwanese semiconductor industry, China presents both a major challenge and a major opportunity. The

[56]Data from Taiwan's Industrial Technology Research Institute cited in Table 5, Chang and Tsai (2002).

[57]Breznitz (2005).

[58]"Trends in SOC design unthaw at SOC 2004," *EDN*, December 9, 2004.

[59]SoC Technology Center interview, March 2005. "SoC" is a common industry acronym for "system-on-a-chip" meaning a complex semiconductor. integrating multiple functions.

challenge is competition in the foundry and fabless sectors, especially for low-cost designs using older technology, as well as competition for talented engineers to work in China and bring with them their knowledge of advanced technology in design and manufacturing. The opportunity is the chance to partner with Chinese companies elsewhere in the value chain, enabling Taiwanese companies to provide high-end design services. In addition, Taiwanese companies would have access to China's rapidly growing markets.

So far, political issues have made it difficult for Taiwanese chip companies to develop partnerships and markets in China, even as they lose experienced engineers to Chinese competitors. Taiwan-born engineers are an important force in technology development in China, in much the same way that the United States was an important force in technology development in Taiwan. Although China seems to be benefiting more than Taiwan from the flow of engineers, capital, and business activities between the two countries, this may change over time if the Taiwanese government changes its policy.

China

China appears to be following a similar pattern—government sponsorship, local access to system firms (such as Haier, Huawei, and TCL) that are increasingly engaged in world markets, and active involvement of expatriates returning from the United States or experienced engineers relocating from Taiwan.[60] In little more than a decade, with the help of foreign companies (as investors or as technology licensors) and the Chinese government, Chinese firms have developed impressive fabrication capability.

Table 15 shows the main chip fabs in China, based primarily in Shanghai. The most striking feature is that they are all foundries working under contract rather than companies that design and manufacture their own products. U.S.-based chip companies have few high-profile deals with Chinese foundries—the major exception being Texas Instruments, which began working with Semiconductor Manufacturing International Corp (SMIC) in 2002 and added a deal to co-develop SMIC's 90 nm process in 2004.[61] Executives with U.S. experience have also played key roles. For example, the CEOs of ASMC and HHNEC previously worked at AMD.[62]

Apart from SMIC, China's foundries have adopted modest growth plans, especially compared to the headline-grabbing predictions of two or three years ago.[63] But chip fabrication is now firmly established in China and will gradually expand. Although China's fabs pose a growing challenge to Taiwanese foundries, from the perspective of U.S. chip firms they add welcome competition to the market for wafer processing.

TABLE 15 Major Fabs in China, 2006

Company	Fab Location	First Year of Production	Capacity (wafers per month, 8-inch equivalent)
Advanced Semiconductor Manufacturing Corp (ASMC)	Shanghai	1995	25,000
Shanghai Hua Hong NEC Electronics (HHNEC)	Shanghai	1999	50,000
Semiconductor Manufacturing International Corp (SMIC)	Shanghai, Tianjin, and Beijing	2001	150,000
Grace Semiconductor Manufacturing Corp (GSMC)	Shanghai	2003	27,000
He Jian Technology	Suzhou	2003	42,000
Taiwan Semiconductor Manufacturing Co (TSMC)	Shanghai	2004	15,000; (40,000 planned)

Source: iSuppli data, reported in Cage Chao and Esther Lam, "Despite China-based foundries reporting full utilization rates in 1Q, Taiwan players not overly impressed," *Digitimes.com*, March 22, 2006.

A potentially more worrisome development for U.S. firms is the emergence of a fabless design sector in China. Since 2003, China has claimed to have more than 400 chip-design firms. Many are small, poorly managed companies that deplete their seed money before they can bring a product to market. Others offer design services rather than their own products.[64] One interviewee, echoed by others, claimed that many, if not most, firms outside the top 10 are engaged in various types of reverse engineering, which is often illegal.[65] Foreign firms are generally reluctant to bring lawsuits, however, for fear of displeasing the authorities and the likelihood of losing in Chinese courts. But at least two U.S. companies are suing Chinese rivals in export markets for intellectual property violations.[66]

China's top 10 chip-design firms in 2005 had total revenues of more than $1 billion, $400 million of which was from Hong Kong-based Solomon Systec, a designer of LCD

[60]Saxenian (2002).

[61]Mark LaPedus, "TI, SMIC sign deal to develop 90-nm technology by Q1 '05," Silicon Strategies, Oct.28, 2004.

[62]Chintay Shih, "Experience on developing Taiwan high-tech cluster," presentation at 4th ITEC International Forum, Doshisha University, June 17, 2006.

[63]Mike Clendenin, "Deflated expectations in China's IC biz," EE Times, August 28, 2006.

[64]Assessment of Byron Wu, iSuppli analyst, reported in "Analyst: China's IC design houses struggling for survival," EE Times, May 20, 2004.

[65]Interview with a European chip executive, conducted by Elena Obukhova in Shanghai, December 2003.

[66]See "An offshore test of IP rights," Electronic Business, May 2004; and "SigmaTel Sues Chinese Chipmaker over IP," Electronic News, January 6, 2005.

drivers that was spun off from Motorola in 1999.[67] The next largest firms (Actions [media player chips], $150 million, and Vimicro [PC camera image processors], $95 million) had IPOs on NASDAQ in 2005.

China's large, growing domestic market provides opportunities for China's chip design companies to grow and become profitable, and in the future Chinese companies may be able to design products for the global marketplace. The local systems firms provide a sizable market for local fabless start-ups. The best chip design work is being done by local systems firms and a few world-class start-ups headed by U.S. returnees.

The Chinese government has taken many steps to support chip-design firms, some of the largest of which are state owned. These measures include tax reductions, venture investing, incubators in seven major cities, and special government projects.[68] A value-added tax preference for domestically designed chips was phased out under U.S. pressure and will reportedly be replaced by a WTO-friendly R&D fund, although this had not been announced as of this writing (September 2006).[69]

The return of Chinese nationals with education and work experience has been an important part of China's recent technology development.[70] Returnees provide valuable management experience and connectivity to global networks that tend to accelerate the development of China's chip sector.[71] According to government statistics on student returnees, in 2003, of the 580,000 students reported to have gone abroad since 1978, 150,000 had returned.[72] The returnees had started 5,000 businesses, including more than 2,000 IT companies in Beijing's Zhongguancun Science Park (one-sixth the park total).[73] China is working to attract more high-tech returnees with a range of specially targeted incentives and infrastructure.[74]

China is not yet an important destination for design offshoring by U.S. firms. Of the top 20 U.S. semiconductor companies, only a handful had opened design centers in China (compared to 18 in India) as of June 2006. Most of these design centers are targeting the local market for the time being, and, according to press reports, some are engaged in software or system design rather than chip design per se. Concerns about intellectual property protection appear to pose a greater barrier to foreign design activity in China than in India.[75]

Chip design in China is at an early stage, but the relatively young Chinese chip-design engineers will steadily build their experience. One factor that favors the development of local design companies is that Chinese engineers prefer to work for domestic start-ups and domestic companies rather than MNCs. Many young Chinese engineers, especially returnees, are willing to risk working for emerging companies that may earn them great wealth. Some companies, particularly those whose founders include expatriates with foreign education and experience, are likely to begin to impact global markets by the end of the decade. It is still too early to predict the future relative importance of domestically owned and foreign-owned chip-design activities, or to predict whether domestic firms will be involved mostly with contract services or with creating and selling chips.

The education of semiconductor engineers in China is also at an early stage. As discussed above, the quality of Chinese engineering graduates varies widely, and few have the knowledge and skills necessary to work on advanced technology or for MNCs. However, MNCs, including chip and EDA firms, have been involved in improving engineering education in China, and the government has been actively recruiting world-class engineering professors to Chinese universities. Over time we expect semiconductor engineering education, especially at the graduate level, to continue improving. For now, returnees from the United States and experienced engineers from Taiwan will continue to play an important role in transferring technology to China.

India

The semiconductor industry in India presents a very different picture. India faces benign neglect by the government, a lack of manufacturing for chips and systems, and fewer returnees from the United States.[76] Unlike Taiwan and China, India has no high-volume chip manufacturing, although as many as five proposals to build foundries are in various stages of negotiation.[77]

India is estimated to have 120 chip-design firms, and revenues from chip design in 2005 were estimated to be

---

[67]Chinese government data cited in Mcallight Liu, "China's Semiconductor Market: IC Design and Applications," Semiconductor Insights: Asia (FSA), Issue 1, 2006 and iSuppli data in Mark LaPedus, "iSuppli lists China's top fabless IC rankings," EE Times, April 21, 2006.

[68]"Synopsys Teams with China's Ministry of Science and Technology, SMIC," Nikkei Electronics Asia, March 21, 2003; "An Uneven Playing Field," Electronic News, July 3, 2003; "China nurtures home-grown semiconductor industry," EBN, December 8, 2003; "China government to support Solomon Systech, Actions and Silan," DigiTimes, April 14, 2005.

[69]"China to form R&D fund to replace VAT rebate, says report," EE Times, April 15, 2005.

[70]Saxenian (2002).

[71]"Story behind the Story: Design in China is growing, but not exploding," audiocast by Bill Roberts, Electronic Business, September 1, 2006, *http://www.edn.com/article/CA6368425.html?text=%22design+in+china%22#*.

[72]"More overseas Chinese students returning home to find opportunities," November 16, 2003, *http://www.china-embassy.org/eng/gyzg/t42338.htm*.

[73]"More overseas Chinese students return home," January 1, 2004, *http://www.china-embassy.org/eng/gyzg/t57364.htm*.

[74]Mike Clendenin, "China starting to lure back its best brains," EE Times, January 3, 2002.

[75]"SIA Pushes Steps to Better IP Protection in China," Electronic News, November 17, 2004.

[76]Saxenian (2002).

[77]Russ Arensman, "Move over, China," Electronic Business, March 2006.

$583 million.[78] Most chip design is taking place in MNC subsidiaries, including most of the top 20 U.S. companies and many European companies. The flow of semiconductor engineering talent to MNCs has slowed the diffusion of technology to local firms, and India has no major fabless companies designing chips for sale under their own brand. Domestic chip-design companies with varied capabilities mainly provide design services. According to a study by the India Semiconductor Association, local design companies use a time- and material-based pricing method by which specific tasks are allocated to be carried out within set time lines.[79] These companies tend to develop simple subsystems based on customer specifications.

Larger independent design-services firms are much more sophisticated. They use a fixed-price method, are able to provide end-to-end solutions that incorporate in-house proprietary intellectual property, and offer design services across the VLSI design flow. The government is developing policies to support domestic chip-design firms.

In our fieldwork we found that Indian engineers prefer MNCs to local start-ups, which are perceived as risky by engineers and their family members. This is a contrast with China, where engineers are relatively eager to join start-ups, which often receive some government support.

Foreign chip companies have been attracted by Indian engineers' knowledge of English and the successful Indian software sector. Many early investments by chip companies were focused on software, the writing of microcode that becomes part of a chip. Over time, Indian affiliates have taken on a bigger role, eventually extending to complete chip designs from specification to physical layout. This transition sometimes happens quickly. Intel, for example, opened a software center in Bangalore in 1999 and began building a design team for 32-bit microprocessors in 2002.[80]

Since most domestically trained engineers lack knowledge of the technology being transferred, the necessary management skills, and knowledge of the entire product cycle, American MNCs are highly dependent on returnees with advanced degrees from the United States to develop new projects in India. So far there have been few instances of design engineers in India leaving MNCs to start their own companies, as often happens in the United States. However, we heard of at least two cases in the past two years at one U.S. subsidiary. We also heard that leaving an MNC to start a company is becoming more acceptable among Indian engineers, many of whom are motivated to help India develop rather than to accumulate great wealth.[81]

Foreign subsidiaries face formidable problems in their Indian operations, including a very tight labor market and inadequate infrastructure. As in China, the quality of Indian engineering graduates varies greatly. This problem is exacerbated in India because most engineers there want to study computer science rather than electronics, and many are not aware of the job opportunities in semiconductors. Graduate education in EE is in its infancy, and doctoral education in the seven major technical universities is not up to U.S. standards. The very low wages paid to professors, the lack of expensive and constantly changing EDA tools, and the difficulty and expense of having sample chips fabricated, all contribute to problems in the development of world-class graduate education.

In addition, India has not attracted nearly as many returnees as China. The low flow of new domestic graduates and returnees into the EE labor supply, coupled with the need for at least three to five years of experience for fully productive chip designers, has meant that the supply of design engineers has not kept pace with increasing demand. As a result, wages for chip designers have been rising rapidly, both at the entry level and during the first five years. As mentioned above, salaries for engineers with five years of experience are double entry-level salaries.

Inadequate infrastructure, especially in Bangalore, also poses serious problems for chip-design centers. Because of the lack of a stable energy supply and lack of office space, foreign subsidiaries must make substantial investments to provide both offices and electricity. Bangalore, the country's primary city for high-tech, is plagued by narrow, pothole-filled roads that are often gridlocked, forcing employees to spend long hours commuting. In addition, high-tech companies are spread throughout the city, making commuting between companies, or even between company locations, very time consuming.

In addition, the housing stock in Bangalore has not kept up with growth, and housing prices and rents have been rising rapidly. Many employees are faced with a choice of living in inadequate housing or living far from work. The housing and schooling problems are especially severe for returnees from the United States, who want to replicate the quality of U.S. housing and schools their families know. Several executives told us that their cost of living in Bangalore was almost as high as in the United States because of the high cost of housing and international schools.[82]

The shortage of engineering talent and weak infrastructure have constrained the rate of growth in the semiconductor design industry, both for foreign subsidiaries and for local companies, in India generally, and in Bangalore particularly. Some companies have been moving operations to areas that have better infrastructure and are less expensive than Bangalore. However, the talent shortage remains, especially for experienced engineers with advanced degrees.

---

[78]Data from Frost & Sullivan, in Chitra Giridhar, "India design firms as product innovators," Electronic Business, July 18, 2006.

[79]"Study: Indian design firms prefer time and material model," EE Times, Sept 22, 2006.

[80]"Intel, TSMC Set Up Camps In Developing Asian Markets," *WSJ.com*, August 30, 2002.

[81]Personal communications in Bangalore, November 2005.

[82]Personal communications in Bangalore, November 2005.

OUTLOOK AND CONCLUSION

The United States remains the world leader in the semiconductor industry in terms of market share, development of successful new companies, supply of experienced engineers, and graduate engineering education. Moreover, the United States is the leading location for system design, the stage at which most semiconductor purchase decisions are made.[83] Our competitors, especially Japan, Korea, Taiwan, and the European Union, look to the United States for lessons on encouraging innovation and start-ups in the semiconductor industry. Nevertheless, competition from low-cost countries, especially China and India, which have rapidly growing and potentially large markets, may pose competitive threats to U.S. companies and engineers in the future.

Outlook for U.S. Engineers

The job market for U.S. semiconductor engineers shows there is some strength in employment and earnings growth, but also shows evidence of labor market problems, especially for older engineers and for the bottom 10 percent at all educational levels. We also observed signs of a decline in the earnings premium for graduate degrees (M.S./Ph.D. compared to a B.S.), and low returns-to-experience for engineers with graduate degrees. The situation is especially difficult for older engineers whose skills can rapidly become obsolete. Experienced design engineers are often forced to work on mature technologies, which pay less and may present fewer interesting problems. For example, according to a salary survey in 2004 by *EE Times*, the average annual salary for U.S. and European engineers skilled at designing for the latest chip-process technology was $107,000, whereas engineers designing for more mature analog technology averaged $87,000.[84]

Results of a regional survey of Silicon Valley, considered the cradle and creative font of the semiconductor industry, reveal that the recent job climate there is difficult. Overall the number of jobs in Silicon Valley has continually decreased since 2001, and jobs in the semiconductor and semiconductor-equipment industries declined 23 percent between 2002 and 2005, although the average wage rose 12 percent during the same period. Thus the survey paints a mixed picture of the health of the industry.[85]

Not surprisingly, industry participants disagree about the significance of offshoring for the U.S. job market. A 2004 survey by *EE Times* of more than 1,453 chip- and board-design engineers and managers showed that about half believed that foreign outsourcing would lead to a reduction in head count. Qualitative opinions were also divided, with optimists noting that reduced costs have strengthened companies and increased job security, and pessimists bemoaning downward pressure on wages and employment as well as the possible loss of intellectual property and, in the long run, industry leadership.[86]

We have observed that some movement of design jobs is related to the business cycle. There was a wave of design offshoring at the height of the dot-com bubble. Then, when the cascading effect of the subsequent downturn reached the semiconductor industry, chip companies began cutting staff at home. Now that the recovery requires the expansion of design operations, chip companies appear to be expanding design operations abroad faster than at home.[87] It is too early to predict where this relative shift in the geographic distribution of employment will find its new equilibrium.

Even experts disagree about whether or not the United States is educating too few engineers and scientists and is facing a shortage.[88] This is partly because economists find it hard to believe there can be a shortage in a labor market when real earnings across the board are stagnant. This is partly a reflection of government policies that affect the immigration and education of high-tech engineers.

Policy Issues

The industry's offshoring has gone well beyond the point at which blunt instruments such as trade policy can help engineers without harming companies. Taxes or quotas on traded activities or goods would raise costs for the many companies that have already invested offshore in a wide array of design and manufacturing activities for both the foreign and domestic chip markets. Policy changes are thus unlikely to improve the demand side of the labor market.

Industry has, however, been actively lobbying for changes on the supply side in the form of changes to educational and immigration policies that increase the supply of high-tech workers. The winter 2005 newsletter of the Semiconductor Industry Association includes two articles on the subject, "Maintaining Leadership as Global Competition Intensifies" by the organization's president and "America Must Choose to Compete" by the outgoing CEO of Intel.

One of the main targets of industry analyses is education. Higher education policies, which reflect both university decisions and government funding, determine the number and country of origin of students at all levels, but especially at the graduate level. Foreign nationals in our M.S. and Ph.D. programs in science and engineering have a direct impact on the supply of knowledge workers, both in the United States and in China and India. Foreign graduates of U.S. universi-

[83]iSuppli data reported in Dylan McGrath, "U.S. still top design influencer; China, India rising fast," EE Times, September 28, 2006.

[84]"After 10-year surge, salaries level off at $89k," EE Times, August 28, 2003.

[85]Joint Venture: Silicon Valley Network, "2006 Index of Silicon Valley," available online at *http://www.jointventure.org/PDF/Index%202006.pdf*. The data are from state unemployment insurance data, which is the basis for the Census data.

[86]"It's an outsourced world, EEs acknowledge," EE Times, August 27, 2004.

[87]See, for example, "The perfect storm brews offshore," Electronic Business, March 2004.

[88]See, for example, Freeman (2003, 2005); Task Force on the Future of American Innovation (2005); NRC (2000, 2001); Butz et al. (2004).

ties must obtain temporary visas, usually H1-B visas, before they can work in the United States after graduation. Legislation is under consideration to provide permanent residency status to foreigners educated in the United States. We are hopeful that this policy will be implemented soon.

Government policies regulating immigration, especially the issuance of H-1B (Non-Immigrant Professional) and L-1 (Intra-Company Transfer) visas, also have a significant impact on the number of foreign engineers engaged in semiconductor and software work. In a delayed response to the recession, changes in policy that took effect in 2004 set severe limits on the number of visas for foreign workers. When the number of H-1B visas was thus reduced, many U.S. companies used the opportunity to send foreign nationals with U.S. education and experience back to India and China to help build operations there.

An area of policy that has received less attention is compensation for engineers who are harmed by offshoring. As a result of the offshoring of chip design, consumers have benefited from lower prices and new products (although much of that benefit is received outside the United States). Some of the short-term cost of offshoring, however, is being borne by engineers in particular companies or industry sectors in which companies are restructuring globally. Currently, white-collar workers like chip designers do not qualify for trade-adjustment assistance from the government when their jobs are sent abroad. It would make sense to help these highly-skilled workers with retraining and other forms of assistance to enable them to remain productive. As Federal Reserve Chair Bernanke remarked, "The challenge for policy makers is to ensure that the benefits of global economic integration are sufficiently wide-shared—for example, by helping displaced workers get the necessary training to take advantage of new opportunities—that a consensus for welfare-enhancing change can be obtained."[89]

Finally, we need more and better data. As researchers in other industries have noted, more labor market data, both for the United States and for our trading partners, are necessary for proper assessments of the effects of offshoring.[90] In the meantime, national policies affecting education, labor markets, and innovation will continue to be based upon informed speculation.

How Should U.S. Engineers Respond?

American engineers are naturally responding to the impact of the changing labor market on their careers. The highly rewarded career path of working for one company for an entire career is no longer an option. Most engineers today must expect to work for several firms. In fact, changing jobs is now the most effective way for them to advance their careers, both in terms of improving pay and learning new technologies and skills. Networking with colleagues from one's alma mater and former companies as well as through professional associations is an excellent way of keeping up with job opportunities as well as learning about new technologies.

Our advice to semiconductor engineers is to embrace the mobile labor market and look to job changes as a way of advancing. Each job should be chosen carefully to improve skills and take advantage of previous job experience. Engineers must continually stay in touch with their networks and share knowledge with their colleagues about what is happening in the field and about job opportunities. In short, engineers today must be in charge of their careers; they can no longer depend on employers to provide them with the training they need to keep up their skills.

Foreign nationals working for U.S. companies can use their networks to develop careers both in the United States and in their home countries. Returnees who are willing to return home for short- or long-term stints can bargain for good salary packages from U.S. employers. U.S. nationals should also go abroad to develop contacts and expertise in specific cultures and regional markets.

Semiconductor engineers are known for their flexibility and ability to solve challenging problems and to learn new technologies. The semiconductor industry is likely to continue to undergo constant crisis and change, and chip engineers should use these industry characteristics to their advantage in planning their careers by seeking jobs where they can learn about new technologies and new markets. To be successful in the industry, an engineer must see change as an opportunity rather than a problem.

Lessons Learned

In its short history, the semiconductor industry has faced continual challenges and has done an extraordinary job of overcoming them, often in innovative ways that were not anticipated. The industry has also continually experienced large swings in demand and prices, and we expect the cyclical nature of the industry to continue, even as the long-term trend moves upward. Our predictions for the future of the industry and recommendations for setting policy must not extrapolate from conditions in the short run, especially during a downturn. We must look to the long-term history of the industry to ensure that policy decisions, either by governments or by companies, are made on a solid foundation. Macro-policies that ensure a strong economy with steady growth are critical to the development of the semiconductor industry, which is negatively affected by national recessions and high interest rates.

Government support for higher education, especially graduate education, should be the cornerstone of public policy to support innovation. A strong university system with state-of-the-art graduate training and strong links to

[89]Edmund L. Andrews, "Fed Chief Sees Faster Pace for Globalization," New York Times, August 25, 2006.

[90]See the excellent study by Tim Sturgeon et al. (2006).

companies is critical for innovation in the semiconductor industry. U.S. universities are essential to educating Ph.D.-level engineers, who are as likely to be from Asia as from the United States. Social networks, such as workers' contacts at their former universities and former employers, are important adjuncts to a company's formal knowledge base. Company awareness of this is critical to ensuring that employees' knowledge is recognized and used rather than flowing outward into these networks.

Conclusion

The semiconductor industry is in the intermediate stages of the complex, dynamic process of globalization. At this point it is hard to predict the impact of offshoring on the competitive position of the U.S. semiconductor industry and on the earnings and employment of domestic engineers, and whether the new equilibrium will be acceptable. Thus policy interventions must be flexible.

Offshoring is an important step in the integration of India and China into the global economy. These countries appear to be pursuing different roles vis-à-vis the United States, with China's chip industry acting more as a competitor (e.g., fabless start-ups) and India's playing a more complementary role (e.g., design services). Both countries will certainly become more important in high-tech industries, both as markets and suppliers. However, their ability to move up the semiconductor technology curve is constrained at the moment by a lack of graduate education, undeveloped financial systems, and inadequate intellectual property protection, as well as severe problems facing their political systems.

We expect that the United States will maintain its leadership position in the semiconductor industry, and we expect the industry and its resourceful engineers to continue to find ways to overcome challenges. For now, modifications in government policies affecting universities, immigration, and workers affected by trade would alleviate some of the labor market problems we have described.

ACKNOWLEDGMENTS

Clair Brown is professor of economics and director of the Center for Work, Technology, and Society (IIR) at University of California, Berkeley; Greg Linden is senior research fellow at the center. Yongwook Paik provided excellent research assistance. The authors would like to thank the Alfred P. Sloan Foundation, the Institute for Industrial Relations at UC Berkeley, and the Institute for Technology, Enterprise and Competitiveness (ITEC/COE) and Omron Fellowship at Doshisha University, Japan, for funding. Bob Doering and Bill Spencer provided detailed and helpful comments on the workshop version of this paper. We are also grateful to Ben Campbell, David Ferrell, Michael Flynn, Gartner Dataquest, Ron Hira, Dave Hodges, Rob Leachman, Daya Nadamuni, Elena Obukhova, Devadas Pillai, Semiconductor Industry Association, Chintay Shih, Gary Smith, Strategic Marketing Associates, Yea-Huey Su, Tim Tredwell, and C-K Wang for their valuable contributions. Melissa Appleyard, Hank Chesbrough, Jason Dedrick, Rafiq Dossani, Richard Freeman, Deepak Gupta, Bradford Jensen, Ken Kraemer, Frank Levy, B. Lindsay Lowell, Jeff Macher, Dave Mowery, Tom Murtha, Tim Sturgeon, Michael Teitelbaum, and Eiichi Yamaguchi, as well as participants at the NAE Workshop on the Offshoring of Engineering, the 2005 Brookings Trade Forum on Offshoring of White-Collar Work, the Berkeley Innovation Seminar, and the Doshisha ITEC seminar series provided thoughtful discussions that improved the paper. We are especially grateful to Gail Pesyna at the Sloan Foundation for her long-running support of, and input into, our research. The authors are responsible for any errors.

REFERENCES

Breznitz, D. 2005. Development, flexibility, and R&D performance in the Taiwanese IT industry—capability creation and the effects of state-industry co-evolution. Industrial and Corporate Change 14(1): 153–187.

Brown, C., and B. Campbell. 2001. Technical change, wages, and employment in semiconductor manufacturing. Industrial and Labor Relations Review 54(2A): 450–465.

Brown, C., J. Haltiwanger, and J. Lane. 2006. Economic Turbulence: Is a Volatile Economy Good for America? Chicago, Ill.: University of Chicago Press.

Brown, C., and G. Linden. 2006. Offshoring in the Semiconductor Industry: A Historical Perspective. Pp. 279–333 in Brookings Trade Forum 2005: Offshoring White-Collar Work—The Issues and the Implications, L. Brainard and S.M. Collins, editors. Washington D.C.: Brookings Institution.

Butz, W., T. Kelly, D. Adamson, G. Bloom, D. Fossum, and M. Gross. 2004. Will the Scientific and Technical Workforce Meet the Requirements of the Federal Government? Santa Monica, Calif.: RAND Corporation.

Chang, P.-L., and C.-T. Tsai. 2002. Finding the niche position—competition strategy of Taiwan's IC design industry. Technovation 22(2): 101–111.

DHS (U.S. Department of Homeland Security). 2004. Characteristics of Specialty Occupation Workers (H-1B): Fiscal Year 2003. Washington D.C.: DHS.

Freeman, R.B. 2003. Trade Wars: The Exaggerated Impact of Trade in Economic Debate. NBER Working Paper W10000. Cambridge, Mass.: National Bureau of Economic Research.

Freeman, R.B. 2005. Does Globalization of the Scientific/Engineering Workforce Threaten U.S. Economic Leadership? NBER Working Paper 11457, June. Cambridge, Mass.: National Bureau of Economic Research.

FSA (Fabless Semiconductor Association). 2005. Global Fabless Fundings & Financials Report: CY Q4 2004/Year End 2004. Dallas, Tex.: FSA.

GAO (Government Accountability Office). 2003. H-1B Foreign Workers: Better Tracking Needed to Help Determine H-1B Program's Effects on U.S. Workforce. GAO Report 03-883. Washington, D.C.: U.S. Government Printing Office. Available online at *http://www.gao.gov/new.items/d03883.pdf*.

Gereffi, G., and V. Wadhwa. 2005. Framing the Engineering Outsourcing Debate: Placing the United States on a Level Playing Field with China and India. Engineering Management Program, Duke University.

Gwartney, J. and R. Lawson, with W. Easterly. 2006. Economic Freedom of the World: 2006 Annual Report. Vancouver, B.C.: Fraser Institute.

Kumagai, J. 2003. It's Cold Out There. IEEE Spectrum 40(7): 13–15.

Macher, J.T., D.C. Mowery, and D.A. Hodges. 1998. Back to Dominance? U.S. Resurgence in the Global Semiconductor Industry. California Management Review 41(1): 107–136.

McKinsey Global Institute. 2005. The Emerging Global Labor Market. San Francisco, Calif.: McKinsey & Company.

NRC (National Research Council). 2000. Forecasting Demand and Supply of Doctoral Scientists and Engineers. Washington, D.C.: National Academy Press.

NRC. 2001. Building a Workforce for the Information Economy. Washington, D.C.: National Academy Press.

NSF (National Science Foundation). Various years. Science and Engineering Doctorate Awards. Arlington, Va.: NSF Division of Science Resources Statistics.

PriceWaterhouseCoopers. 2004. China's Impact on the Semiconductor Industry, TC-02-04. San Jose, Calif.: PriceWaterhouseCoopers.

Saxenian, A. 2002. Transnational communities and the evolution of global production networks: the cases of Taiwan, China and India. Industry and Innovation 9(3): 183–202.

Sturgeon, T.J. 2006. Services Offshoring Working Group: Final report. September. MIT Industrial Performance Center, Cambridge, Mass.

Task Force on the Future of American Innovation. 2005. The Knowledge Economy: Is the United States Losing Its Competitive Edge? Available online at *www.futureofinnovation.org/* as of April 20, 2005.

ONLINE NEWS SOURCES USED

The following list provides Internet addresses for the online publications cited in our footnotes. In some cases, the original Internet site we used to access the article has been folded into another site. The one listed is the current site. Some of the articles are also archived at third-party sites such as *findarticles.com*, so it is generally simplest to search for the article title as an exact phrase. Articles older than five years may no longer be available online.

DigiTimes *http://digitimes.com/*
EDN *http://www.edn.com*
Electronic Business *http://www.edn.com* (formerly separate)
Electronic News *http://www.edn.com* (formerly separate)
Express Computer (India) *http://www.expresscomputeronline.com*
The Register *http://www.theregister.co.uk/*
Silicon Strategies *http://www.eetimes.com* (formerly separate)
Electronic News *http://www.edn.com* (formerly separate)
New York Times *http://www.nytimes.com*
Nikkei Electronics Asia *http://techon.nikkeibp.co.jp*
Semiconductor Business News *http://www.eetimes.com* (formerly separate)
Semiconductor International *http://www.reed-electronics.com/semiconductor/*
Taipei Times *http://www.taipeitimes.com*
WSJ.com *http://online.wsj.com*

ADDITIONAL READING

Borrus, M. 1988. Competing for Control: America's Stake in Microelectronics. Cambridge, Mass.: Ballinger Publishing.

Ernst, D. 2004. Internationalisation of Innovation: Why is Chip Design Moving to Asia? Working Paper 64 (rev. March 2004). Honolulu, Hi.: East-West Center.

Flamm, K.S. 1985. Internationalization in the Semiconductor Industry. Pp. 38–136 in The Global Factory: Foreign Assembly in International Trade, J. Grunwald and K.S. Flamm, editors. Washington, D.C.: Brookings Institution.

Garner, A.C. 2004. Offshoring in the service sector: economic impact and policy issues. Economic Review (Federal Reserve Bank of Kansas City), Third Quarter: 5–37.

Groshen, E.L., B. Hobijn, and M.M. McConnell. 2005. U.S. jobs gained and lost through trade: a net measure. Current Issues in Economics and Finance 11 (8): 1–7.

Gupta, D. 2005. Ties That Bind: An Econometric Analysis of the Return Home of Foreign-Born U.S. Ph.D.s. Working Paper 2005-1. Amrita School of Business, Ettimadai, India.

Ham, R.M., G. Linden, and M.M. Appleyard. 1998. The evolving role of semiconductor consortia in the U.S. and Japan. California Management Review 41(1): 137–163.

Hemani, A. 2004. Charting the EDA roadmap. IEEE Circuits & Devices 20(6): 5–10.

Henderson, J.W. 1989. The Globalisation of High Technology Production: Society, Space, and Semiconductors in the Restructuring of the Modern World. New York: Routledge.

Henisz, W.J., and J.T. Macher. 2004. Firm- and country-level trade-offs and contingencies in the evaluation of foreign investment. Organization Science 15(5): 537–554.

Hira, R. 2004. U.S. immigration regulations and India's information technology industry. Technological Forecasting & Social Change 71: 837–854.

Hira, R. 2005. Impacts and Trends of Offshoring Engineering Tasks and Jobs. The Bridge 35(3): 22–27. Available online at *http://www.nae.edu/NAE/bridgecom.nsf/weblinks/NAEW-6GEK5J?OpenDocument*.

Howell, T.R., B.L. Bartlett, W.A. Noellert, and R. Howe. 2003. China's Emerging Semiconductor Industry: The Impact of China's Preferential Value-Added Tax on Current Investment Trends. San Jose, Calif.: Semiconductor Industry Association.

International Business Strategies. 2002. Analysis of the Relationship between EDA Expenditures and Competitive Positioning of IC Vendors: A Custom Study for EDA Consortium. Available online at *www.edac.org/downloads/resources/profitability/HandelJonesReport.pdf*. Accessed April 15, 2005.

Leachman, R.C., and C.H. Leachman. 2004. Globalization of Semiconductors: Do Real Men Have Fabs, or Virtual Fabs? Pp. 203–231 in Locating Global Advantage: Industry Dynamics in the International Economy, edited by M. Kenney with R. Florida. Stanford, Calif.: Stanford University Press.

Linden, G., C. Brown, and M. Appleyard. 2004. The Net World Order's Influence on Global Leadership in the Semiconductor Industry. Pp. 232–257 in Locating Global Advantage, edited by M. Kenney with R. Florida. Stanford, Calif.: Stanford University Press.

Linden, G., and D. Somaya. 2003. System-on-a-chip integration in the semiconductor industry: industry structure and firm strategies. Industrial and Corporate Change 12(3): 545–576.

MAEI (Malaysian-American Electronics Industry). 1995. Malaysian-American Electronics Industry: Annual Survey 1994/1995. Mimeo. Kuala Lumpur, Malaysia: MAEI.

McKendrick, D.G., R.F. Doner, and S. Haggard. 2000. From Silicon Valley to Singapore : Location and Competitive Advantage in the Hard Disk Drive Industry. Stanford, Calif.: Stanford University Press.

Semiconductor Industry Association. 2003. International Technology Roadmap For Semiconductors: Design." Available online at *public.itrs.net/Files/2003ITRS/Home2003.htm as of April 19, 2005.*

Tilton, J.E. 1991. International Diffusion of Technology: The Case of Semiconductors. Washington D.C.: Brookings Institution.

Workshop Presentations

Implications of Offshoring for Engineering Management and Engineering Education

Anne Stevens

An initial hypothesis of the study group was that offshoring is a huge risk and a major issue for all of us in engineering in the United States. But, from another point of view, it is not as big a risk as all the hype makes it out to be. Which is it?

In terms of engineering, how much is actually being offshored? If we look back at history, we have been offshoring engineering, distribution, marketing, and selling of products all over the world since the early 1900s. Ford Motor Company has been operating in Argentina, Brazil, and Mexico for more than 90 years. General Motors, too, has been participating in business, including offshoring engineering, for just as long.

But what are the facts? One problem in understanding the issue is some confusion about both the binning of the data and the accuracy of the data. But what is important is what we do with what we know. What are the implications for engineering management? What are the issues? the opportunities? What should we do as an academy, as leaders in academia and industry and government?

What is the role of engineering managers? Does that role have to change? Should engineering managers be spending their time differently? Do they require new skills? If so, what are they? What is our plan to give engineering managers, in their 20s, 30s, 40s, 50s, or 60s, those skills?

The first issue is that in some countries, such as India, there is a "hire-ability" measure for one-year engineers, three-year engineers, and 10-year engineers. It was noted that 10-year engineers really wouldn't be considered for hiring because they would be too hierarchical. I do believe that in some of our industries—such as Ford and General Motors and some others—in some academic areas, and in the government, we do have an issue with hierarchical structure. To be competitive in the new world, we must know what is going on in the global economy. Old hierarchies cost money. They cost the morale of young employees. And they are inefficient in terms of bringing products or services to the marketplace or education to students. What is really important is how we transition these hierarchies into more nimble structures. There are already many nimble structures in existence today. Just look at firms like Google.

The second issue is workforce transition and adaptability. A 50-year-old engineer who is in or out of the market today for various reasons needs reentry plans. Another issue is the middle class. Companies like Ford Motor Company and General Motors are displacing tens of thousands of hourly workers, who were the foundation of the middle class in the United States. Their ancestors were farmers who came off the land and into the factory. Companies started paying $5.00 a day, which grew the domestic automotive market. The issue is about transitioning people in the prime of their lives who have amassed much education and experience. How can they—as well as other segments of the population—be mainstreamed back and contribute value to our economy or our educational system?

The third issue is allocating the right level of resources to the right areas of R&D. Where, what, when, who, how much? Bob Galvin, a keynote speaker for this workshop and retired Motorola CEO, is still very active and is working to realize important engineering activities. There are

---

Anne Stevens is chairman, president, and CEO of Carpenter Technology.

many more Galvins out there. The question is how we use their brainpower and their energy to address some of these issues.

As for engineering management, we have to retool, but not just retool. We must change our attitude in several ways. It is critical that we tap into the skills and inherent creativity and innovation in the millennial generation, the group born between 1977 and 1995, which is going to be bigger than the baby boomer generation. Estimates range from 60 to 74 million. So we must look at the millennial generation. These are the customers of the future, the politicians of the future, the students of today and of the future. Engineering management cannot expect to lead this generation with the same set of skills that were used to lead many of us.

The millennial generation will be a huge economic and social force in our world. Their perceptions and attitudes are different from those of previous generations. They are a connected generation. The Net is their primary source of news; the next closest source is radio. Compare that with older generations, which strongly prefer TV. Technology has always been part of their lives. Whereas prior generations see the Net as something to connect to, millennials see the Net as a way to connect to each other and the world.

Here are a few personal examples of the millennial generation.

First story: I have a granddaughter, Courtney Anne. She's nine years old, but the story I am going to tell you took place when she was four. Courtney Anne's other grandfather, my daughter's father-in-law, has a Ph.D. He is a very, very educated, savvy professional. He is retired now. Courtney Anne was sitting with him at the computer. I don't know what he was saying, but he was pretty upset. She looked at him and said, "What are you trying to do, Pop-pop?" He told her, and she said—at four years old—"let me show you how."

The second story took place when I was with Ford. A young man sent a letter to Jack Nasser who was then CEO of Ford Motor Company. "Jack, let me tell you something. I'm 13 years old, but I know I need to start thinking about things now. I know for sure in the future that what I want to be is CEO of Ford Motor Company. The thing I don't know is what I should be studying. What is my field when I go to get a university degree? Where do I start in Ford Motor Company when I come in, knowing that I want your job?"

Public Affairs was fascinated with this young gentleman. I got a phone call asking if I would please meet with him when he came into the company, which I did. He came in with his mother. They flew in from California. We set up a very interesting day for this young man. We put him in the design studio with the designers. We took him into the manufacturing facility. The individual engineers who met with him at the end of the day were absolutely shocked, because in the design area, this young man knew more about future design trends from what he had read than many of the designers did sitting at the tubes. When he was taken around the manufacturing plant by a superintendent, he asked more questions about the power train of the trucks than the superintendent could answer. All this information was learned from the Web!

These are the kinds of individuals we are going to be dealing with, as customers, employees, and students. They are different, and the rules have changed.

The last example: I was privileged to give the commencement speech this year for the engineering students at the University of Michigan. Before my address I had lunch with several of the best and brightest in the class. They had a lot of questions for me, and I had one question for them. My question was what don't I know about them, what is it that people in my generation don't understand. These very savvy, brilliant, top-of-the-class engineers all looked at me and said one thing: Facebook.

Technology like that can really network in the virtual world, as we offshore engineering. But if we bring these young people into organizations and try to evaluate their worth and their performance by whether they are in at 7:00 in the morning and whether they are in their seats until 7:00 at night—if we use those rules and many other current rules—we are not going to be able, as engineering management, to tap into their creativity and innovation. As leaders in engineering management, we need to increase the appeal of science, engineering, and technology to the millennial generation. Galvin's road map, with his grandson leading it in terms of defining technology management and areas of engineering and science for the future, is right on and brilliant.

And what about those 50-year-olds who have been displaced and the middle-class hourly workers or workers in other fields? Many of our institutions were really responsible for their predicament. In the Depression era people were encouraged to go to school, join a company like AT&T, Bell Telephone, or Ford and GM, and stay there until they retired. Today that model of job security has changed!

Many people in the generation that have been displaced basically thought they had signed a parental contract, with employers and with governments. "We are going to take care of you. In exchange for your loyalty, you'll have a retirement benefit that's going to be there, health care, and a job for 30 years." The rules have changed. Generation Xers know it, and the millennials know it. This is not going to be an issue for them. They have their skills, their abilities, their capability, and their networks. We need to figure out how to take these lessons from the Generation Xers and millennials and re-teach 50-year-olds and the middle class and hourly population—how to generate value back. All of us have some responsibility there.

Going back to the appeal of science, engineering, and technology, Galvin's example is one that works. Maybe it's "hairy, audacious goals," as put forth by Collins. Maybe it's higher salaries, as some speakers have said. Maybe it's figuring out the next version of the space program to reenergize youth, or maybe it's MacGyver on steroids on the Web. But whatever it is, we know we need to reengage the population to give us the talent we need for the future.

In summary, what about offshoring? First of all, it differs from industry to industry. It is a very exciting issue, and, at the end of the day, we should be optimistic. Offshoring for us is an opportunity. But we have to get at the roots of who we are as people living in the United States of America and bring forward what we have always been able to do best. The biggest risk we face is complacency. But, culturally, as a country, we have what it takes to succeed.

America was, is, and always will be the lion. America knows how to eat. The gazelles are out there, and we are going to figure out how to feed ourselves. The key to that is some of the things summarized during this conference. Academia, industry, and government all have major roles to play.

An Academic Perspective on the Globalization of Engineering

Charles M. Vest

When I was asked to speak at the beginning of this session, I pointed out that I do not know a great deal about the topic of offshoring and that everyone else in the room probably knows more. So, my purpose today is to provide some context along with my personal views before you begin your deep exploration of the topic.

My main message this morning is that I wish you well in sorting out, as the workshop subtitle says, the "facts" from the "myths," and in coming to a deeper understanding of the nature of globalization, particularly for engineers and engineering work. This understanding is very badly needed. Above all, we need guidance on how we as a nation can stop thinking about globalization as a set of awful problems and begin thinking of it as a set of opportunities for America and, indeed, for the world.

THE CURRENT SITUATION

Let's start with the basics. Where is the expertise going to be in the future? We know that natural shifts and changes are occurring in where engineers and scientists are being educated, although there is some debate about the accuracy and meaning of the statistics. Asia now accounts for a growing share of first science and engineering degrees (NSB, 2006). However, if we look at doctoral degrees, the picture is quite different, with Europe, as a collection of nations, ahead of both Asia and North America.

If we look at first degrees in science and engineering—the bachelor's level—country by country, the United States has a relatively constant production in natural science and engineering, and China's production is rising rapidly. If you look behind those facts and separate science from engineering, you see that the United States continues to lead in science degrees, but not in engineering degrees (Figures 1 and 2).

These figures have generated a great deal of debate over the last few years. I learned many years ago that if you want to write a paper that gains a high rank in the science citation indexes, you should make a very obvious error so that everybody will write papers correcting it, thereby driving up the ranking. The first draft of *Rising Above the Gathering Storm* (COSEPUP, 2007) quoted inaccurate statistics on Chinese and Indian degrees, contributing not only to debate over the report, but to a feeding frenzy of pundits focusing on this error. Vivek Wadhwa, who is speaking later today, more substantively pointed out the problems with those statistics (Wadhwa et al., 2005).

Indeed, I do not disagree very much with what Vivek has to say, which can be summarized in four major points. First, all degrees are not created equal. That is absolutely true. There is a significant disparity in quality among degrees from various institutions in the United States, and far greater disparity among degrees from institutions in places where the higher education system is developing very rapidly, such as China and India.

Second, in proportion to population, there is no obvious imbalance in the numbers of engineers being produced by the United States, China, and India. After all, the United States has less than 5 percent of the world's population, a percentage that is expected to drop going forward, so why should we expect our absolute number of engineering graduates to be as large as those in much larger countries?

Charles M. Vest is president of the National Academy of Engineering and President Emeritus of the Massachusetts Institute of Technology.

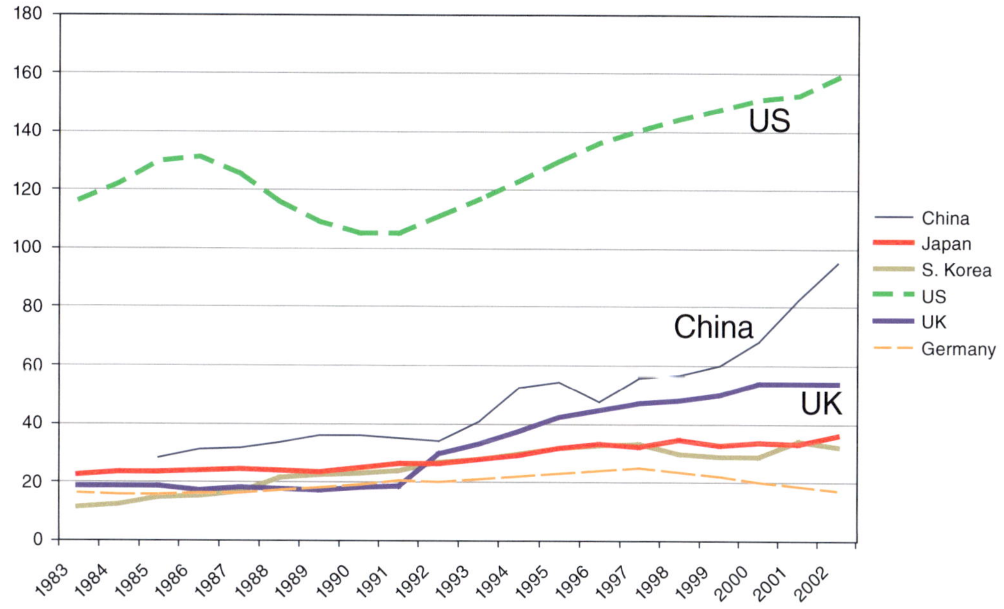

FIGURE 1 First natural science degrees. Source: NSB, 2006.

Third, salary trends and other labor market information indicate that there is no shortage of engineers in the United States. This may be true today, but here I must raise a caveat. Everyone I know who has looked at current labor market conditions and predicted what they mean for the future, especially in engineering, has fallen on their sword. I claim no particular wisdom about the "right" number of engineers we should be graduating. But I do think we have to be very careful about basing decisions on today's marketplace conditions. We really should focus on the future.

Finally, the fourth point is that our universities are better than those of China and India. I agree with that. I don't know if that situation is fleeting or will last forever. But I believe we should aim at making it last forever. In fact we currently have major advantages over the rest of the world in the way most of our institutions educate most of our engineers.

No matter how you look at it, there are mixed messages out there. Earlier this month, on October 12, 2006, *The New York Times* carried a story with the headline "Profit Rises 53% at Infosys, a Top Indian Outsourcing Company" (Rai, 2006). A mere five days later, on October 17, there was another headline in the *Times*, "Skills Gap Hurts Technology

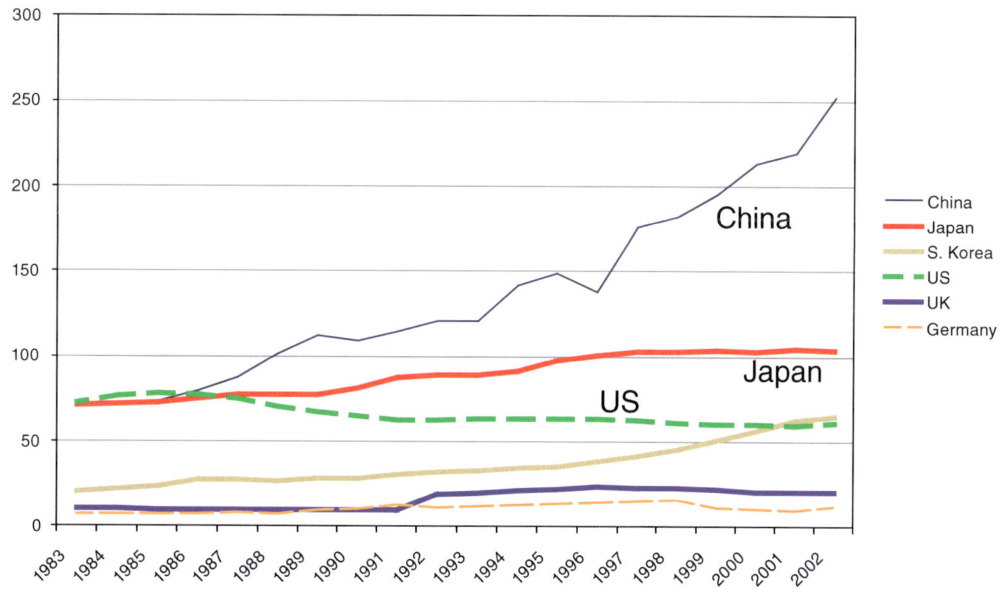

FIGURE 2 First engineering degrees. Source: NSB, 2006.

Boom in India" (Sengupta, 2006). One important point in the second story was that many software and service companies in India report that the quality of engineering and computer science education is sufficiently bad in their country that they consider only about one of every four engineering graduates employable.

Notwithstanding these mixed signals, I believe the broad trend in graduation numbers matters and should give us pause, for two very simple reasons. First, we must compete in the global economy, while simultaneously maintaining our American standard of living. This is a daunting challenge. Second, I believe that prospering in the Knowledge Age requires people with knowledge, much of it relatively deep in the areas of science and engineering. So we must closely monitor trends and not base decisions just on our current position or where we have been. The important question is what happens to the next generation. What should we do now to prepare young people for their personal and professional lives in the future?

THE IMPORTANCE OF LOCATION

Much of the debate in the popular press and in politics, and certainly in our own profession, has to do with location. For example, if we think very broadly about industrial R&D and innovation and the importance of location, there appear to be two camps out there, as one might expect. The first says, fundamentally, that location does not matter any more and is going to matter even less in the future. This view has been popularized and communicated extremely effectively in *The World Is Flat* by Thomas Friedman (2005). His basic view is that the Berlin Wall came down in 1989, but Microsoft Windows went up; and one day we woke up and found that $1.5 trillion worth of optical fiber had connected all of us around the world. The interesting story behind that, of course, is that most of the businesses that laid the fiber failed, and some of their managers are sitting in jail. Nonetheless, we ended up connected in a way that we could never have imagined. Friedman goes on to say that globalization has accidentally made Beijing, Bangalore, and Bethesda next-door neighbors, and many jobs are now just a mouse click away from anywhere.

There is another camp that tends to take the position that location does matter. While I am not sure Michael Porter would appreciate me viewing him as a representative of one end of this discussion, because he is a very broad and thoughtful person, he has built a very powerful case over the years about the importance of regional innovation clusters in the United States and elsewhere. These clusters are groupings of industries related to one another, and their proximity and interaction leads to an accumulation of human capital, expertise, synergy, communication, and so forth.

The importance of proximity to universities for small companies and corporate laboratories is well established. Not only do universities often spawn new enterprises, but they also tend to play a very important role in bringing people together, in effect forming a centroid for the boiling and perking that leads to the development of small, technology-based companies.

Another factor that is not mentioned so much is that venture capitalists often prefer working in a small region where they know everybody and can stay close to the companies they invest in as they build their networks.

Manufacturing Migration

My own view is strictly middle of the road, namely, that both camps are correct in that there are some aspects of globalization that make location less important and some that make it more important. I will start with one obvious trend, what I will call "manufacturing migration," or the idea that many industries, particularly industries that manufacture products, may first develop in the United States, but then migrate to, say, Taiwan and then, perhaps, to Korea, to China, to Vietnam—and who knows where next? One of the questions before you is whether this migration is inevitable. What are its pluses and minuses?

Whether or not migration is inevitable—and I suspect that it is—it is serious business. Just a few factoids here (COSEPUP, 2007; Palmisano, 2006):

- Between 2000 and 2003, foreign firms are estimated to have built 60,000 manufacturing plants in China.
- In 2004, chemical companies closed 70 facilities in the United States and tagged 40 more for shutdown.
- Of the 120 major chemical plants currently under construction, at least as of about two years ago, one was in the United States, and 50 were in China.

So good, bad, or indifferent, manufacturing migration is happening.

What does this mean for the quality and quantity of jobs in the United States? What are we really losing and gaining? In the brief period from the beginning of 2000 to the end of 2002, it is estimated that about 400,000 jobs in IT manufacturing were lost in the United States (PCAST, 2004). While overall employment in U.S. manufacturing declined by 6 percent between 1997 and 2001, employment in computer manufacturing declined by 20 percent.

Changes in Innovation and R&D

Consider the evolution of U.S. corporate innovation and R&D over the past several decades. We might think of the 1970s as the golden age of corporate research laboratories, some of which still exist, generally in rather different forms. The key point here is that corporate R&D labs of that era not only generated new ideas for their own companies, but also contributed enormously to the science and engineering commons by virtue of publications, participation in meetings, and collaborations with universities and each other.

In the 1980s, due to the near-death experiences of many segments of U.S. industry, the R&D function was dramatically transformed and was largely absorbed into product development. This was necessary, because it enabled a number of our companies not only to survive, but also to prosper, at least for a period of time. But this trend did represent a change in the U.S. innovation landscape, as corporate labs became less active as sources of non-proprietary ideas.

In the 1990s, of course, having largely turned away from longer term R&D, although there are obvious exceptions here and there, many of our large companies began acquiring their innovation rather than carrying it onboard, by, for example, purchasing high-tech start-ups.

One indicator of this trend is provided by Robert Lucky, who plotted the affiliations of authors of papers published in the *IEEE Transactions on Communications* by percentage (Figure 3). In 1970, the vast majority of papers were actually written by computer scientists and engineers working for U.S. companies, with only a small percentage authored by academics. This has shifted and changed in two directions. The percentage of authors from both U.S. and non-U.S. industry, at least in this field, has declined to almost nothing these days. This has been accompanied by a rise in authorship by academics, with academics outside the United States now slightly ahead of U.S. academics.

So, with migration between countries and shifts in the roles of companies and universities, the innovation landscape is changing. The question is why. There are some very basic reasons—economics and wage rates, the availability of the Internet and the World Wide Web, and tax and trade policies. However, the fundamental reason is that innovators and the innovation system are just reacting to the increased speed and complexity of business, technology, and markets.

Figure 4 makes this point. If you go back to the introduction of the automobile around the beginning of the twentieth century, it took essentially a lifetime from the first marketing of this product to the point at which it had reached 25 percent of the U.S. population. For the telephone and then the radio, it took something on the order of a professional career for a similar diffusion. In the case of the World Wide Web, it took, astoundingly, only about eight years to reach a quarter of the U.S. population. So things certainly are speeding up.

Stuart Feldman at IBM put together a chart confirming something we all probably know (Figure 5). In 1800, virtually everybody in the United States worked in agriculture, a sector that now accounts for an almost immeasurably small share of our employment. Manufacturing rose but has now declined, being replaced very rapidly by services, particularly services based on information technology.

Population and development are also shifting regularly. A paper from Goldman Sachs a year or two ago estimated, in just one decade, about 80 percent of the world's middle-income consumers will be living in nations that we currently consider to be outside the industrialized world (Wilson and Purushothaman, 2003).

In addition, consider two facts. First, people everywhere in the world are smart and capable, and when given the opportunity, they will do amazing things. Second, the Internet and the World Wide Web are major democratizing forces that have opened up opportunities and possibilities for people who may not have had them before.

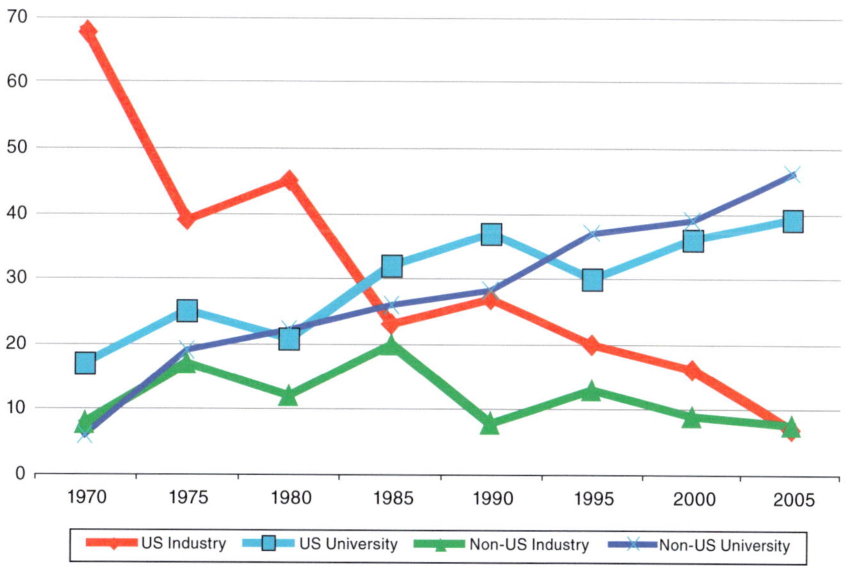

FIGURE 3 Percent authorship of papers in *IEEE Transactions on Communications*. Source: Lucky, 2006. Reprinted with permission.

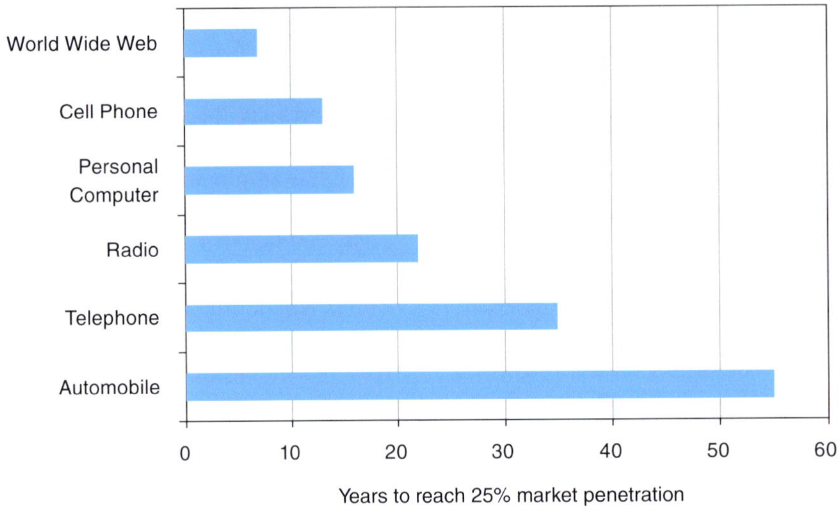

FIGURE 4 Why everyone is in a hurry. Source: Charles M. Vest (compiled from NSB, 2006).

New Business Models

If we add all these trends together, we can see why people are thinking about new models of conducting business. One example is the concept of open innovation that has been popularized by Henry Chesbrough of the Harvard Business School. He points out that companies increasingly find they have to reach beyond their own boundaries—perhaps beyond their own countries, perhaps even into competing organizations—to find the people who do particular things best, where the best ideas originate. Companies have to reach out, grab those people, and somehow bring them together. This has stimulated debate in the business world because of issues about licensing, partnering, joint-venturing, and so forth. But, clearly, some form of openness is developing in our innovation system.

More recently—and more radically—Sam Palmisano, the CEO of IBM, traces the history of corporations over the last two centuries and asserts that we are now shifting away from the model of the multinational corporation to what he calls the "globally integrated enterprise." That is increasingly the way his company and many others are being run. Globally integrated enterprises are driven by globally shared technologies and standards and linked by information technology, and their focus is shifting from products to production. New borderless strategies, management, and operations for integrated production and value delivery are being developed.

So life, and innovation, today are not simple. Take the recent example of Sony and Toshiba in Japan, which excel at conceiving, designing, and building computer games for young people. IBM, based in the United States, excels at designing and manufacturing sophisticated chips. Those companies got together and, in Austin, Texas, developed new processors designed to drive the next generation of computer games. Then, a few weeks ago, Los Alamos National Laboratory ordered what will probably be the world's largest supercomputer based on these chips, which were designed for the gaming industry.

Those of you who have a few gray hairs will remember the furor a decade or two ago in this country when someone

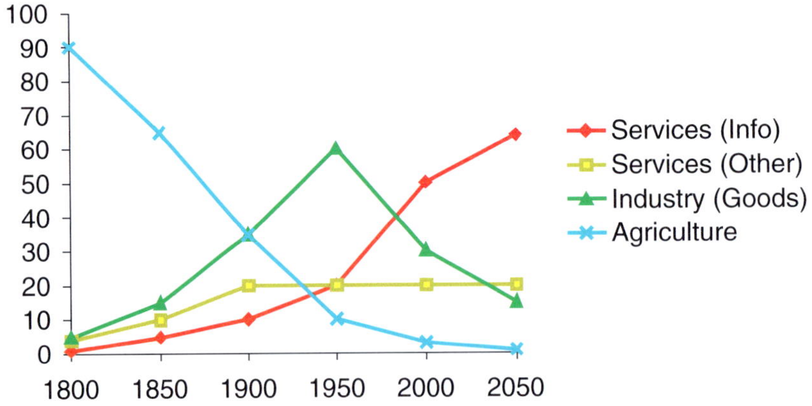

FIGURE 5 Percentage of U.S. employment by sector, history and projection. Source: Stuart Feldman, 2005. Reprinted with permission.

attempted to buy a Japanese Fujitsu supercomputer to use for government-sponsored research. We have come a long way when the Japanese game-chip industry ends up driving Los Alamos's most advanced computer.

THE GLOBALIZATION OF HIGHER EDUCATION

What does this all this mean for education? Let's look at how the research university has evolved over time and how it has globalized. We begin in the nineteenth century with Humboldt University in Germany, which developed the model of the research university as we know it. That model was transplanted in the United States with Humboldt rather directly inspiring Johns Hopkins University. In the second half of the nineteenth century and first half of the twentieth century, Berkeley, Stanford, Michigan, Illinois, and others began to adopt, and adapt, the research university model. Institutions such as MIT, RPI, Caltech, and so forth took the model in a somewhat different direction.

Then, in the 1960s and 1970s, this model was literally transplanted into India through the founding of the IITs (Indian Institutes of Technology). In my view, the development of IITs over the past 50 years is one of the most amazing success stories in the world. In 2006, the European Union started to establish EIT, the European Institute of Technology. We do not know how the EIT will develop, but it is interesting to note that Germany and, indeed, Europe are in the process of re-importing the very research university model they first sent to us.

So that was Research University Globalization, Part I. Part II encompasses three trends. First, some institutions are establishing a physical presence in other countries. Many U.S. universities are either opening or have opened campuses abroad, primarily to give their students a different perspective and different experiences. In addition, laboratories, research facilities, medical schools, and other operations are being opened in Singapore, the Middle East, and elsewhere. Some of these are already being dissolved!

A second trend is that strategic alliances are being built among universities around the world. This is an old tradition in basic sciences, such as physics, but somewhat newer in a lot of other areas, including engineering. The Cambridge-MIT Alliance is a good example.

The third, and perhaps most interesting and exciting trend, is virtual presence, which tends to take two different forms. There is a big argument going on about which is best, although there is probably room for both. One is distance education, both synchronous education—for example, the MIT-Singapore link using Internet2 to conduct classes that we conceive of as occurring in a big room, half of which is in Cambridge and half in Singapore—and asynchronous education through various Web-based tools.

The other form of virtual presence is the open-content movement, which I believe represents the emergence of a new meta-university, a platform on which institutions all around the world can share teaching materials, information, methodologies, and so forth. Educators can pick and choose and shape the best material from everywhere and integrate it in ways that fit the local context. In addition, there is a growing number of experiments out there in telepresence, the ability to operate laboratories from a distance, particularly from poorer parts of the world, running expensive educational laboratory equipment in wealthier states.

The next development I will call Research University Globalization, Part III. A lot of groups are beginning to work together and think about the best way to educate and prepare our engineers for the coming century. One example is a study sponsored by Continental AG that has been going on for about a year now called Global Engineering Excellence: Educating Engineers for the 21st Century. The study involves faculty members from ETH Zurich, Georgia Tech, MIT, Shanghai Jiao Tong University, Technical University of Darmstadt, Tsinghua University, E.P.U. Sao Paulo, and the University of Tokyo. Thus excellent minds representing several continents, several approaches, and some of the best engineering schools in the world can think together about the nature of the curriculum and the experience we owe our students.

To summarize, a number of things are going on in the globalization of higher education. I do not think it is a matter of which approach wins, but we will see which approaches succeed—propagation/emulation, overseas campuses and facilities, multinational alliances, distance education, the meta-university (or, as those in industry prefer to call it, "digital convergence"), or plain old-fashioned redefining of our curricula and goals for globalization.

LOOMING PROBLEMS

I leave you with the thought of some real policy clouds looming over globalization. First, there are serious unresolved issues about the control of "deemed exports." A deemed export, of course, means that an export license is required when sensitive information is shared with a non-U.S. citizen in a research context. We have to resolve this issue. Second, although great improvements have been made in visa policies, there are still problems with the issuance of visas by the U.S. government, particularly for short-term visitors, scholars, participants in joint research and technical meetings, and so forth.

A third looming issue that must be approached carefully is overdependence on foreign graduate students. I think one of the greatest absolute strengths of this country is that wonderful, bright young men and women come to us from all over the world, and I am fully behind as much openness as we can have. At the same time, we must educate more U.S. citizens in science and engineering and encourage them to contribute at advanced levels.

CONCLUSION

I believe we are the most innovative nation on the planet, and we still have the best research universities in the world. We are still the king of the hill in R&D in most fields. We have these comparative advantages—a strong science and technology base and a free-market economy built on a substrate of democracy and freedom.

But I leave you with this paranoid thought. The enemy I fear most is complacency. We have work to do in this country. I very much look forward to hearing your thoughts and what you can learn and teach us all about the real, evolving nature of globalization.

REFERENCES

COSEPUP (Committee on Science, Engineering, and Public Policy). 2007. Rising Above the Gathering Storm: Energizing and Employing America for a Brighter Economic Future. Washington, D.C.: The National Academies Press.

Feldman, S. 2005. Presentation at Carnegie-Mellon University, June 29, 2005. IBM Research.

Friedman, T. 2005. The World Is Flat: A Brief History of the Twenty-First Century. New York: Farrar, Straus and Giroux.

Lucky, R.C. 2006. Personal Communication.

NSB (National Science Board). 2006. Science and Engineering Indicators. Arlington, Va.: National Science Foundation.

Palmisano, S.J. 2006. The globally integrated enterprise. Foreign Affairs 85(3): 127–136.

PCAST (President's Council of Advisors on Science and Technology). 2004. Sustaining the Nation's Innovation Ecosystems: Report on Information Technology Manufacturing and Competitiveness. Available online at *http://www.ostp.gov/PCAST/FINALPCASTITManuf%20Report Package.pdf*.

Rai, S. 2006. Profit Rises 53% at Infosys, a Top Indian Outsourcing Company. New York Times, October 12.

Sengupta, S. 2006. Skills Gap Hurts Technology Boom in India. New York Times, October 17.

Wadhwa, V., G. Gereffi, B. Rissing, K. Kalakuntla, S, Cheong, Q. Weng, and N. Lingamneni. 2005. Framing the Engineering Outsourcing Debate: Placing the United States on a Level Playing Field with China and India. Pratt School of Engineering, Duke University. Available online at *http://memp.pratt.duke.edu/downloads/duke_outsourcing_2005.pdf*.

Wilson, D., and R. Purushothaman. 2003. Dreaming with BRICS: The Path to 2050. Global Economics Paper 99. New York: Goldman Sachs. Available online at *http://www2.goldmansachs.com/insight/research/reports/99.pdf*.

Keynote Talk on the Globalization of Engineering

Robert Galvin

I have with me *Mayflower: A Story of Courage, Community, and War*, a book by Nathaniel Philbrick that takes us back 350 years. It's a pleasant book if you like reading light history. In it, Philbrick tells a story about our predecessors that was almost prescient about the reason for this meeting. Here are the three sentences most relevant to our subject:

> Governor Bradford was disheartened when he learned that Brewster, Winslow, Myles Standish, and John Alden had left Plymouth and moved further north. He was particularly disheartened because the new towns being established there were doing better than Plymouth, which had fallen on hard times. Bradford noted that the problem was mainly the shallow anchorage in Plymouth harbor, which doomed it to eventually becoming the poorest of the New England colonies.

New England has been at the center of our industrial history. Historically, it was the center of great development, and it still is. People like Chuck Vest and others have continually renewed that community and will probably do so for the next 100 years. But look what's happened in the country over the last 75 to 100 years. Industry that was centered in New England moved, first a little bit, to the Midwest, then farther west, some to the Southwest, some to the South. With each move, there were dislocations—disruptions to our comfort zone—and many people were terribly upset. Why couldn't everyone just stay where they were and things remain as they used to be?

In effect, the Pilgrims began "onshoring" by moving north and inland from Plymouth. Offshoring—a fascinating new term—is a significant word, for it represents opportunity and movement. But it does not spell "inevitability." My message is, among other things, that we can still do things of great significance in the 50 states of the United States. Chuck, in effect, challenged us to do so at the end of his excellent talk. Change is an old story in the United States that will surely continue. But our mobility alone still offers expansive possibilities and will continue to be a vitalizing phenomenon.

We can do things to affect the offshoring situation. I started in business in 1940 and worked for exactly 50 years in our company. During that time, in an honorable way, we changed the rules of the game we played, its governance and industry affairs, in all kinds of ways, thereby influencing our neighbors. My message today is that we can do many things, if we are courageous enough, if our backbone is strong enough, to change the rules as we go along. In particular, we can establish rules for the new things we should do, which I will illustrate in just a moment. We can influence policy on foreign trade—what goes out and what comes in. I spent a lot of time on that in the past, and we showed that we could increase our sales of U.S.-made products in overseas markets. One of the largest producers of cell phones is our plant in Arlington Heights, Illinois, which is as competitive as any plant in the world.

We can, and should, invest *integratively*. All the things Chuck talked about had a component of significant investment. If we put something in Scotland, it should be in harmony with what we have in Austin. It is very satisfying for me to observe that, when the brilliant people at SEMATECH encountered problems, they could find a solution in one of our laboratories—in Scotland or Toulouse or Angers or Germany or China.

People who are multinational in their living are coopera-

Robert Galvin is Chairman Emeritus of Motorola Inc.

tive. They do things in a centralized way with family institutions. Motorola began offshoring to improve its service to customers, and most of our thinking and planning has to do with serving our customers. With that emphasis, we come up with some rather conventional but very bright observations. Our goal then becomes assembling a team of people to serve that customer. If that means we need a factory in Toulouse, France, then that's where we go. For the factory in Toulouse, France, to be successful, we need a hometown boy (someone from Toulouse, or at least from France) to run it. There are a lot of fundamentals involved.

To ensure that we become a very significant servant to our customers, we do what we have to do anywhere in the world. There are places that are still untouched and at least two continents in this world that are virtually unsettled. There is no middle class in South America, Africa, or most of China. Nine hundred million people live way below the level of the middle class. We have an anthropological responsibility over the next 50 to 100 years to change human relations so that there will be significant middle classes in the other half of the world. When they are tuned in to the opportunities available, they will become, first, very significant customers, and then, very good servants to customers (i.e., competitors).

Here at home we can do things better, too. We shouldn't be afraid to move around in our own country. I remember when Mark Shepherd from Texas Instruments called me and said, "I hear you're coming to Austin. Don't you realize all the problems are here? You don't want to do that." But I knew Austin was a good place to go, that it would be best for our people. It was where they wanted to live, where their families could prosper, where the team worked best together. With enough flexibility, you can move your operations around in this country—not just offshore. We must move to create change, if that's the best way to pull our people together.

I am now going to take advantage of this distinguished audience and tell you about three things I'm doing privately that are about to become public, and I think they will have significant consequences in our country. Chuck has said we have to be bold, make changes, do things that will make a big difference. This means not just designing next year's product line. I offer three examples where teams of experts like you might do even better than I.

The Galvin Electricity Project

The first is a project to revolutionize and "re-found" the electric power industry. We waste 40 percent of our energy just delivering electricity by wire around the country. Most of these systems break down somewhere two or three times a year. But this can be changed. I have felt for some years that there should never be an outage, that we should never be disadvantaged because things go wrong in the electric power system. Over the last couple of years I have met with hundreds of experts like you and put together a plan that will be made public sometime after the first of the year. The plan is already on the Web, but it will be publicized more effectively in the first part of next year.

The plan is essentially a distributed system that involves mostly onsite generation, thus making delivery unnecessary. I'm not going to describe it in detail, but it has already been through an extensive review and assessment process, and we are now moving to prototyping. This very significant digital system—with automation, instrumentation, self-correction, new forms of storage, et cetera—is part of the Galvin Electricity Project.

As a matter of fact, this is going to be a business of interest to those of you who are entrepreneurs. It's not my business, and I'm not investing in it. I am investing in the ideas and then opening the business to everyone on the open-market. People can start a business in their region or their town or go national if they want. We already have quite a few active thinkers and investors ready to move ahead.

Making this kind of change will take a couple of decades to become manifest in the country. With the Galvin Electricity Project, we are well on the way to completely changing the way the electricity industry provides power. The change will require that many new engineers do many different things—in the United States. This low-cost system will bring great benefits to our citizens and increase the efficiencies of manufacturing and of services.

The Galvin Project on Eliminating Congestion

The second project, which is called the Galvin Project on Eliminating Congestion—and I do mean "eliminate"—is also moving ahead. This operation was born of my personal conviction that all cities will die by 2050 unless we make drastic changes. The project is not public yet, in the sense of having government step in and help us, but I have convinced a large number of experts in the traffic-management field—technical, business, model systems, et cetera, and some public officials—of this. It's a cardiovascular problem. The arteries are clogging up and will be clogged up completely soon, creating total gridlock. This may sound heretical, but many experts now agree and will be publicizing this prediction.

As a consequence of congestion, property values will be severely degraded. Things being built in Chicago's downtown or around the Chicago region will be worthless in 45 years because people won't be able to get to them. There will be no accessibility. Every ordinary citizen knows this, although some experts say, "Oh, no, it won't be that way." But ask your neighbor's wife. She knows it. And your neighbor who has trouble getting to work knows it.

We can avoid this tragedy through a surgical process. People are already thinking about and designing what I call "Lego sets," that is, overpasses that can be installed, in just a few weeks, in very congested intersections with difficult traffic patterns, enabling traffic to pass over the congested area.

This will require significant new engineering contributions from the construction industry. Our cities will be networks of tunnels. Tunnels will crisscross Chicago, New York, Los Angeles, Albuquerque, Beijing, et cetera.

We are not making a lot of public fuss about how we are going to "popularize" this concept. We are going to convince our friends in China first, because the Chinese have the authority to do this in their cities. That authority is not as readily available in our democratic society. My son will be giving a presentation in a few weeks when he goes to China for a meeting at a university where he is a trustee. So, we will be publishing our first document in Mandarin and giving it to the Chinese before Christmas. The document will explain how they can eliminate congestion in about 120 cities with large numbers of tunnels.

The business model will be a toll business, and we expect there will be tremendous competition internationally to win, in effect, the right to collect tolls in a given section or a given city. I imagine that there will also be a dramatic number of technical achievements as people learn how to build these overpasses and tunnels, much like what happened with the introduction of the cellular telephone. When we announced the cellular telephone about 25 years ago, AT&T wasn't ready for the change. Neither were the Japanese. Almost every husband we ever talked to said, "Well, that's a very nice thing to have. I think I might need one in my business. But I won't let my wife have one, and I'm not going to let my children have them." But who has cellular telephones today, at almost no cost?

The people in this room, and your engineering associates, have a great talent—the ability to take the essence of an idea and refine it. In the process, costs will go way down, and services will become remarkably reliable. By 2030, a new transportation system will be evident, a system that was formed well before that. I assure you that if it's not done by 2035, the new Trump Building in Chicago will begin losing value. But I think we can convince the American people, and the American leadership, that they must go to a radically new system to prevent the death of cities.

Science Road Maps

Finally, I want to bring up an issue I have talked about often but have never been able to sell, although I think the concept is fundamental. About 35 or 40 years ago, after one of many days per week spent in our laboratories—I frequently spent time with our bright, young people, who were always giving me ideas that had never gotten to the top of their divisions—I said, "We will have road maps." And I drew an XY chart and put some lines on it. Even our brightest, top people who happened to be sitting in that room that afternoon couldn't grasp what I had in mind.

I didn't have a clear idea of how to present my idea, but I knew what the end objective was. I told them I'd be back in six months for the first meeting on road maps and that they had better have a damn good story to tell about their plans for the future, in immense technical detail, or there would be a radical change in the organization. Three people picked up on the idea and designed engineering road maps for our company that led to dazzling results in our product-development programs for more than three decades.

We discussed the idea of industry road maps with Ian Ross, who was then heading a commission in D.C. studying the semiconductor industry. Finally, I convinced him to support the concept of engineering road maps for that industry. We worked together to develop road maps on pre-competitive ideas, all the ideas that engineers could come up with. Today, I think we are in the 9th or 10th edition of biannual technology road maps for the industry that have done a giant job, particularly at IBM, which was one of the companies that helped us develop the road maps.

Road maps for technical management are far more useful than many science and engineering people realize. I know some top science people rather intimately, which gave me insights as to what they were thinking. I told them there should be science road maps—a chemistry road map, a physics road map, and so forth. About 10 years ago I saw Dan Goldin at a party one evening, and I asked him, "What does NASA think about road maps?" He said, "We've got the most distinguished road map on biology you can imagine." I asked, "Why biology?" He said, "We have to figure out where we're going. We have to know it to the essence." So I sent our team down to see what the NASA biology road map looked like.

But I have failed to convince laboratories, universities, this distinguished institution, and the overall National Academies, to adopt and promote science road maps. A few people have tried them, but, like many new ideas, they get lost if they are not directed by an enthusiastic head person. I was able to do it in my company, where I was at every technology road map meeting for 10 years.

This time around I'm going to succeed, and I'll tell you how. I have discussed road maps extensively at home with one of my grandsons, a sophomore at Harvard studying physics. After he and Leon Lederman (a wonderful man, very intelligent, who looks down his nose at my ideas on science road maps) and I had spent a number of hours together talking six or eight weeks prior to this meeting, my grandson came to me and said, "Grandpa, we've started these conversations by you saying the first thing we must do is talk about how we think. We have to know how to think in a process way about creativity, and we must never think negatively about an idea until time for judgment comes. I have an idea, and I expect you to accept it." I said, "I do." He said, "How would you like me to lead the science road map parade?" I said, "William, that's a statement of genius." I called Leon Lederman and asked him what he thought of the idea. He said, "I'll work with him." Now when I talk to people who run great institutions, they say, "Oh, my God, we have to get a couple of our kids on this road map committee."

To write a road map, you have to bring together 100 or

150 people in a big room, a big ballroom someplace, and for two days, just put out ideas. One idea begets the next idea, and so forth, and off we go. So starting from his position as a "matriculator," William is going to start recruiting a friend at Caltech. We have contacts at Texas, and I wrote to Donna Shalala at Miami a couple of days ago, because she's strong on women scientists. We are going to recruit 40 or 50 young people at the regular college level—we are not going for postdocs yet—and let them start to write physics road maps. I think they will have a pretty good idea as to what that road map should be by the end of this year.

We are not thinking in terms of urgency. I see this as a program that will grow gradually over 10 years. About two or three years from now, the students who are active in my road map program will be learning more from their road mapping experience than their courses for general matriculation.

I have also taken the idea to Jiang Zemin in China and to top people in Israel. I said, "Why don't you embarrass the United States? Why don't you write the road map?" But the Israelis have been muddling around with the idea. But now, through our youth, we are going to excite a science road map program, eventually with international membership in our road map workshops.

Conclusion

We are going to accomplish all three of my blockbuster changes. We are on the cusp of taking on the first two, changes in energy distribution and the elimination of congestion. In three or four years, someone like a Chuck Vest will be saying, "Let me tell you about that program with the college kids writing road maps. They're actually making some progress."

We will draw the geniuses back in. I talked to Jim Cronin, a Nobel scientist working in Argentina, eventually in Utah, on the Pierre Auger Project, cosmic rays, et cetera. He said, "I don't understand what that's all about, but I wish I was under 75. I'd like to be a part of that team." So I think we are going to excite people about science road maps also.

As I said before, these are things we can do in America. You don't have to go offshore, but the things you do will have a tremendous impact offshore. Great things can be done with your next project. I just have three ideas, and I'm pursuing them on my own, recruiting people to meet for extended periods of time to come up with practical ways to make them happen. Where are your ideas, the fourth or the fifth or the ninth? They would be so welcome! You have the technical talent to lead the way. I respectfully suggest that tremendous things could be done here.

Let me end with one odd comment that's not obviously related to this agenda. Bill Spencer and a few marvelous academics were on a committee that put together the Galvin Report, at the invitation of the federal government, on how the U.S. Department of Energy laboratories could be more effective. I came up with a heretical idea that government laboratories should be privatized (none of the other members was too keen on it, but I had the authorship, so I got it into the report).

The details are simple and not worthy of comment here, but that's the kind of thing that has to happen to bring America back to greatness. We have to privatize the laboratories. IBM can't afford a total laboratory. Nor can AT&T. But we could figure out a way to privatize those 10 government laboratories. The idea is still being talked about in Washington, but the current Congress hasn't got the stomach for those kinds of things.

Nevertheless, these are the kinds of things I have been changing for over 50 years. I changed the constitution of Ireland and the economy of Israel and moved them away from socialism. I gave Jiang Zemin an idea that had to be implemented in China to keep it from failing. He bought it, and brought my company in as a private-sector investor. For a long time, we were the largest foreign investor in China. With Bob Strauss and Akio Morita, I opened the Japanese market. It takes only two or three people to do these things. The minority always has to push things through.

The people in our industries can think great things and do great things.

Software-Related Offshoring[1]

Alfred Z. Spector

I'll begin with four rather simple observations about software. Although they are simple, I believe they are important to any discussion of software-related offshoring:

1. There is a *global* leveling of opportunity in the software field.
2. The software field is very large, and its subcomponents include many diverse practices and skills.
3. Terrific opportunities for innovation in software remain, and demand for software should increase as prices decline and innovation continues. Economists would say there is both high-price and high-innovation elasticity of demand.
4. As applications of information technology continue to increase rapidly, they provide increasing opportunities and reinforce the centrality of software to science, engineering, and, indeed, society at large.

To illustrate some of these points, I will continue with a brief description of some offshoring activities by IBM, where I was recently vice president of strategy and technology for the IBM Software Group. Then I will talk about the implications of offshoring for growth in the software field, the impact of diversity in the software industry on the potential competitiveness of different populations, the importance of the software industry to our society, and—finally—that the United States cannot expect to dominate the software industry they way it once did.

I will then propose three possible scenarios for the future. In Scenario 1, although offshoring continues, the United States retains its dominant role in many segments of the software field as a result of the differentiation I mentioned above. In Scenario 3, I present a worrisome picture of accelerating migration of software jobs overseas as talent in the United States dries up, perhaps the result of an expectation-driven downward spiral. In Scenario 2, I describe an intermediate situation in which many traditional programming-related jobs migrate, but high-value growth both in the field and around its periphery is sufficient to sustain the industry in the United States.

Finally, my primary conclusion is that we must attend to the talent available in the U.S. labor pool. It seems self-evident that unless we have a sufficient number of enormously talented individuals, whether U.S.-born or immigrants, who have been given the best training in the world, we will gradually drift toward Scenario 3. I also briefly mention that we must vigilantly protect the laws and economic structures that encourage continuing investment in both new research and novel businesses that generate the opportunities on which we will depend.

MESSAGE 1: WORLDWIDE LEVELING

To introduce my first observation about the global leveling of opportunity, I'll tell you a personal story. In the summer of 1973, at the end of my freshman year at Harvard College, I got a programming job at Harvard. I recall vividly one particular morning, after I had pried myself out of

Alfred A. Spector is a consultant and former vice president of strategy and technology of IBM Software Group.

[1]The views expressed herein are the author's and not those of institutions with which he is affiliated.

bed early in the morning and headed off to the then-extant Harvard Aiken Computation Laboratory to write and debug code, I found myself the sole user of the Harvard PDP-10 research time-sharing computer. This was my reward for getting up before 5:00 a.m.

The PDP-10 ran at roughly 400,000 instructions per second. It had less than half a megabyte of memory and a few megabytes of disk storage; we used small magnetic tapes, called DEC tapes, for all long-term storage. Even in that early era, the Harvard PDP-10 was networked with a few other computers via the ARPANET, the predecessor of the Internet. In fact, in my work I regularly accessed computers at MIT and Carnegie Mellon. At the time, I didn't know how much the PDP-10 cost. However, I did a little research for this presentation, and I believe that it would have cost about $2 million in today's dollars.

In addition to pursuing my debugging work, I remember my mind wandering and contemplating my career options. At the time, I was considering going into economics and journalism, but I was also thinking about computer science. That early morning I do remember explicitly thinking about the comparative advantage I had as a student at a U.S. university, capitalized by *my very own PDP-10* (at least at 5:00 a.m.), and thinking about all the folks in Europe who had minimal computational access (note that Europeans used to do much less hands-on computer science because of this). At the time, I never even considered India and China as having any software capabilities. I believe my unquestionable comparative advantages impacted my career choice.

Almost 35 years later, the contrast is clear. Modern computation and networking are four to five orders of magnitude better, cheaper, and more ubiquitous than they were then. And most necessary information is on the Web. Take just one example: MIT's plan to put most of its instructional materials on the Worldwide Web. Even machine translation is making some progress making information available in multiple languages. *Thus leveling of opportunity is undeniable.*

This leads me back to my first observation. A U.S. student going into the field of computer science today does not have as great a comparative advantage as a student even 10 years ago. This is not a reason to avoid computer science and software, but it is important to recognize that the U.S. advantage has decreased.

MESSAGE 2: A VERY DIVERSE FIELD OF ENDEAVOR

The second point I want to make is that "software" or "information technology" is not one large, coherent, aggregated profession, but is instead a very diverse field. This is partly because it is a very big field—more than a trillion dollars are spent on software worldwide (in aggregate). To illustrate this diversity we can look at four different "cuts" across the variety of activities in and around software (Figure 1).

The first cut considers software from the vantage point of software production. Here are some aspects of the process, although not all of the elements I've listed are applicable to all software production:

- conceptual work as a basis for deciding what can and should be done
- competitive analysis to determine how to succeed in the market
- work on requirements as a basis for making a formal decision about what a program must do
- various perspectives for considering the design of a system:
  –the human interface
  –the security of operation
  –the robustness of operation in the presence of faults
  –other factors
- development of the high-level design of the major modules and information structure of a program
- the low-level design of individual modules
- coding
- porting to alternative platforms
- formal and informal verification
- testing of components, modules, and systems
- evaluation and tuning of performance
- intellectual property protection and licensing
- development of documentation/information and national language support
- packaging and delivery
- project management

Undoubtedly, many more activities could be added to this list.

The second cut across the field, the application domain of software, influences development processes in many ways. Systems software (e.g., operating systems, database management systems, server infrastructure, middleware) that run continuously have different requirements, such as robustness and scalability, than tools that are executed and re-executed periodically. Packaged applications that are sold to numerous customers have different requirements (e.g., significant expertise requirements in the huge number of potential application domains) from programming tools, although they must still be of use to a variety of customers within a particular industry or problem domain. Custom applications for one or a few uses or customers may be considerably easier to develop because they require less generality, and there is, therefore, less of the combinatorial explosion that makes packaged software so expensive. These different types of applications also require significantly different production methodologies.

Even in each of these application areas, there are many approaches to developing software:

- The traditional waterfall method is a common baseline,

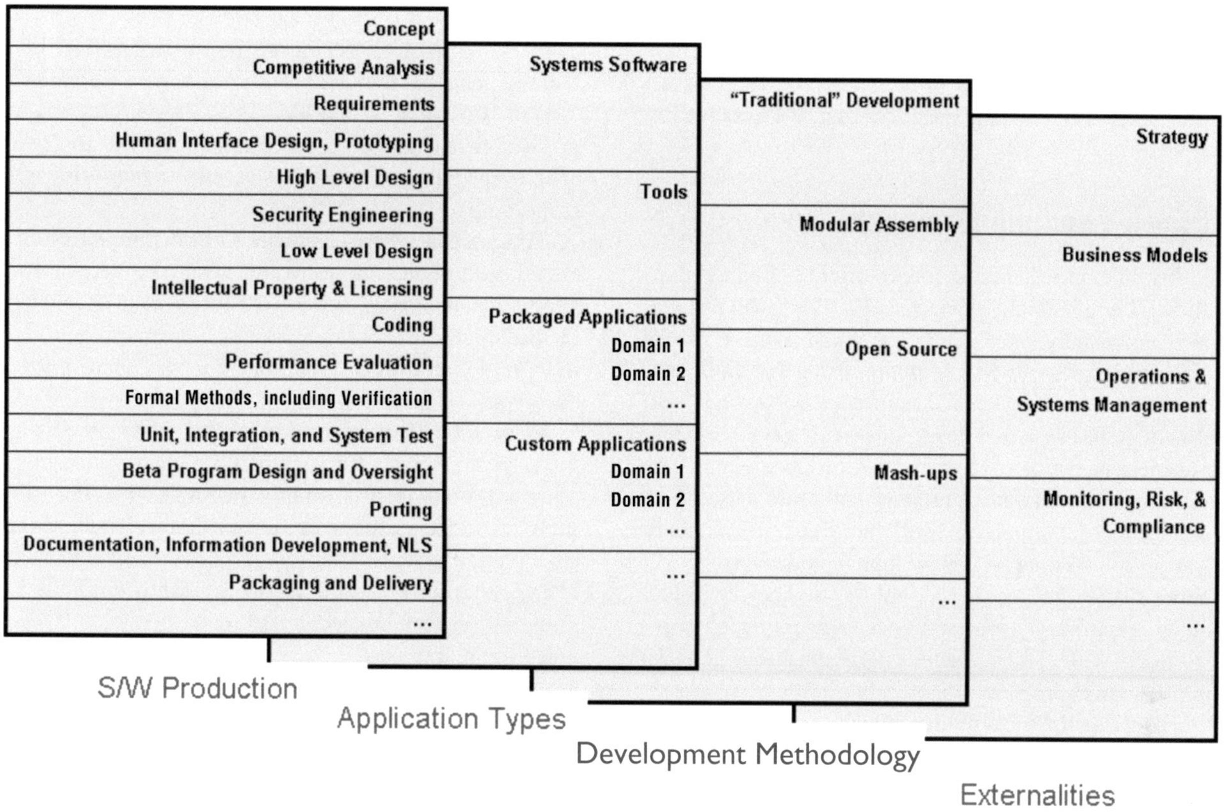

FIGURE 1 The diversity of software activities.

but as requirements and designs are refined, every step might need to be revisited requiring cycles in development processes.

- Interest in modular assembly (e.g., web service-based development or previously object-oriented techniques), with its greatly reduced emphasis on new coding, is increasing. The newest incarnations are "mash ups," connected groupings of reusable components that provide a new function, often intended for a modest-sized audience. Simplicity of assembly is the focus, and success is based on the existence of a massive, society-wide capital plant of increasingly modular components, such as maps, calendars, group bulletin boards and editors, etc.
- Open-source techniques have been remarkably effective for creating good software. To the amazement of some, volunteer groups in modest organizational structures, often using many preexisting software components, are proving adept at developing quality software.

There is no agreed-upon standard methodology for creating software (and there may never be); so software development is not amenable to rigid standardization. Creating software-development organizations is not like designing a semiconductor fab, for which one can create a design that can be cloned in whatever locations have the most favorable cost or regulatory structures. Software development is too variable for that.

Finally, the last cut attempts to capture the important interactions with the world around software including other items that go into the software life cycle. For example, one cannot undertake the automation of a medical procedure without understanding the impacts of failure, FDA requirements for proof of safety and efficacy, and much more. Automation strategies, the creation of business models for supporting software, an understanding of the management of holistic systems in which software will operate, and an increasing focus on risk and compliance management must all be considered part and parcel of the software field.

I suspect that with more time and thought we could fill out and enlarge this multidimensional matrix in each dimension.

We can conclude, however, that software is a very diverse field. Thus, when we consider offshoring, we must remember—and this is my second message—that *there is great variability in software objectives, job types, and practices around the world.* Thus, even if a population somewhere becomes very, very good at one aspect of software, the field is so diverse that it is unlikely that population, or any population, will be very, very good at everything. If, when we think "software" we think only "coding," we miss the big picture.

MESSAGE 3: UNBOUNDED OPPORTUNITY

My third message is that software offers unbounded opportunities. I emphatically disagree with those who say that software opportunities are fading away and that the bloom is off the rose. Software is a synthetic discipline for creating the logic to encode virtually anything! Software is also operational in the sense that it is a constructive synthesis that generates useful entities that work and produce value. The target domain for software is broader and more varied than for other fields of engineering. Software has applications in all areas of human interest and all human endeavors.

I believe software people are ambitious. They (we) feel that those magnificent computational engines, called computers, along with their storage, communication, and I/O capabilities, are capable of vast, nearly infinite brilliance.

First, consider the most traditional space of software: enterprise computing. Even the most mundane application system in any corporation has a backlog as far as you can see for improvements, endless requirements that have not been met. This necessitates a continuing prioritization process. When I managed software products at IBM, we always(!) had much more to do than we could do with available resources. So we are not running out of even the most traditional work.

More important, consider the plethora of uses for computing that have been postulated. Very gradually, we are getting to some of these, but we have a huge backlog. I submit that most of the "fantasies" about computer applications will eventually come true. Whether playing chess, supporting autonomous robots, providing universal access to information, answering questions, or you name it, these and many other uses for computing just keeping coming.

Some argue that Moore's law will meet its limits in the near term and slow innovation. Even if this were true, we have massive underutilized capacity today. Even if we hit a brick wall tomorrow, it would not have a great impact.

Second, although nothing grows to the stars, frequency and density scaling are continuing (particularly the latter), and this growth will continue for a while, for many reasons.

Third, given the very low manufacturing cost of silicon devices, we can have as much processing power as we want as long as we are willing to embrace parallelism. The world's fastest computers are already made up of tens of thousands of processing units, and there are no limits to their feasible expansion. Although exploiting parallelism is sometimes challenging, the challenge itself opens up fascinating opportunities.

So what are the factors that could limit growth? First, design and engineering costs are the primary reasons we can't implement all of the requirements and make rapid enough progress to meet grand challenges. Apparently, it's not cost-effective to tackle some of these, or, presumably, rational firms would do so. Clearly, it's not the marginal cost of production (duplication or transmission) that is stopping us.

There are also management and operational costs associated with deploying software. When we software people generate new software, we often forget to take into account the life-cycle costs of maintaining and managing it. Management and operational costs are important factors that limit the impact of software.

Finally, a lack of innovation is always an impediment. For many years, people think something cannot be done. Then a breakthrough occurs, and we begin to do it. For example, there was once a general consensus that we could not do "artificial intelligence." Then, all of a sudden, we had a winning chess program and Internet-scale search engines. Maybe neither is perfect, but both are exceedingly good. If the initial entry in an application domain is successful, market forces stimulate iterative enhancements that generate a long stream of continuing advances.

Thus my third message is that software is a field that offers tremendous opportunities for the future. The level of opportunity has not reached its peak. There is a tremendous amount still to be done, and given that the major stumbling blocks are not the lack of opportunity, but limited cost-effective resources and talent (innovation), I submit lower prices and greater innovation will spur more demand for people—that is, there is high price and innovation elasticity of demand for software people.

MESSAGE 4: THE EXPANDING SPHERE OF SOFTWARE

My fourth and last point is that the field of software is not just concerned with refining past achievements. The field is expanding (Figure 2).

In November 2004, I gave a talk entitled "Research on the Edge of the Expanding Sphere" at Harvard's Center for Research on Computation and Society. The message of that talk was that the software field clearly has a core, say, inside a sphere. Today, that core includes, for example, the study of algorithms, compilers, operating systems, distributed computing, et cetera. However, every year, the sphere gets bigger.

For example, when I was a Ph.D. student in the 1970s at Stanford, there were six elements to the field of computer science: algorithms, complexity theory, software, artificial intelligence, numerical methods, and architecture. Since then, the field has grown incredibly in two ways:

1. The density of elements inside the sphere has increased. There is more in operating systems today than

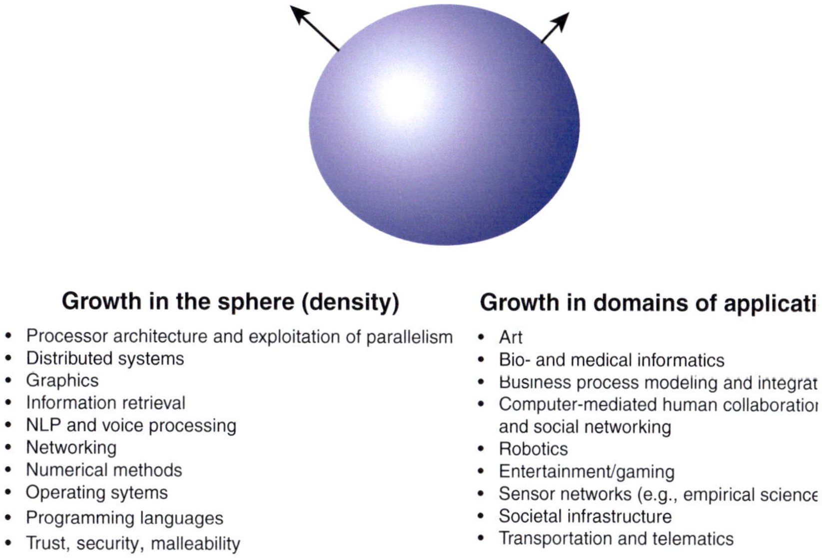

FIGURE 2 The software sphere is simultaneously denser and expanding.

there was 30 years. There is a lot more in programming. There is a lot more in artificial intelligence, the study of algorithms, and so on.

2. The sphere has also expanded into new domains that were once unrelated to software. Examples include e-commerce, social networking, bio-informatics, e-voting, and very many more.

So there are immense opportunities in what I refer to as $CS + X_i$, for many values of X_i. The X_i could be art, computational biology, medical informatics, entertainment and gaming, sensor networks, and so on. $CS + X_i$ has an impact on all aspects of the economy and on the way we conduct science as massive growth in these hybrid software-related activities continues.

For example, our ability to measure all sorts of natural phenomena with very low-cost sensor networks will continue to revolutionize some aspects of engineering, but also of science. Take another example, something in which all of us have an interest, the relationship of software to health care, a circa $2 trillion industry in the United States alone. Most people believe that many hundreds of billions of dollars are wasted because of a lack of good information technology support, and that health care suffers as well. This topic, because of its scale and urgent need for creative solutions, will undoubtedly generate significant incremental demand for software. My list of examples could continue.

IBM AND GLOBALIZATION

As the recent vice president for technology and strategy of the IBM Software Group, I should say a bit about IBM's recent moves toward globalization. IBM's chairman and CEO, Sam Palmisano, has explained the many reasons IBM has created an overseas technical presence. These include proximity to markets, the capability of understanding overseas markets, the availability of talent, lower costs, and so on.

In the past 10 years, IBM's presence in India has increased dramatically, especially in the last few years. Mr. Palmisano reported in June 2006 that IBM had 43,000 Indian employees out of a total of 340,000 employees.[2] Thus India now has the second-largest IBM employee population in the world.

A tremendous variety of jobs are filled by Indians. Most are in IBM's vast services business and involve custom application development, systems management, and call-center automation. However, about 5 percent of the Indian employees produce packaged software for the circa $15 billion IBM software business; this percentage is somewhat low because of the particular skill/experience requirements of that business. So, as I mentioned earlier, it is much more difficult to move certain software-related jobs overseas than others.

IBM has a somewhat larger number of software-group employees in China, interestingly enough, which is a little bit inconsistent with the prevailing wisdom that India is the developing IT powerhouse. I'm not sure of the reason for this, but quality English is not as much a requirement in the development of packaged software products as it is in service-related jobs that involve direct interaction with customers.

Based on my experience in the software business, I can explain the factors that influence a decision about where to locate employees:

[2]In early 2007, after this talk was given, the number had risen to 53,000.

- Talent, experience, and maturity of the teams. Software has sufficiently high margins that talent, quality, experience, and maturity can mean more than costs per hour.
- Organizational capability, including managerial leadership and—importantly—technical leadership. The lack of leadership tends to be the most difficult impediment to growing teams in new locales.
- Capability in a wide range of software activities, including interfacing with project management, customers, sales teams, and finance teams. These activities are not ancillary to software; they are a core part of the business.
- Co-location with a market. This is related to the previous point but bears repetition.
- And, very important, lower labor costs.

Let me describe one situation in which IBM moved some software development from England to India fairly recently. IBM had a few tens of people doing somewhat repetitive, but still high-skill, high-profile, Java-related work in England. It took about a year-and-a-half to make the transition to India, during which time the English and Indian teams had to work together closely. The move was successful, with much of the work now being done in India. To the best of my knowledge, the English team was not unhappy because members felt there would be new, more exciting work to replace what they had been doing. That is, the move freed up the talent in England to do things that would generate more revenue growth and employment.

Overall, in the position I held at IBM, my biggest worry was always about leadership. The same need for talented leadership was also important in offshore software research—and was a persistent problem throughout the decade or so when I visited IBM's newer research sites.

IMPLICATIONS

I have discussed four observations relating to (1) the global leveling of opportunity in software, (2) the great variety of objectives, job types, and practices in software, (3) the high elasticity (price and innovation) of demand, and (4) the interaction and mutual impact of software and computer science on more and more fields of human endeavor (my shorthand for this last point is $X_i\,(CS + X_i)$).

The most important implication is that there are vast opportunities in software. The technology provides sufficient benefits to ensure employment for many populations, with no obvious limits. This has not been true in other areas or other U.S. industries, where there has been significant degradation. Perhaps, for example, only so much innovation occurred in the steel industry over the years, and there is only so much demand. I do not believe there are similar limits in software (Message 3).

Second, because of the variability in the field, some populations have comparative advantages. One can differentiate to gain comparative advantage in many ways—talent, experience, capitalization, location, trust, risk, and so on. Take just one example, the need for trustworthy systems as software moves into life-critical domains.

I believe the application of software to other fields and vice versa will be increasingly important to opportunities for differentiated innovation. A situation may require not just software talent, but also multidisciplinary critical mass (Messages 2 and 4).

Just because of its centrality in so many fields, computer science and software are important. The ability to lead in IT development and IT applications continues to be important for our security and our economy (Message 4).

Finally, global leveling means that Americans cannot take software leadership for granted (Message 1).

SCENARIOS FOR THE FUTURE

To crystallize my, and perhaps your, thinking on the impact of offshoring, I've developed three admittedly overly simplistic scenarios.

Scenario 1

Certain activities, such as testing, integration testing, internationalization (to make software ready for use in many countries), and coding are much less expensive because of offshoring. Nevertheless, elasticity of demand is still high, so lots of opportunities remain for talent in the United States. Dollars saved by the reduced costs of offshoring of certain activities are available for higher value activities that encourage growth in overall output and employment. U.S. innovation, employment, and economic contributions increase.

I think this scenario is not only possible, but is also the most likely to be realized. When I was at IBM, if development had been more cost effective, more development would have been done, much of it naturally in the United States.

To make this scenario even more comforting for the long term, certain coding, testing, and design activities would remain in the United States to ensure that American universities, labs, and corporations retained sufficient skill and training capabilities and to prevent insidious "technical hollowing out." The United States is likely to retain some jobs across the spectrum for two reasons: (1) if all members of a team are co-located, the work goes faster; and (2) overseas cost benefits tend to decrease as workforces there gain skills and experience (note the significant wage inflation for talented Indian software professionals).

Scenario 2

In this scenario, more and more employment in the central sphere moves offshore. But, the software field continues to change fast enough to generate new subdisciplines or, if you will, "superdisciplines" based on hybrids of software and other endeavors. Although the United States may have lost competitiveness in significant aspects of the core of information technology, the country's attention has turned to topics related to $CS+X_i$ thereby providing continuing opportunities. As long as high value is created in these hybrid activities, this is a good outcome.

An analogy is to think of our jobs arranged in a pyramid. As certain jobs at the bottom of the pyramid migrate overseas, the outcome is fine as long as we can move to the top of the pyramid, which of course keeps growing higher. Scenario 2 is optimistic and possible with a field as open ended as software and $CS+X_i$.

Scenario 3

This scenario is pessimistic. When certain activities move offshore, our students and funding agencies take this to mean that opportunities in software have dried up. As a result, U.S. talent dries up, creating a downward spiral. Although there may still be elasticity of demand for innovation, we no longer have the capability to innovate. Given the centrality of software to everything in our lives, this has profound, negative implications throughout the country. I consider this scenario a risk.

CONCLUSION

Given that software is central to so much in our lives, I believe IT is a crucial fulcrum for American prosperity. I think leadership in aspects of software, particularly the most innovative aspects, is important for the United States. This does not mean we must dominate all elements of software, which is fortunate, because we cannot dominate software as completely as we did in the past. I also do not think software leadership is incompatible with significant offshoring. However, we must remain strong in areas of differentiated value.

As I consider what we should do, my obvious conclusion is that we should attend to our future workforce. We must have a creative workforce that has high value compared to others around the world and that can keep us on the leading edge of high-value opportunities.

Ensuring that we have this workforce will require both in-depth and interdisciplinary education. I think we don't yet fully understand the requirements and advantages of interdisciplinary education. With our flexible institutions, the United States may be better than most at "interdisciplinarity," but, when teaching people about computer software and the fields in which we need software applications, interdisciplinary education will require a great deal of careful thought and planning.

It is one thing to argue for a better educated populace. It is something far different to suggest exactly what we should do: incentives, curricula, organizational structures, ethnic and geographical diversity, and so on. It will take some very deep thinking to get these things right. I believe it is time we revisited these topics in far greater depth than we have so far!

When we think of our future workforce, we must also think about immigration. There have been discussions in academic circles about how difficulties in getting into the United States have reduced the immigrant graduate population. Although we want native-born Americans to go into science and engineering, we cannot afford to lose the creative, entrepreneurial immigrants who are integral to our talent pool—and who have done so much for our country throughout our history.

Finally, we must retain economic incentives to encourage people to pursue an education, to work hard and be creative, and to accomplish great things. We must have the right laws to enforce business ethics and honesty, but we must not go overboard in a way the drives the locus of industry off shore.

Implications of Offshoring for the Engineering Workforce and Profession

Ralph Wyndrum

I am pleased to speak today on the implications of offshore outsourcing for engineering. I speak not only as a representative of the engineering profession, but also as an engineer with high-level management experience at AT&T/Bell Labs, as an entrepreneur, and as a consultant. The views I express today are my own, but they are based on my experience in the global engineering-services marketplace, my interactions with the engineering profession as 2006 president of IEEE-USA, and various studies and analyses of offshoring.

THE BIG PICTURE

Let me start with some observations about the large picture as I see it:

- Offshore outsourcing of engineering services is an almost inevitable outcome of the globalization trends created by the basic economic forces of shareholder value, efficiency, productivity enhancement, and the free flow of capital. These trends are enabled by the very technologies engineers created and are continually improving, such as broadband communications and the Internet.
- Offshore outsourcing occurs for a number of reasons, all of which are grounded in basic business logic. Much emphasis has been put on wage differentials and labor arbitrage as the principal driving forces behind offshoring, and labor costs are undoubtedly the major factor at present. But offshoring is much more complex than that, and business decisions are also motivated by other considerations, such as market access and market development, access to talent, the cost of capital, government economic incentives, special or lower cost infrastructures and capabilities (e.g., subsidized telecommunications or Internet), access to universities and research centers, government regulations, and a host of other factors. Therefore, even if labor-cost margins can be narrowed, there will still be strong incentives for offshoring.
- The engineering profession in the United States is not monolithic. Thus offshoring does not affect all engineering disciplines in the same way, at the same pace, or to the same degree. New opportunities for engineers are constantly being created by challenges arising from circumstances, such as hurricanes Katrina and Rita, increases in oil prices, and military operations abroad. Technologies mature and become obsolete, along with the academic disciplines that rose up around them. Although some electrical and computer engineering disciplines are still maturing and in transition in many ways, new disciplines are emerging and other disciplines, such as bioengineering and nanotechnology, are experiencing growth and creating new opportunities.
- If the United States maintains its leadership in emerging technology fields, the U.S. engineering profession will continue to create new engineering opportunities and be somewhat insulated from offshoring. Even then, however, engineers in affected disciplines will continue to struggle as individuals to make career transitions.

Ralph Wyndrum is CEO, Executive Engineering Consultants, and the 2006 president of IEEE-USA.

Prior to the emergence of offshoring, the U.S. engineering profession was already wrestling with significant challenges, not the least of which were the dot.com and telecom busts, which led to major contractions (estimated at a half-million jobs or more) between 2001 and 2003 in the high-tech sector, particularly electrical engineering. These busts came on the heels of the major downturn in the U.S. aerospace industry after 1998, another engineering-intensive sector. Other structural issues in the profession are contributing to the problem:

- The post-WWII/Cold War technology boom that fueled America's high standard of living was based on amazing improvements in productivity that drove the nation's economic growth, while at the same time automating and streamlining many engineering-intensive tasks. Engineers joke, and with good reason, that they are the only professionals who work hard to put themselves out of a job. That translates into professionals who are, by necessity, highly mobile, moving from assignment to assignment and employer to employer.
- Engineering is a profession whose members are challenged to keep up with the latest developments in technology, and continuing education has become critical for engineers in mid/late career. At the same time, employers are becoming less and less likely to invest in training or to support time off for professional activities. Electrical and computer engineers increasingly face early obsolescence (as early as their mid-30s or early 40s) unless they continually reinvent themselves. With most engineering Ph.D.s leaving school in their early 30s, the productive lifespan of a research or design engineer is shorter than ever before, making the opportunity-cost calculation less than compelling for bright students weighing their career options.
- The educational barriers to entry in the engineering profession are constantly getting higher and more expensive as more and more content is squeezed into traditional four-year degree programs, which typically take nearly five years to complete. Recently, the National Council of Examiners for Engineering and Surveying voted to amend the model state engineering licensing law to require "30 credits of acceptable upper-level undergraduate or graduate level coursework from approved course providers" in addition to a B.S. degree as a prerequisite for licensure. The change would not take effect until 2010 at least. The additional work, however, does not seem to be paying off in terms of future compensation. According to the National Association of Colleges and Employers, beginning salary offers for electrical and computer engineers at both the B.S. and M.S. levels were flat, or actually fell, between 2001 and 2005. In other engineering disciplines during the same period salaries varied. Some underwent seesaw fluctuations, some remained flat, and some experienced modest growth.
- We are also facing a demographic issue. As the U.S. engineering workforce ages, a high percentage of baby-boom-generation engineers will reach retirement age in the next 10 to 15 years. The losses will be felt most strongly in mature engineering sectors, such as aerospace and power. The National Science Foundation's most recent *Science and Engineering Indicators* reports that 29 percent of all science and engineering (S&E) degree holders and 44 percent of all S&E doctorate holders in the workforce are now 50 or older. Among S&E doctorate holders in the labor force, 44 percent are over 50. We see the same demographic trend in IEEE, where the average age is now 47 for regular members (up from 44 in 1997). Employers taking the long-term view are looking to secure labor resources to meet future needs (hence their interest in tapping the global services market), as well as to shed pension and other overhead costs that make it difficult for them to compete.

As engineering labor becomes more and more of a commodity, the fundamental relationship between engineers and employers is changing. As a consequence, a significant percentage of the U.S. engineering workforce is becoming increasingly apprehensive about their careers and the future of the profession. Some feel they have been used and discarded. Many want or need to keep working in their later years but feel the environment is neither receptive nor enabling. A small percentage is challenging apparent discrimination in employment.

Against this somewhat troubled backdrop, the offshore outsourcing trend gained high-profile attention after 2001, on a par with the related trends of guest workers and domestic outsourcing. Many companies have reduced their engineering payrolls and moved engineering work to services firms, thus creating new jobs in those services firms, but often at lower pay, with fewer benefits, and with less job security. Some of those firms rely almost exclusively on in-sourced guest labor (with H1-B and L-1 visas) as their business model, using labor arbitrage to gain a competitive edge. In many instances, in-sourcing has been used to facilitate planned offshoring of business operations; in other cases, it had that consequence as in-sourced managers used their business contacts to offshore engineering services. Nine of the top 10 engineering-services firms that use L-1 visas to bring foreign high-tech workers to the United States are also engaged in offshore outsourcing.

In the three years since offshoring in the information-technology (IT) services sector began in earnest, the whole IT industry has been transformed. Virtually all bids for commercial work now include an offshore component, and the "global delivery model" has become THE business model in the IT sector. The potential for equally rapid and far-reaching transformations exist in IT engineering as well.

In this environment, engineering jobs tied to manufacturing (which was already moving overseas) and lower level service work have become ripe for offshoring. Many displaced engineers are considered too expensive or not qualified for the new, often more limited opportunities that are available. Those who cannot find new jobs often turn to consulting or contract work or made transitions to non-engineering jobs. A few are chronically unemployed. As president of IEEE-USA, I hear from all of them with some frequency.

For a while there was almost irrational exuberance about the anticipated benefits of offshoring. Now, companies have learned a few hard lessons and are much more thoughtful in implementing their global strategies and in communicating their plans. Offshoring remains a business priority, especially for smaller entrepreneurial start-ups for which investment capitalists and Wall Street require that business plans include an offshore component. Although the public rhetoric has softened, the pace of offshore outsourcing appears to be steady or growing.

IMPACTS ON ENGINEERS, THE ENGINEERING PROFESSION, AND OUR NATION

An article of faith for many proponents of offshoring in the public dialogue is that only low-level service-sector jobs, such as call centers and business-process support, will be offshored. IEEE-USA disagrees and notes that there is already considerable evidence that high-level research and design work are also moving overseas.

The Commerce Department's 2004 report on workforce globalization concluded that "long-term trends in the structure of the (semiconductor) industry suggest that employment in manufacturing by U.S. semiconductor companies will decline, both in the United States and abroad, and employment in research and development (R&D) and design work will increase at a faster rate outside the United States."

Innovation, R&D and Offshoring, a report by Ashok Deo Bardhan and Dwight Jaffee (for the Fisher Center at the University of California, Berkeley) published in fall 2005, based on a survey of industry R&D offshoring practices, concludes that "the emerging situation with offshoring of R&D related activity is going to pose a series of challenges to white collar workers, engineers, designers and scientists, to U.S. firms, as well as to policy makers. It is possible that the future of R&D offshoring will include continued innovation and R&D in the U.S . . . leading to a win-win situation where the U.S. develops/markets the "new" good, and the now "routinized" goods and services are offshored. On the other hand, there exists the distinct possibility of major innovations originating abroad."

In a February 2006 report, the Association for Computing Machinery found that "globalization of, and offshoring within, the software industry are deeply connected and both will continue to grow." The report goes on to note, "one example of a higher-skill area now subject to global competition is computing research. Historically, the bulk of this research was carried out in only a few countries . . . this situation is changing rapidly and the trend looks inexorable."

Innovation Offshoring: Asia's Emerging Role in Global Innovation Networks, a July 2006 report by the East-West Center, notes that "it is time to correct earlier claims that only low-level service jobs will move offshore and that there is little 'evidence' of a major push by American companies to set up research operations in the developing world. Innovation offshoring goes far beyond the migration of relatively routine services like call centers, software programming, and business process support . . . beyond adaptation, innovation offshoring in Asia now also encompasses the creation of new products and processes."

The National Academies 2005 report, *Globalization of Materials Research and Development*, cautions that globalization in the materials area could threaten U.S. access to advances in materials science and engineering (MSE). The report notes that the effects of globalization on U.S. leadership in MSE R&D vary by field and subfield and warns that the emergence of new centers of high-value research around the globe is challenging the ability of the United States to attract top research talent.

In another recent report on offshoring implications for the U.S. semiconductor and software industries, the Government Accountability Office concluded that "recently U.S. firms have offshored more complex research and design activities; they have also sought to take advantage of Asian engineering talent and to target the rapidly growing Asian market." The report adds that "as firms experienced cost savings and observed high-quality work in these offshore locations, they expanded offshore operations to include more advanced operations, such as software design and systems integration."

The Insight 2005 study of U.S. technology innovators conducted by McClenahanBruer Communications, CMP, and *Electronic Engineering Times*, reported that 64 percent of respondents "worry about the future of the engineering profession in the U.S. because of the impact of outsourcing." Of the survey respondents, 46 percent indicated that their companies have sent electronics design work overseas; 70 percent was at the low end of software development, hardware design, or manufacturing; and 30 percent was characterized as high-end software or hardware design.

Last May, Booz Allen Hamilton and Insead surveyed 186 companies operating in 19 countries and 17 industry sectors to assess trends in the dispersion of innovation in R&D. They found that companies are increasingly siting R&D operations outside their headquarters market (45 percent in 1975 and 66 percent in 2004). Foreign R&D sitings have shifted toward China and India and away from the United States and Western Europe. The survey respondents suggested that the pace of offshoring will increase, with 77 percent of new R&D sites planned through 2007 slated for either China or India. By the end of 2007, China and India's share of global R&D staff is projected to jump from 19 percent to 31 percent, replacing Europe as the most important location for foreign R&D for U.S. companies.

A more recent study by Booz Allen Hamilton conducted for the National Association of Software and Service Companies in India highlights growing demand for engineering services. The study estimates that $10 to 15 billion of engineering services is currently being offshored, with projected growth to $150 to 225 billion by 2020.

This summer, *Electronic Engineering Times* conducted a survey of its electrical-engineering readers to gauge their thoughts on offshoring. What they found was described as a "grim acknowledgement" of the trend and a sense that the United States has been complacent. The authors concluded that "American EEs [electrical engineers] fear that U.S. companies are looking for equally smart, but cheaper, engineers in developing markets who can be future stars once they gain experience. Moreover, they wonder if America is trading away its future industrial leadership for short-term gains in the bottom line."

If both low-level and high-level engineering work is being offshored, what are the prospects for U.S. engineers in the future?

- Engineering jobs tied to creating and maintaining geographical infrastructures will clearly still be in demand.
- Large companies will retain some level of R&D and design work close to their U.S. markets and manufacturing enterprises even as they shift their investment priorities to opportunities abroad.
- Engineers with entrepreneurial sensibilities and bright ideas will create their own opportunities.
- Higher level research jobs will remain around federal laboratories and academic research centers as long as federal R&D dollars continue to flow.
- For the foreseeable future, it seems likely that job opportunities that involve sensitive or classified work will remain in the defense and homeland security sectors.
- As new and emerging technologies are commercialized, they could also drive job creation in the United States.
- According to most macroeconomic projections, the overall size of the U.S. engineering workforce will increase in the short term, keeping pace with the growth of the U.S. economy. It is not clear, however, how the U.S. engineering workforce will fare if the United States is unable to retain its leadership position in technology innovation over the longer term.

POLICY IMPLICATIONS

In a talk in October 2003 reported by *Forbes*, Andy Grove, then chairman of Intel, described the cost benefits driving offshore outsourcing and acknowledged that he was torn between his responsibility to shareholders to cut costs and increase profits and his responsibility to U.S. workers. He concluded that the government must help establish a proper balance between the two. Otherwise, he cautioned, companies will revert to their obligation to increase shareholder value. So far, government has not risen to that challenge, and, in effect, Grove's cautionary note is increasingly becoming the reality.

IEEE-USA believes that offshoring is inextricably tied to the broader issue of preserving our national competitiveness and technological leadership in an increasingly global economy. IEEE-USA also believes we need a coordinated national strategy to sustain U.S. technological leadership and promote job creation in response to the concerted efforts of other countries to capture U.S. industries, jobs, and markets.

Rising Above the Gathering Storm, a recent National Academies report, draws attention to the competitiveness challenges facing the nation, challenges that are inextricably linked to engineering and the engineering profession and to offshoring and other trends. In response to the report, a number of advocacy coalitions have been formed, more than a dozen bills have been introduced in Congress, and the president has announced the American Competitiveness Initiative focused on reprioritizing federal R&D appropriations. For all this talk, however, relatively little has actually been accomplished so far in the policy sphere.

Some common points of consensus are being advanced, however, most of them supported by IEEE-USA and other professional engineering societies, as well as by industry and other groups. We collectively endorse the following points:

- a renewed federal commitment to support front-end research and development to encourage innovation
- permanent extension of the federal R&D tax credit
- programs or tax incentives to assist in the development of human capital and worker training
- improvements in K–12 science, technology, engineering, and math education in the United States to ensure the availability of a technologically literate workforce

I believe these are necessary policy responses, but not nearly sufficient to the challenges of the situation. It may be that how effectively expenditures are made, rather than how much we spend on R&D, makes the real difference. R&D geared toward product/process improvement helps drive incremental innovations that fuel commercialization and promote prosperity in the short term. But to remain competitive over the long term, the United States (both in the public and private sectors) must invest more of its resources in both exploratory and applied research in the physical sciences, particularly in high-risk areas of new and emerging technology that can lead to new technology-based industries.

We must also find ways to capture the benefits of that research, protect the intellectual property, and commercialize it so that we create high-value jobs here in the United States that will drive our economic prosperity and sustain our national standard of living. As for K–12 math/science education, we must advance technological literacy and expand the pool of prospective scientists and engineers who can fill the gaps when the engineers in the baby-boom generation retire. At the same time, we must be wary of shortsighted reactions that encourage individuals to enter the engineering "pipeline" in numbers disproportionate to realistic projections of workforce demand. Our policies must align the employment opportunities and career prospects for those individuals when they reach the end of the pipeline.

Some have suggested that we should aggressively promote engineering training as a pathway to nontechnical careers. Their argument is based on data suggesting that 24 to 40 percent of recent engineering graduates end up in nontechnical fields, such as investment banking, law, medicine, and management consulting. I endorse the view that a degree in engineering can lead to a variety of careers, but I'm not convinced that engineering is likely to become a popular degree choice for entry into nontechnical professions because of the high threshold requirements and comparative difficulty of obtaining an engineering degree combined with the current career outlook for professionals in our field. I worry that engineering graduates are opting out of technical careers because of the financial incentives to go elsewhere or the perceived lack of opportunities in their preferred fields.

We must be wary of the potential for "hollowing out" the profession if the flow of jobs overseas translates into fewer entry-level jobs here that will enable new engineering graduates to gain the experience necessary for them to move to higher level jobs. With the probable exception of new Ph.D.s, who have research backgrounds, most newly graduated engineers are not equipped to apply their academic backgrounds to innovative solutions to engineering problems. We are already starting to read about U.S.-born electrical and computer engineering graduates going to India to build up their resumes.

IEEE-USA'S POSITION ON OFFSHORING

IEEE-USA's position on offshoring rests on several specific proposals for action:

- Prudent steps should be taken to determine the implications of offshoring for the nation and the engineering profession. The federal government must collect and publish reliable statistics on the kinds and numbers of manufacturing, R&D, and service jobs that are being moved offshore. IEEE-USA was pleased to work with Congressman Frank Wolf in supporting the appropriation for the National Academy of Public Administration's series of studies on offshoring. We also worked to secure the release of the Commerce Department's offshoring study to the House Science Committee. Although these reports are useful, they are essentially snapshots of trends taken at particular points in time. Only a thoughtful, continuous examination of offshoring and its implications for the engineering profession and for the national interest will provide a basis for a strategic approach to national policy making.
- New U.S. workforce assistance programs should be created to help displaced high-tech workers find productive employment and ensure that employed workers can acquire the knowledge and skills they need to remain competitive. This is an extremely challenging and potentially costly problem, compounded by the fact that employers can no longer be counted on to invest in their technical workforce.
- New, or more effective, incentives are necessary to help engineers and other professionals tackle the challenges of mid-career education.
- It is appropriate for government procurement rules to favor engineering work done in the United States, absent compelling reasons to do it elsewhere. Government often purchases products and services that "stretch the envelope" of the market, and firms that win those contracts accumulate knowledge and capabilities that give them competitive advantages.
- Policy makers should take a systematic look at U.S. immigration policy and its implications for the global trade in services. To meet the competitiveness challenges of the future, the United States will benefit more from an open, competitive labor market that encourages the permanent immigration of the best and brightest individuals than from increasing reliance on the in-sourcing of guest workers through a regulatory system that suppresses wages, limits opportunities, and then sends those same guest workers home to use what they have learned here to benefit our competitors overseas. This problem must be resolved before large numbers of baby-boomer-generation engineers retire to avoid creating another incentive for offshoring engineering services.

Additional Considerations

I want to emphasize this last point because it has become increasingly apparent to me that Congress is so caught up in the politics of immigration policy that it is not thinking carefully about the consequences of its proposals, particularly for skilled workers. As a case in point, IEEE-USA commissioned a study by Dr. Lindsay Lowell of the Georgetown

University Center for International Migration. Dr. Lowell concluded that current legislative proposals would, conservatively, allow the entry of 1.88 million high-tech workers to fill the 1.25 million new computer and engineering jobs (as projected by the Bureau of Labor Statistics) that will be created in the next 10 years. Therefore, I would add three points to the IEEE-USA list:

- Although I endorse immigration as a positive means of building our talent pool, I believe that as a nation we need to do more than "poach" the world's best and brightest engineers. We need to look more closely at providing incentives for qualified American students to pursue technical careers. The offshoring trend is now a disincentive for many of those students that is prompting an undetermined number of engineers/parents to actively discourage their children from following in their footsteps. Consider this quote from James Finkel, an engineering manager for B.E. Wallace Products, from "Engineering Becomes a Perilous Career Choice," an editorial in the *Wall Street Journal* on April 29, 2006. When asked about recommending engineering as a career option, Finkel responded, "Given the time and effort of becoming an engineer, who wants to be unemployed every few years? . . . why choose your lifetime salary the day you graduate from college?"
- The global labor market in engineering services ought to work both ways. As a nation, we need a better understanding of the barriers facing U.S. engineers seeking work abroad so we can prepare them to work in the global engineering-services market.
- One final critical point. Engineers, as individuals and as a profession, must be more effective and more proactive participants in the public policy process and in public discourse about technology-related issues. A little bit of active citizenship will go a long way toward ensuring that public policy is better informed and more responsive to the competitiveness challenges we face.

POSITIONS TAKEN BY OTHER ENGINEERING SOCIETIES

Other engineering societies have emphasized different points in their offshoring position statements:

- The American Society of Mechanical Engineers rightly points to the need to secure America's job-intensive manufacturing base. According to one estimate, nearly 48 percent of American engineers work in the manufacturing sector, which also currently accounts for 62 percent of the total U.S. R&D investment. Because the prevailing management practice is to locate R&D as close to manufacturing production as possible, and because manufacturing is increasingly moving overseas, engineering design and R&D will inevitably follow.
- The American Society of Civil Engineers frames offshoring as a homeland-security issue because non-U.S. architects and engineers are increasingly gaining access to information about U.S. facilities and infrastructure.
- The National Society of Professional Engineers notes the difficulties offshore engineering raises for administering the engineering licensing system used by states to protect the public safety.

CLOSING THOUGHTS

In closing, I would observe that there are limits to what policy can or should do when it comes to the free market. Much of what needs to be done is the responsibility of engineers and the engineering profession. We must attend to our own needs and our own best interests.

Professional engineering societies, including IEEE, will quickly become irrelevant unless we enable our members to thrive in their profession and provide them with the necessary tools and direction to deal with the challenges posed by globalization. We must be ready to respond to members who ask how they can be more "innovative," what it means to be "entrepreneurial," and which technologies they must master to remain competitive for the next five years. We must also break down our disciplinary barriers and expose our members to the intersections of technology, where innovation is most likely to occur.

This is why during my tenure this year as IEEE-USA president I have pressed our Board of Directors to shift the focus to increasing our value to members, emphasizing mid-career education and the importance of lifelong continuing education, providing innovative leadership, and enlisting engineers to support K–12 education for future technologists. We are developing new programs, such as a proposed innovation institute, where we can tap the expertise of our members to help promote the profession. Carrying out these policies will require a modicum of "tough love" at times to change the thinking of established engineers about their careers. And it will take time to effect changes and to see the results. But I am convinced we're moving in the correct direction.

The offshoring challenge is real, as is the challenge to continued U.S. technological leadership in the face of growing global competition. We must move beyond simplistic "win-win" rhetoric to a thoughtful and deliberative understanding of the effects of the phenomenon and respond accordingly. We have many advantages as a nation and as an engineering profession, but we do not have a monopoly on bright people, technical know-how, or investment capital. We *will* lose our competitive edge if we are not focused and persistent.

The U.S. engineering profession is in the early stages of a painful transition as it adapts to the hard realities of globalization. There will be some who are unable to make the transition and some who will need help. As a profession, we must be prepared to help all of them rise to the challenge. As a nation, we must find a way to preserve and support a vital domestic engineering capability that can sustain the technological leadership and innovation that underpins America's economic and national security.

My thanks to the National Academy of Engineering for sponsoring this discussion and my appreciation to the United Engineering Foundation for funding it. I hope this is the start of an ongoing dialogue in the engineering community that will help us reach an actionable consensus on sustaining a strong U.S. engineering profession for the future.

ACKNOWLEDGMENTS

I'd like to thank Dr. Ron Hira, Rochester Institute of Technology and IEEE-USA vice president of career activities, and other members of the IEEE-USA Board of Directors and staff who reviewed my comments and provided valuable input.

Industry Trends in Engineering Offshoring

Vivek Wadhwa

I am wearing two hats—one as a technology entrepreneur and one as an academic. As a tech entrepreneur, I was one of the first to outsource software development to Russia and to build a company dependent upon having its entire development team at the other end of the globe. I was also one of the first to outsource research and development to India.

I started my operations in Russia in 1992, right after the fall of the Iron Curtain. I can tell you stories about how we hired ex-KGB programmers to reengineer code based on skills they had gained from reengineering American systems during the cold war; but that is a different topic. We employed 50 Russian scientists who built technology that led to the establishment of my second start-up company, which created 200 U.S. jobs and helped many businesses improve their operations.

In my first start-up, we also employed many workers on H1-B visas. In the early 1990s, we recruited them from London and Dublin because our British and Irish hires typically cost 40 percent less than Americans with equivalent skills. The fact is that when you hire H1-Bs, they cost a lot less. We built a very successful company as a result, and we didn't take American jobs away. In fact, we created new jobs.

Then I had a heart attack—a career-changing event for me—and I couldn't continue in tech, so I ended up joining Duke University as an executive in residence. My goal was to give something back by mentoring students and sharing my business knowledge. But when I joined Duke, there were some surprises in store for me.

First surprise—I thought I was joining a country club. I thought that academia was pretty laid back—beautiful campuses, easy life styles, and so on—and that this would be a part-time job. It wasn't a part-time job, though. There's no such thing as part time in academia, as you folks know.

Second surprise—having been a tech entrepreneur during the dot-com days when it was hard to hire good talent—especially from universities like Duke—I didn't expect students to ask me what sort of courses they should take to make their jobs "outsourcing proof." After all, I thought, the fact that these kids had made it into Duke University meant that they were highly sought after, top-notch students who would be set in their careers.

I talked to many bachelor's and master's students, even some Ph.D.s, and found that there were two types of students. One type had no clue about what was going on in the world—these were just hard-core engineers. The other type was more business savvy. These kids worried about their careers and were planning their future. They were trying to figure out how they would pay off the student loans they had amassed at Duke.

The third surprise—30 to 40 percent of our students in the Masters of Engineering Management Program were accepting jobs outside the engineering profession. This didn't make sense to me. All of us talk about the shortage of engineers. Yes, I accept that we want engineering education to be widely disseminated and that engineering education can be a foundation for many professions. But these students weren't going to J.P. Morgan or to McKinsey Consulting to leverage their engineering education; they were going because of economic opportunity.

The bottom line was that engineering was not cool, with some exceptions, of course. Biomedical engineering is really

Vivek Wadhwa is executive-in-resident, and adjunct professor, Pratt School of Engineering, Duke University.

cool. The biomedical students I met showed the same passion, the same fire I used to see in my technologists. But many students in civil engineering and technology didn't have that passion. They were the ones who were looking for jobs in investment banking. Top IIT graduates from India would take courses in our Fuqua School of Business just to position themselves for jobs in investment banking.

I couldn't answer the question my students had asked, so as an academic, I decided to research the topic, and I asked Professor Gary Gereffi, a professor of sociology at Duke, to help. We thought we'd start by assessing the facts in the outsourcing debate, but we couldn't find many. Other than three or four academic papers, including one by Harvard professor Richard Freeman, there wasn't much research on the subject of outsourcing and its impact on the engineering profession. And I didn't give much credence to reports by industry analyst groups because, as a tech executive, I knew that you could often pay an analyst group to produce a report that would support your point of view.

The facts—the numbers commonly cited about the United States graduating 70,000 engineers a year and China and India graduating a million a year—didn't make sense to me. I had worked in India, and I knew how weak education in India was. I didn't believe that India was graduating 350,000 engineers a year, as the media often reported. And, as a board member and advisor to several companies doing business in China, I didn't believe that China was graduating 600,000 engineers either. So the first question we asked was where these data were coming from. None of it made sense.

We decided to start by researching this issue, so we enlisted some of our brightest students to investigate the statistics. Here is what we found (Figure 1).

The statistics in common use were wrong. We were comparing four-year degrees in the United States with three- and four-year degrees in China and two-, three-, and four-year degrees in India. I have to add a caveat here—the Chinese numbers are suspect. In India, independent bodies track graduation rates. In China, provinces report to the central government, and they tell the government what it wants to hear.

The problem is that when you have the wrong information, you reach the wrong conclusions. But when you focus on a single metric, like the number of engineering degrees, there seems to be a simple solution. If the problem is the number of engineers that China and India are graduating is high compared to the number the United States is graduating, then the simple solution for U.S. competitiveness seems to be to for the United States to graduate more. Yet there is no indication that we need more engineering graduates. If we do graduate more, all we will be doing is helping McKinsey, J.P. Morgan, and First Boston with their recruiting because more of our engineering students will have to find employment there.

Recently, thanks to the Sloan Foundation, we expanded our research. We went to India and China and met with academics, business executives, and Communist Party officials to get a better understanding of the situation. Almost everyone agreed with our conclusions—that the numbers for India and China were questionable and that the quality of the graduates was questionable. The vast majority of engineers that graduate in India and China are low quality.

In China, we met with executives of about a dozen companies, each of whom had a list of as many as 10 universities they would hire from. They all said the rest of the graduates were unemployable by multinationals. If you put the lists of universities together, there are probably 20 in the whole of China (about 5 percent of the engineering schools) from which multinationals or start-up companies can recruit.

We learned that the Chinese government created this situation deliberately. About 8 or 10 years ago, they realized

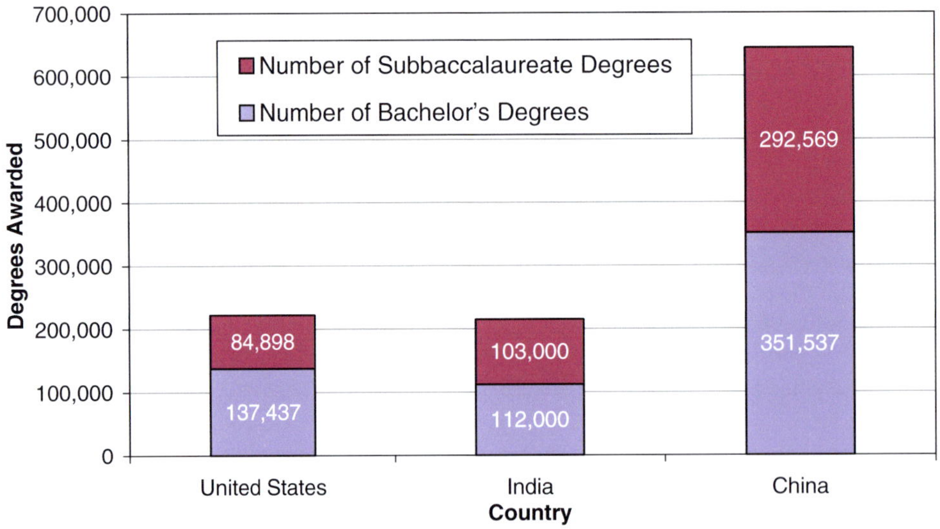

FIGURE 1 Engineering, computer science, and information technology degrees awarded in 2004.

they needed more engineers, so they decided to flood the market with engineering graduates. They told the provinces to increase engineering graduation rates, and the provinces complied, as they always do. But universities like Fudan and Tsinghua—the top universities—resisted. They were able to show the government how quality dropped when graduation rates increased over one or two years. The government then gave the top universities special permission to reduce graduation rates to maintain quality.

The China National Reform Commission issued a report about four months ago that said 60 percent of the graduating class of 2006 would not be able to find employment. About two years ago, the Chinese government decided to slow down engineering graduation rates. So you will see two years from now that the engineering graduation numbers drop off again.

India is facing different challenges. In India, the country seems to succeed *despite* the government, while in China the country succeeds *because* of the government. In India, private industry compensates for the weakness of the government. Right now, in Indian newspapers debates are raging about quotas (more than 50 percent of the seats in all universities are reserved for so-called "scheduled castes"). In fact, India may be messing up its own educational system because politicians get more votes that way. That's a problem with a democracy.

But India also has its own self-defense mechanisms. It is a relatively open, democratic country, and private colleges are beginning to provide high-quality education. Multinationals in India told us they could hire the top 5 to 10 percent of graduates from almost any college in India. Private companies like NIIT provide a "finishing school" for graduates from bad universities and give them enough training to meet the needs of multinationals.

We then took our research further and conducted a survey of U.S. companies that outsource engineering jobs. We interviewed 78 presidents, division heads, CEOs, and senior HR representatives from 58 companies. For more information see *http://memp.pratt.duke.edu/downloads/Duke_Industry_Trends_in_Engineering_Offshoring_10_24_06.pdf.*

Our previous research had raised questions about whether companies really hire individuals with two- or three-year diplomas who are graduating en masse from Indian and Chinese universities. I thought they didn't, but the survey proved me wrong. We found that 40 percent of the companies we interviewed gave us an unqualified yes—that they do hire two- and three-year diploma holders. Seventeen percent said maybe—depending on what kind of additional experience an individual has. We asked companies what additional training they would like the engineers they hire to have. The answers were more communication and presentation skills, internships, computer-related skills, and so on.

We hear a lot about the shortages of people with engineering skills, and there was a serious shortage of programmers during the days of the dot-com boom and Y2K. In our survey, we asked companies a series of questions to determine the extent of the shortages in engineering skills today. Eighty percent of the companies said their acceptance rates were greater than 40 percent. In other words, about half the people they offered jobs accepted them. Eighty percent of the companies said that acceptance rates had remained constant or increased over the last few years. Most of these companies don't offer sign-up bonuses, which are offered when a company is eager to hire people and they are not accepting offers. Today, engineering jobs are being filled in less than four months. This doesn't look like a skill shortage to me.

We asked about what has changed over the last three to five years, and we left the question open ended because we didn't want to bait our respondents. Most said that the engineers they have hired in the last three to five years generally have better technology skills, better communication skills, and a broader global outlook. Some said there was no change.

When asked about the advantages of U.S. engineers, respondents said they understand the market, the business, and communication, they have better interpersonal skills, they are creative, they are good at problem solving, and so on. Thirty-seven percent said U.S. engineers were more productive; 24 percent said equal; 9 percent said overseas engineers were more productive. Thirty-eight percent said U.S. engineering employees produce higher quality work, and 40 percent said equal.

When we asked where companies are sending their jobs, India was number one, China number two, Mexico number three, and then a long list of other countries. Here's where things got really interesting. We found that a very wide variety of jobs where being shipped overseas. When we asked companies to compare jobs overseas to jobs in the United States, 44 percent said U.S. jobs were more technical; only 1 percent said that offshore jobs were more technical; and 33 percent said the jobs were more or less equivalent. When we asked what they gained by offshoring, the responses included access to new markets, culture, co-location, 24/7 development cycle, salary savings, and so on.

When we asked companies to compare the availability of engineers in the United States, China, and India, I was astonished at the responses. Seventy-five percent said that India has a large to adequate supply of well qualified entry-level engineers; 59 percent said the United States did; and 54 percent said China did. I didn't expect this. I thought India would have a greater shortage of engineers than the United States, but the respondents we surveyed said they could hire entry-level graduates more easily in India than in the United States or China.

We asked about the strengths and weaknesses of each workforce. For the United States, the weaknesses were salary demands—not a big surprise, lack of industry experience, unwillingness to relocate, and poor work ethics. In China, they were communication skills, visa restrictions, proximity, lack of loyalty, cultural differences, IP theft, lack of a big-picture mindset. In India, it was communication skills, lack of industry knowledge, proximity/visa, poor project management, high turnover, and cultural differences.

I thought turnover would be at the top of the list for India, but it was mentioned somewhere in the middle of the survey. This was a big surprise because many articles are about massive turnover. Executives in India said turnover is a big issue, but in our survey it was just a passing point. Turnover didn't seem to faze these companies.

We asked companies about the relative advantages of engineers from each country. For the United States, advantages were communication skills, understanding of industry, superior business acumen, better education/training, a sense of creativity, desire to challenge the status quo. For China, advantages were cost followed by work ethics and willingness to work long hours. For India they were cost, technical knowledge, knowledge of English, education, ability to learn quickly, and work ethics.

Cost was cited as the most important reason most companies go overseas. When we asked companies what lies ahead, most said they expected the offshoring trend to continue and to expand. Only 5 percent said it would diminish over time.

To draw some general conclusions, I will put my tech CEO hat back on. When I was a tech executive, I learned that you must always fear your competition. You have to be alert and awake. You have to know your competition's strengths and weaknesses. And you have to be ruthless in crushing the enemy. That's the way to compete. You learn to take advantage of your strengths, the things that make you what you are. You have to do those things better than the competition, and you have to battle the competition on your turf.

In this debate, we have been focusing on the strengths of our competition and competing on their turf. India and China will always have an advantage in numbers, and there is no way we will ever catch up. They graduate more engineers, more dentists, and more shopkeepers. Who cares?

We should focus on what makes us what we are. American workers are creative, hardworking, innovative, and can think outside the box, and American universities excel in basic and applied research. The quality of education in America is not just a little better than in the rest of the world; it is miles ahead. I acknowledge that K–12 education can be improved and that we should teach our kids more math and science. But we must start by focusing on our key strengths and doing what we do better.

Offshoring in the U.S. Telecommunications Industry

Theodore S. Rappaport

I have been asked to discuss network systems, another way of describing the telecommunications industry. From other speakers at this workshop, you can get a sense of the rapid growth of global telecommunications markets and the massive adoption of wireless and telecommunications technologies—with surprisingly little job growth, and sometimes job losses, in the United States.

I have spent the last 18 years as a professor in the telecommunications field and an entrepreneur who has started and sold two businesses to publicly traded companies. I just returned from a wonderful stint in industry, at Motorola, and am pleased to be back in academia. Motorola is headed by a pioneering chairman and world leader in telecommunications. In his keynote address, Bob exhorted us to remain hopeful, and I agree.

The massive bubble in telecommunications of the late 1990s, followed by the telecom crash in 2001–2003, were the most dramatic events I have seen in my professional career. I do not think I can overstate just how dramatic, and devastating, the dot-com implosion has been for the telecommunications research community in the United States.

I will show data indicating that the crash of the telecom industry, combined with a lack of public policy to bring together industry and academic institutions, have created a crisis in the United States that must be addressed if our country wishes to maintain its technical superiority and product development/job creation capabilities in the communications field. Some of these data are anecdotal, and some are compiled from public records.

EFFECTS OF THE DOT-COM CRASH

The fallout from the dot-com crash of 2001–2003 persists, which has greatly impacted the behavior of large telecommunications companies, and, I contend, has affected, in turn, the behavior of students entering engineering undergraduate and graduate schools. Compared to other countries—particularly nations that are emerging rapidly, either through emerging markets or through technological innovations in selected technologies that have become national initiatives—the United States has "lost" its way.

Look at Korea, for example, the most wired country in the world. Korea had a focused, well funded government initiative involving universities, major corporations, and carriers of Korea to bring fiber to the home. As a result, more than 30 megabits per second of data are available in more than 95 percent of Korean homes, and new applications and services are a major part of Korea's technology future. Of course, Korea is a compact country where laying fiber is much more affordable than in the United States. But this effort was supported by a big push from government, which helped bring together corporate and university leaders to create new technologies.

That push continues with wireless technology. In fact, Korea has stated, as a national policy, that the country wants to be the exporter of the next-great revolution in telecom, which they believe will be broadband and wireless. Today, companies like Samsung and carriers like KT Freetel and SK Telecom, with government support and subsidies, are build-

Theodore S. Rappaport is the William and Bettye Nowlin Chair in Engineering and Founding Director, Wireless Networking and Communications Group, University of Texas at Austin.

ing future technologies, which they are exporting already and will be exporting more aggressively in the future. Korean companies are encouraged to send their brightest researchers to U.S. graduate programs, usually subsidized at least partly by government.

The Chinese government has provided significant focus and funding for the broadband build-out in the rapidly growing Chinese economy through investment and subsidies for capital infrastructure projects, start-up ventures, wireless spectrum licensing, and technology standards. China is most likely to insist that its billion-plus population have wireless technology standards based on Chinese intellectual property. New roads, built in very remote parts of China, are being installed with massive fiber-optic capacity, even where the nearest community is dozens of miles away. The Chinese government understands that communications connectivity will be vital for the developing knowledge-based capabilities in rural areas. The government also supports R&D in many areas of telecommunications and sends tens of thousands of China's brightest students to attend graduate school in the United States each year.

India has a truly fascinating engineering culture. I recommend Tom Friedman's book, *The World Is Flat* (2005), as an excellent read on this subject. One of the interesting things I have learned from talking to businessmen in India is that the tax structure for outsourcing and for IT and telecom companies is so favorable that it's bad business *not* to invest in India. Under the Indian tax code, IT, software, and telecom types of companies pay virtually no income taxes. In fact, just today, I turned on *Bloomberg*, and Azim Premji, the founder of Wipro, was talking about how his tax breaks will end in 2009 and that for the next decade, there will be a 0.5 percent tax (maybe) on IT companies.

The tax structure in India requires that employees pay personal income tax, but companies do not pay corporate taxes. This is very, very lucrative for business and another example of a government focusing on engineering. In India, government policy is designed to bolster an industry or a capability to increase national competitiveness. The professions of "engineer" and "medical doctor" are two of the highest callings in Indian society. When I ask Indian graduate students to describe their culture, they remark how very different it is from the culture of the United States, where engineering is in decline among our own citizens and where capitation and insurance/drug policies are making things difficult for the U.S. medical profession. Not so in India where government policies encourage and reward people for pursuing careers in these fields and encourage businesses to establish centers there.

In the European Union, as many of you may know, the Cooperative Arrangement with Science and Technology (COST), which began in the mid-1980s, has funded telecom research at much higher levels than anything in the United States. COST in Europe created the GSM cellular standard (the most widely used cellular standard in the world with 85 percent world market share). COST, which has brought industry and academic researchers together for the last 20 years, provides strong government funding and matching corporate funding, which leads to a great deal of cooperation between industry and academia. Research expenditures are in the hundreds of millions of dollars per year.

The United States has traditionally relied on the Defense Advanced Research Projects Agency (DARPA), the National Science Foundation (NSF), and the major telecommunications laboratories to fund research in the United States. But since the mid-1990s, when DARPA adopted a "problem of the week" mentality, the agency has moved quickly away from basic research. Compare that to the late 1980s, when DARPA program managers were instructed to fund the best academic minds in the United States and let them create great technologies. Today, DARPA requires strict deliverables and quarterly project reviews, as if academic labs were for-profit contractors.

The Industry-Academia Chasm

Although accountability is a good thing, there are likely to be fewer companies like Broadcom, Atheros, SUN, FORE, and others that spun out of universities with DARPA-funded research. DARPA used to empower highly motivated and entrepreneurial faculty. Now it is difficult even to find entrepreneurial faculty on U.S. campuses, because industry and academia are separated by a chasm created by the dot-com bust.

NSF's proposed funding rates are in the small single digits, making it very difficult for faculty to obtain funding for basic research. A "follow-the-herd" mentality has developed, and only esoteric, far-removed projects, which are of great cerebral academic interest but have little industrial relevance, are selected for funding by NSF peer reviewers, who are also removed from industry's competitive needs.

The chasm between industry and academia in the United States, which developed in the wake of the dot-com bust, has become worse over time. Another victim has been the major corporate R&D laboratories that supported applied research and brokered activities between academia and mainstream U.S. corporations. The United States no longer has Bell Laboratories or Xerox PARC (as they once existed). A huge part of the telecom research community has been demolished.

With the dot-com crash, stock prices dropped by orders of magnitude in the telecommunications field. QUALCOMM dropped from $200 to $20 a share; Lucent dropped from $80 to $2 a share. The Telecom Act of 1996 had opened the floodgates of competition, and thus capital, between competitive local exchange carriers and incumbents. Huge amounts of money had flooded in, and maybe expectations were too high. Yes, clearly they were. But the fall of these companies and the layoffs in telecom hit the American psyche harder than anything else I have seen in my professional career.

Renewing U.S. Telecommunications Research, a report published by the National Research Council in June 2006, documents, in very real terms, some of the issues facing the U.S. telecommunications industry and engineering and telecommunications in the United States vis-à-vis the global economy. The study committee for this report, of which I was a member and which included many national leaders, had difficulty agreeing on the reasons for the dire situation of telecom in the United States. We all agreed we were in trouble, but we did not agree on how we got there. When we got to how we could make up for lost ground, we had trouble finding a strategy that would enable us to "save" ourselves.

One thing we all agreed upon was that there is now a tremendous chasm between U.S. industry and U.S. academic programs. Before the dot-com bust, corporations were active on campuses, funding research and scholarships and investing in a dialogue. After the dot-com bust, telecom companies went into hiding. They were (and many still are) in survival mode. The companies that did survive saw their market capitalization decrease by factors of 10 or even 100. According to some sources, more than a million people in the telecom engineering sector lost their jobs. One million engineering jobs in one sector is a lot of jobs to lose.

Imagine the dinner-table conversations of grade-school children or high-school children whose moms or dads were laid off from Lucent or MCI or WorldCom or Global Crossing. Imagine them seeing their parents lose their jobs and their parents' employers going bankrupt or their companies being sold.

The dot-com bust dealt a severe psychological blow to the telecommunications engineering profession. I have seen this first-hand in students who enroll in undergraduate programs at the University of Texas. Something must be done to change the situation if we are to have a reservoir of future technical experts who are U.S. citizens in the communications field.

Since 2002–2003 when U.S. companies went into survival mode, they have been trying to make quarter-by-quarter results for Wall Street. At the same time, U.S. funding agencies have moved farther from the needs of the wounded telecom industry. In the 1980s, DARPA funded entrepreneurial faculty and students who created technologies that fundamentally changed the telecom landscape. DARPA's funding supported advances in fiber optics, the Internet, even cellular technology. However, in the last 10 years, and especially in the last few years, DARPA has moved away from supporting long-term research on academic campuses. In fact, it is now hard to find students who can qualify for DARPA funding, because recipients must be U.S. citizens, and they are becoming exceedingly rare in U.S. graduate programs.

Although NSF continues to do its part, and is a sponsor of the present study, it has not been able to fund relevant, long-term, industry-captivating R&D with its small grants. And telecom has not been included in NSF's bold engineering research center (ERC) program, which has captivated the minds of American universities and industry and promoted interdisciplinary research.

STRATEGIES FOR THE FUTURE

As the study committee of *Renewing U.S. Telecommunications Research* argued, we need a U.S. initiative, a national policy that can repair the industry-academy chasm in the United States, particularly to address the huge number of corporate research jobs that were lost in the dot-com implosion. The industry-academy chasm will not be closed by market forces, which, in reality, encourage telecom companies to invest in emerging markets outside the United States where government subsidies and billions of potential customers await. If the United States hopes to invent the next Internet or remain a leader in telecommunications, if only to meet our own security needs, we must have a national policy that reverses the decline of U.S. citizens in graduate programs and ensures that they can pursue careers and research in communications.

Anecdotal evidence based on conversations with department heads, colleagues, and others involved with electrical and computer engineering departments at various universities reveals these changes dramatically. When you and I were in engineering school, a large majority of the students were U.S. citizens. Today, at the University of Texas, Purdue University, and the University of Florida—three schools I picked at random—two-thirds to three-fourths of the undergraduate students are U.S. citizens, and one-quarter to one-third are from other countries.

In graduate programs today, U.S. students account for 12 to 15 percent and are greatly outnumbered. In some schools, the number is as low as 8 percent; in others, it's, perhaps, 20 percent. In short, U.S. students in graduate programs in electrical and computer engineering are a small minority. In undergraduate programs, they are still a majority, but just barely.

This isn't necessarily a bad thing, except that recently I have noticed that my own students, and other students I talk to, want to go home after they graduate. They see huge economic opportunities in their homelands where multinational companies are investing. They see that the U.S. telecom industry is still in "hunker-down" survival mode and investing its precious capital in the emerging markets where future customers will come from.

So we have a problem. On the supply side, we have a problem attracting U.S. citizens at the undergraduate level. We must figure out how to attract students whose parents, and others in their parents' generation, lost their jobs. If we don't attract them they will feel more and more like strangers in their own land, as they become a smaller and smaller minority in engineering. As for demand, it is clearly in emerging markets and not in the United States.

Another aspect of the slippage in the U.S. position is a precipitous drop in the number of conference papers authored by researchers from industry. In fact, industry is now all but missing in action at the IEEE Global Communications Conference (Globecom) and other international communications conferences. In 2005, of all the companies in the world, only eight published more than one conference paper. In the mid-1980s, Globecom was dominated by presenters from Bell Labs, Xerox PARC, IBM, and Motorola. Today, however, U.S. industry participation has dropped by more than an order of magnitude to 1 in 12 papers presented at international conferences. Industry is no longer a participant in the research dialogue. In effect, professors now present papers to other professors.

For the purposes of this talk, I collected some data on how multinational companies view funding for R&D. I asked one of my research assistants to look up financial records and press releases at the largest telecommunications companies for the last four years. She scoured thousands upon thousands of Web announcements by large multinational companies to find out where they are making R&D investments. We then correlated the results for location and types of R&D investments being made and estimated the expenditures. The results, not surprising, perhaps, underscore that major corporations are not investing in the United States but are turning instead to the most promising emerging markets.

Figure 1 shows the companies we studied and their annual sales for their most recent fiscal years (in U.S. dollars). These companies, which are household names, have the bulk of the worldwide market capital of telecommunications companies—network systems, software, and so on. You could say that Texas Instruments should be on the list (Intel is there as a proxy) or other companies. But keep in mind this is just a sample of major companies that have presence around the world.

As Figure 1 shows, the annual revenues of these 16 companies is about $0.5 trillion—a lot of infrastructure, software, and handsets. Some companies, like Samsung and Siemens, are not limited to telecommunications, but also make refrigerators and power turbines. I included Microsoft as a proxy for software companies. A few companies from China, Huawei, UTStarcom, and ZTE, that have emerged rapidly on the telecommunications scene are also on the list. Although their annual revenues are only $3 billion to $6 billion, they have grown rapidly in the last few years.

The five major focus areas of R&D, based on all of the company press releases we collected, will give you a sense of the needs being met by the $0.5 trillion of products being sold around the globe each year:

- Subscriber devices or premise equipment—cell phones, videophones, voice over IP—purchased by end users is a major component of telecommunications revenues.
- Infrastructure equipment and services are the base stations and switching stations, the large infrastructure that connects carrier-grade telephone, Internet, and wireless systems.
- Switching and routing equipment is the technology/equipment used to connect large hubs with an enterprise or premise.
- Integrated circuits are essential to the chips in all equipment.
- Software and applications, a more and more vital part of the value chain, create the features, adaptability, and upgradability of all products.

The corporate research themes of global telecom companies (listed below) can scarcely be found among research initiatives at NSF or DARPA:

- the development of Internet protocol (IP) to replace circuit switching (IP has a double meaning; Internet protocol and intellectual property [a creator/extractor of value], a key theme in corporate America)

- Alcatel: $15 billion
- Cisco: $28 billion
- Ericsson: $19 billion
- Huawei: $6 billion
- Intel: $39 billion
- LG: $23 billion
- Lucent: $10 billion
- Microsoft: $44 billion
- Motorola: $37 billion
- NEC: $46 billion
- Nokia: $40 billion
- Nortel: $10 billion
- Samsung: $79 billion
- Siemens: $91 billion
- UTStarcom: $3 billion
- ZTE: $3 billion

FIGURE 1 Companies and annual revenues (USD). Source: Compiled from data on various company web sites.

- the convergence of wired and wireless networks throughout an enterprise and throughout the home
- the expansion of intelligence and massive bandwidths to the edge of the Internet and into the home or enterprise
- the development of multimedia data transfer for multiple providers (cable, telephone, wireless converging, competing, and offering content)
- the development of low-cost devices and low-cost infrastructure for emerging economies that have different price points and applications than in the mature U.S. market
- the ongoing development of software and middleware and increasing reusability

The absence of these research themes in U.S. academic R&D shows the complete disconnect between what companies see in their future (e.g., the corporate vision or road map) and what professors and graduate students in the United States are working on. I fear that professors are becoming increasingly isolated from "customers."

R&D Initiatives by Major Companies

The discussion that follows briefly covers some of the major R&D initiatives of the corporations listed in Figure 1.[1] In 2002, Alcatel made R&D investments in Canada, Australia, and China; in 2003, the company invested in R&D initiatives in Australia, France, and China. In fact, China appears on every company's R&D investment list. Alcatel did not invest in the United States until 2007, when it purchased Lucent.

Only a small number of R&D projects by Cisco have been publicly announced. Cisco uses an open IETF model in developing much of its technology, but the company has opened centers in Japan, India, and Vietnam. Instead of funding basic R&D, Cisco typically buys 10 to 15 small companies per year. Ericsson, like most telecom companies, struggled for survival during the dot-com bust. Since then, it has managed to open an R&D facility in China.

Huawei has made investments in China (its home country), India, and Malaysia. The Malaysian government offers huge incentives to companies to locate jobs there, especially in telecom and manufacturing. For example, the government often pays the salaries of the first 100 or so engineers to help a company establish a beachhead.

Intel has not publicly announced any investments in R&D centers in the United States. The company has invested in China, Spain, and England. LG, a leading Korean company, established a U.S. facility in 2005, but its primary focus is in China, Korea (its home country), Italy, and France. Microsoft has announced one major U.S. investment, R&D initiatives in England, and a major investment in India. Motorola has been active in China, Brazil, and Denmark and, in 2005, made a major R&D investment in the United States.

NEC and Nokia both opened R&D centers in China. Nortel is investing in China, France, and India. Samsung made major R&D investments in China and Korea. Siemens made an R&D investment in Korea and is partnering with Nokia. UTStarcom, a Chinese company, is investing heavily in India for wireless infrastructure and for IPv6. ZTE, another Chinese company, is investing in its home country.

In summary, major telecommunications companies announced 57 major R&D initiatives in the past few years. Of these, only five were in the United States. Thirty-five, the overwhelming majority, were in Asian countries, where public policy and regulations are much more welcoming and where markets are experiencing higher growth rates. Twelve major R&D investments were made in Europe, more than twice as many as in the United States.

OBSERVATIONS

What can we learn from these data? First, R&D investments are going to high-growth countries. Second, multinational companies are investing in countries that have made telecom a priority, either through tax incentives, research-expenditure incentives, or other government policies. Businesses are going where there is less friction, because it makes good business sense.

We now see many foreign students who come to the United States to get educated in telecom return to their home countries. For example, of the Ph.D.s who graduate from our wireless center at the University of Texas, about half take jobs back in China, Korea, Pakistan, and India—something I haven't seen in my career before.

That trend is not a bad thing in itself, but it points to the fact that the United States has lost its national focus on telecom, which is clearly not on the research agenda of the U.S. government or U.S. industry. When I look 10 to 15 years ahead, I am deeply concerned about what might happen in the United States. As the data clearly show, U.S. companies, rather than investing in the United States, are investing overseas and hiring overseas, for R&D positions.

Under these conditions, I wonder if, today, we could invent the Internet or cell phone technology. DARPA funded the Internet. NSF took over the build-out of the Internet to campuses across the country. This required a very long investment period, 10 years of funding, before there was even a hope of creating an Internet. Yet that stay-the-course, long-term research was a national policy by a government determined to develop a failsafe communications network that could survive and operate in a national emergency.

In the aftermath of the dot-com crash, we must rethink how U.S. corporations should engage with the U.S. government and with academia. We must ask ourselves how long the United States can continue to produce talent at home to build secure networks, defense networks, without the old

[1] For details see *http://users.ece.utexas.edu/~wireless/NAE%20Research.htm.*

DARPA or Bell Labs. And how we will be able to operate secure networks. Perhaps most important, how will the United States be able to compete globally in telecom against countries that are making concerted efforts to develop technology for export? Do we need a government policy to prop up the telecom industry, just as SEMATECH came to the rescue of the semiconductor industry two decades ago?

The United States does not like to pick winners or losers, but considering the dot-com situation and the lack of investment in U.S. R&D, combined with the chasm between academia and industry, I submit this is one time when the government should reach in to encourage and assist industry in reinvesting and engaging with U.S. universities to try to repair the damage done by market forces. If we care about U.S. citizens having expertise and research acumen in the telecommunications field, something must be done to encourage them and to give them a reason to put forth the effort to attain the necessary skills. U.S. taxpayers should also be made aware of the falling enrollments of U.S. citizens on engineering campuses throughout the United States.

TURNING THE TIDE

In this section I suggest some steps that could be taken to turn the tide for telecommunications R&D in the United States. First, we need a federal policy that encourages and rewards U.S. industry for engaging in a new social contract like the one the telecom community had for decades in the United States. That contract meant that companies were actively involved on campuses, and they provided scholarships. Reviving this social contract may require government engagement.

Before the dot-com crash, when I headed the wireless center at Virginia Tech, more than 30 major companies came to Blacksburg, Virginia, to invest in our research center and engage with our students. After the dot-com crash, the number fell to about 10, and two of those were from China and Korea. That is a real drop-off! If we can't bring more companies back to academia, professors will be working on problems that are peer-reviewed by other professors in a vacuum, without regard to market needs and without long-term marketability for the benefit of the U.S. telecommunications industry.

Second, we need a public-private "big picture," a big, hairy, audacious goal (BHAG), that will lead to hope for the future. Bob Galvin gave us some examples of big-picture projects in his keynote address—new architecture, wireless Internet, middleware. The United States has so many opportunities to take the lead, to solve a national problem through technology. But we are not doing that. We need a "man-on-the-moon" kind of mission to rally our industry and engage creative minds in thinking about how we can improve our nation.

I agree with Bob Galvin that, in some instances, public policy needs to help pick future technologies. "Road mapping" is probably too strong a term in the case of telecommunications, but as you can read in *Renewing U.S. Telecommunications Research*, if we don't do something, we are going to continue to lose our lead. As a successful businessman once told me, if you are not going up, you are going down. Unless the United States moves forward with a vision in telecommunications, we are in for a tough time.

Now that Bell Labs is gone and DARPA is no longer doing what it used to, we need a national policy or entity that can bring us together, in telecom. As we just saw, IEEE conferences aren't doing that. Universities are there, but industry is absent on campuses. Industry is doing a lot of internal research, but the results lead to filing for patents instead of publishing papers. Microsoft and Motorola have huge research organizations, both internal. These companies used to engage with the academic community. My concern is that if we don't bring industry and academia together, our relevance on the global telecom stage will be greatly diminished.

CONCLUSION

I will close with this final thought. Let me ask everyone in the room to please raise your hand if you grew up somewhere other than the United States. Raise your hand proudly. So, maybe 10 or 12 percent of the people in this room were born outside the United States. How many of you grew up in the United States and went to college in the United States? The overwhelming majority, 90 percent.

Consider that the data I have presented show that in graduate programs in telecom throughout the United States today, the numbers are diametrically opposed to the numbers in this room. On college campuses today, most of the graduate students in telecommunications and electrical engineering did not grow up in the United States; they did not grow up in the culture of the United States. This presents us with a great opportunity, but also a great challenge.

If we want this room to be filled 20 years from now when the United States faces a new crisis, we must make sure that the students in graduate schools today stay in the United States, can find gainful employment in the United States, and have the necessary support for the United States to remain a leader in telecommunications innovation. If we don't do that, all the creative minds will leave the United States and go to other countries, perhaps their home countries, where they have the opportunities and incentives to make an impact.

REFERENCES

Friedman, T.L. 2005. The World Is Flat: A Brief History of the Twenty-First Century. New York: Farrar, Straus, and Giroux.

NRC (National Research Council). 2006. Renewing U.S. Telecommunications Research. Washington, D.C.: The National Academies Press.

Appendixes

Appendix A

Workshop Agenda

NATIONAL ACADEMY OF ENGINEERING
THE OFFSHORING OF ENGINEERING: FACTS, MYTHS, UNKNOWNS, AND IMPLICATIONS
Auditorium
2100 C Street, N.W.
Washington, D.C.
October 24–25, 2006

AGENDA

Tuesday, October 24, 2006

8:00 a.m. Continental Breakfast

8:30 Welcome
William Spencer, Chairman Emeritus, SEMATECH

8:40 Keynote Talks: The Globalization of Engineering
Charles Vest, President Emeritus, Massachusetts Institute of Technology
Robert Galvin, Chairman Emeritus, Motorola Inc.

9:50 Software
Moderator: *Susan Graham, Pehong Chen Distinguished Professor, University of California, Berkeley*
Speakers:
Rafiq Dossani, Senior Research Scholar, Shorenstein Asia-Pacific Research Center, Stanford University
Alfred Spector, Consultant and Former Vice President of Strategy and Technology, IBM Software Group

11:00 Break

11:15 Autos
Moderator: *Peter Bridenbaugh, Retired Executive Vice President of Science, Technology, Engineering, Environment, Safety and Health, Alcoa*
Speakers:
John Moavenzadeh, Executive Director, International Motor Vehicle Program, Massachusetts Institute of Technology
John Cohoon, Executive Director, Global Engineering Processes, General Motors Corporation

12:15 p.m. Lunch

1:30 Panel: Implications for the Engineering Workforce and Profession
Moderator: *Lori Kletzer, Professor of Economics and Department Chair, University of California, Santa Cruz*
Speakers:
Ralph Wyndrum, President, IEEE-USA, and CEO, Executive Engineering Consultants
Richard Freeman, Herbert S. Ascherman Professor of Economics, Harvard University; Co-Director, Labor and Worklife Program, Harvard Law School; and Director, Labor Studies Program, National Bureau of Economic Research

Vivek Wadhwa, Executive-in-Residence/Adjunct Professor, Pratt School of Engineering, Duke University

3:00 Break

3:30 p.m. Network Systems

Theodore Rappaport, William and Bettye Nowlin Chair in Engineering and Founding Director, Wireless Networking and Communications Group, University of Texas at Austin

4:00 PC manufacturing

Jason Dedrick, Co-Director, Personal Computing Industry Center, and Project Scientist, Center for Research on Information Technology and Organizations, University of California, Irvine

4:30 Pharmaceuticals

Moderator: *Stephen Drew, Retired Vice President of Technical Operations and Engineering, Merck & Co. Inc.*

Speaker: *Charles Cooney, Professor of Chemistry and Biochemical Engineering, Massachusetts Institute of Technology*

5:00 Adjourn to Public Reception

Wednesday, October 25, 2006

8:00 a.m. Continental Breakfast

8:30 a.m. Welcome

Wm A. Wulf, President, National Academy of Engineering

8:40 a.m. Semiconductors

Moderator: *Marie Thursby, Hal and John Smith Chair in Entrepreneurship, Georgia Institute of Technology*

Speakers:

Clair Brown, Director, Center for Work, Technology, and Society, University of California, Berkeley

Robert Doering, Senior Fellow, Silicon Technology Development, Texas Instruments

9:40 Construction Engineering and Services

John Messner, Director, Computer Integrated Construction Research Program, Pennsylvania State University

Jan Tuchman, Editor-in-Chief, Engineering-News Record

10:40 Break

11:00 Implications for Engineering Management and Education

Moderator: *Linda Abriola, Dean of Engineering, Tufts University*

Speakers:

Anne Stevens, Chairman, President, and CEO-designate, Carpenter Technology

James Porter, Chief Engineer and Vice President, DuPont Engineering and Operations

Richard Newton, Dean of Engineering, University of California, Berkeley

12:30 p.m. Chairman's Closing Remarks

12:40 p.m. Adjournment

Appendix B

Workshop Participants

Linda Abriola
Dean of Engineering
Tufts University

Ramakrishna Akella
Professor and Director
Silicon Valley Center & Santa Cruz
University of California

Thomas Arrison
Senior Program Officer
Government-University-Industry Research Roundtable
National Research Council

Eugene Arthurs
Executive Director
SPIE, The International Society for Optical Engineering

Ramakrishnan Balasubramanian
Student Researcher
Duke University

Pritam Banerjee
Graduate Research Assistant
School of Public Policy
George Mason University

Jennifer Baument
Senior Trade Analyst for Labor
U.S. International Trade Commission

Richard Benson
Dean, College of Engineering
Virginia Polytechnic Institute and State University

William Berry
Distinguished Research Fellow
Center for Technology and National Security

Yudhijit Bhattacharjee
Writer
Science Magazine

Peter Blair
Executive Director
Division on Engineering and Physical Sciences
National Research Council

Robert Blankenbaker
Senior International Trade Specialist
U.S. Department of Commerce

Erich Bloch
Principal
Washington Advisory Group, LLC

Dorothea Blouin
International Trade Specialist
U.S. Department of Commerce

Robert Boege
Executive Director
Alliance for Science and Technology Research in America

Stephen Bonk
IEEE/AAAS Congressional Fellow
Office of Rep. Dana Rohrabacher

Maria Borga
Economist
International Investment Division
U.S. Bureau of Economic Analysis

Chris Brantley
Managing Director
IEEE-USA

Peter Bridenbaugh
Retired Executive Vice President, Automotive
Aluminum Company of America

Clair Brown
Professor of Economics and Director
Center for Work, Technology, and Society
University of California, Berkeley

David Bruggeman
Public Policy Analyst
Association for Computing Machinery

Melissa Carl
Government Relations Representative
American Society of Mechanical Engineers

Erran Carmel
Professor, Kogod School of Business
American University

William Chadwick
International Trade Analyst
U.S. International Trade Commission

Tien-Sheng Chang
Visiting Professor
Department of Mechanical Engineering
University of Maryland

David Cheney
Director
Science and Technology Policy Program
SRI International

Michael Chinworth
Director
U.S.-Japan Center
Vanderbilt Institute for Public Policy Studies

Margaret Chu
Office of Research and Development
U.S. Environmental Protection Agency

John Cohoon
Executive Director
Global Engineering Processes
General Motors Corporation

Charles Cooney
Professor of Chemistry and Biochemical Engineering
Massachusetts Institute of Technology

John Damoulakis
Division Director
Information Sciences Institute
University of Southern California

Lance Davis
Executive Officer
National Academy of Engineering

Gordon Day
Director of Government Relations
Optoelectronics Industry Development Association

Jason Dedrick
Staff Research Associate
Center for Research on Information Technology
and Organizations
University of California-Irvine

Robert Doering
Senior Fellow
Silicon Technology Development
Texas Instruments Inc.

Rafiq Dossani
Senior Research Scholar
Shorenstein Asia-Pacific Research Center
Stanford University

Stephen Drew
Science Partners, LLC

Robin Elsham
BPO researcher/writer
Mendota Heights, Minnesota

Kevin Finneran
Editor-in-Chief
Issues in Science and Technology
National Academies

Samuel Florman
Chairman
Kreisler Borg Florman Construction Company

Richard Freeman
Herbert S. Ascherman Professor of Economics
National Bureau of Economic Research

Lisa Frehill
Executive Director
Commission on Professionals in Science and Technology

Robert Galvin
Chairman Emeritus
Motorola Inc.

Mark Gaylord
Executive Technical Assistant to Cohoon
General Motors Corporation

Al Ghorbanpoor
Interim Dean
College of Engineering
University of Wisconsin, Milwaukee

Michael Gibbons
Director of Data Research
American Society for Engineering Education

Susan Graham
Pehong Chen Distinguished Professor
Computer Science Division—EECS
University of California, Berkeley

David Alan Grier
Associate Professor of Science and Technology Policy and Associate Dean
Elliott School of International Affairs
George Washington University

Robert Gustafson
Associate Dean
Ohio State University

Bill Hall
Economist
Bureau of Economic Analysis (IID)

Oscar Harasic
Science and Technology Principal Specialist
Organization of American States

Magnus Harviden
Counselor, Science and Technology
Embassy of Sweden

David Harwell
Assistant Director
American Chemical Society

Richard Heckel
Founder and Technical Director
Engineering Trends

Susan Heller-Zeisler
National Institute of Science and Technology

Jungeun Heo
Center for International Science and Technology Policy
George Washington University

Martin Hight
Senior Manager
Government Relations
American Society of Civil Engineers

Christopher Hill
Professor of Public Policy and Technology
George Mason University

Greg Hill
Managing Editor
Today's Engineer
IEEE-USA

Ron Hira
Vice President, Career Activities
IEEE USA

E. Dan Hirleman
Professor and Interim Director
Global Engineering Programs
Purdue University

Michael Holland
Committee on Science
U.S. House of Representatives

Virginia Hughes
Assistant Director, International
U.S. Government Accountability Office

Yongsoo Hwang
Visiting Scholar
George Washington University

Marilyn Ibarra
Economist
Bureau of Economic Analysis

Mike Jacobsak
Engineering Manager
Teradyne

Roger Jones
President
Franklin International, LLC

Russel Jones
Managing Partner
World Expertise, LLC

Eva Kaplan-Leiserson
Staff Writer
National Society of Professional Engineers

Faraz Khan
Researcher
International Federation of Professional and Technical Engineers (IFPTE)

Lori Kletzer
Professor of Economics and Chair
Department of Economics
University of California, Santa Cruz

Saki Kumagai
Research Analyst
Washington CORE

Theresa Laranang-Mutlu
Assistant Director
American Chemical Society

Rolf Lehming
National Science Foundation

Wil Lepkowski
Science Writer and Editor, *Public Policy*
Reston, Virginia

Nishanth Lingamneni
Graduate Student
Duke University

Fred Looft
Professor of Electrical and Computer Engineering
Worcester Polytechnic Institute

Neil MacDonald
Editor
Federal Technology Watch

Jeffrey Macher
Assistant Professor
Georgetown University

Tom Mahoney
President
WV MEP

Chris McManes
IEEE-USA

Stephen Merrill
Senior Director, Board on Science, Technology, and Economic Policy
National Research Council

John Messner
Assistant Professor of Architectural Engineering
Director, Computer Integrated Construction Research Program
Pennsylvania State University

Scott Midkiff
National Science Foundation

John Moavenzadeh
Executive Director
MIT International Motor Vehicle Program

Christopher Morecroft
Duke University

Robert Morgan
Professor Emeritus of Technology and Human Affairs
Washington University in St. Louis

Donald Mosebrook
Vice President
Lutron Electronics Co. Inc.

Richard Newton
Dean
College of Engineering
University of California, Berkeley

Janet Norwood
Chevy Chase, Maryland

Don O'Neill
President
Center for National Software Studies

Vin O'Neill
Senior Legislative Representative
IEEE-USA

Ryan Ong
Research Associate
Center on Globalization, Governance and Competitiveness

Hilliard Paige
Retired President
General Dynamics Corporation

Greg Pearson
Senior Program Officer
National Academy of Engineering

Gail Pesyna
Program Director
Alfred P. Sloan Foundation

James Porter
Vice President of Engineering and Operations
and Chief Engineer
Dupont

Jan Post
Oklahoma City, Oklahoma

Ricardo Pozo
Director, Electronics Engineering Center
Raytheon Company

Alan Rapoport
Senior Analyst
National Science Foundation

Theodore Rappaport
Director, Wireless Networking and Communications
Group
University of Texas at Austin

Niyanthi Reddy
Graduate Student
Duke University

William Rees
Deputy Under Secretary of Defense, LABS
U.S. Department of Defense

Proctor Reid
Director, Program Office
National Academy of Engineering

Michael Reischman
Deputy Assistant Director
National Science Foundation

Benjamin Rissing
Research Scholar
Duke University

George Robinson
Engineering Management Student
Duke University

James Rudd
Program Officer
National Science Foundation

Jeff Russell
Projector Director
Duke CIBER, Fuqua School of Business
Duke University

Hal Salzman
The Urban Institute

Keith Sandroni
George Washington University

Lyle Schwartz
Retired Director
Air Force Office of Scientific Research

Edgar Seagle
Consulting Engineer
Rockville, Maryland

Omar Serrano
Consultor, Department of Science and Technology
Organization of American States

Charles Smollin
Professional Engineer
Hamilton, New Jersey

Alfred Spector
Senior Advisor
IBM Software Group

William Spencer
Chairman Emeritus
SEMATECH

Anne Stevens
Chairman, President, and CEO-designate
Carpenter Technology

Edward Sullivan
Associate Dean of Engineering
California Polytechnic State University

Batul Tambawalla
Duke University

Nicholas Terrell
Economist
Bureau of Labor Statistics

Marie Thursby
Hal and John Smith Chair in Entrepreneurship
DuPree College of Management
Georgia Institute of Technology

Timothy W. Tong
Dean and Professor of Mechanical Engineering
George Washington University

Charles Trembley
Northeast Area Vice President
IFPTE

Jan Tuchman
Editor-in-Chief
Engineering News-Record

Daniel Ullman
AMS/AAAS Congressional Fellow
Committee on Science
U.S. House of Representatives

Charles Vest
President Emeritus
Massachusetts Institute of Technology

Vivek Wadhwa
Executive in Residence
Pratt School of Engineering
Duke University

Scott Walthour
Associate
Arium AE

Mark Weaver
Student
Duke University

Larry Weber
Program Manager
National Science Foundation

Scott Weidman
Director
Board on Mathematical Sciences
National Research Council

Cameron Wilson
Director of Public Policy
Association For Computing Machinery

Belle Woods
Government Relations and Public Affairs Associate
Council of Graduate Schools

Wm. A. Wulf
President
National Academy of Engineering

Ralph Wyndrum
President
IEEE-USA

Isabelle Xu
Regional Economist
Arlington Economic Development

Zhenyu Yang
Research Assistant
Duke University

Falan Yinug
International Trade Analyst
U.S. International Trade Commission

Wu Zhang
Senior Research Engineer
Brookhaven National Laboratory

Appendix C

Biographical Information

Chairman

WILLIAM J. SPENCER is Chairman Emeritus of SEMATECH and International SEMATECH. Created in 1990, SEMATECH is a consortium of companies whose goal is to improve semiconductor manufacturing technologies. As chief executive officer and president, he refocused the organization's efforts on streamlining the manufacturing process and introduced standardization. Under his guidance, the institution fostered cooperative relationships among competitors, expanded to include non-U.S. members, and transitioned from government support to increased industry funding. Previously, Dr. Spencer held key research positions at Xerox Corporation, Bell Laboratories, and Sandia National Laboratories. He is a member of the National Academy of Engineering and a fellow of IEEE. He received an A.B. from William Jewell College and M.S. and Ph.D. degrees in physics from Kansas State University.

Committee Members

LINDA M. ABRIOLA is dean of engineering at Tufts University. Previously, she was professor of civil and environmental engineering at the University of Michigan. Her research interests relate to the prediction of the transport and fate of organic chemical contaminants in the subsurface. She has a Ph.D. in civil engineering from Princeton University and is a member of NAE.

PETER R. BRIDENBAUGH is retired executive vice president-science, technology, engineering, environment, safety, and health of Alcoa. Dr. Bridenbaugh joined Alcoa in 1968 at the Alcoa Research Laboratories, New Kensington, Pennsylvania. During his career, he has held positions in Alcoa Laboratories Warrick (Indiana) Operations and Tennessee Operations. He has led Alcoa Technical Center since 1983 and was appointed to his present position in 1991. Dr. Bridenbaugh serves on advisory boards at Carnegie Mellon University, University of Pittsburgh, Pennsylvania State University, Stanford University, Massachusetts Institute of Technology (MIT), University of Virginia, Lehigh University, and Northwestern University. He is chair of the Engineering Design Research Center Industrial Planning Committee at Carnegie Mellon University. Dr. Bridenbaugh has a Ph.D. in materials science from MIT. He is a member of NAE.

STEPHEN W. DREW is retired vice president of technical operations and engineering, Merck & Co. Inc. Currently with Science Partners LLC, his technical areas of expertise are chemical, biological, and engineering technology for the bulk chemical manufacture of pharmaceuticals. He was elected to NAE for his work in this area in 1993. He has a Ph.D. in biochemical engineering from MIT.

SAM FLORMAN is a writer as well as a practicing engineer and chairman of Kreisler Borg Florman General Construction Company in Scarsdale, New York. Florman was elected to NAE in 1995. He is the author of six books dealing with the relationship of technology to the general culture and has written more than 250 articles in professional journals and popular magazines. Florman, a lifelong resident of New York City, is a fellow of the American Society of Civil Engineers. He holds a bachelor's degree and a civil engineer's degree from Dartmouth College and an M.A. in English literature from Columbia University.

SUSAN L. GRAHAM is Pehong Chen Distinguished Professor, Computer Science Division—EECS, University of California, Berkeley. Her expertise is in the design and implementation of programming languages; techniques, methodology, tools, and environments for software development; and software support for high-performance computing. She has a Ph.D. in computer science from Stanford University and was elected to NAE in 1993.

LORI KLETZER is a professor and chair of the Economics Department at the University of California, Santa Cruz. Her areas of specialization are labor economics, industrial relations, and applied econometrics. Her current research interests include consequences for the domestic labor market of increasing economic integration; the causes and costs of job displacement; differences in educational attainment, occupation, and earnings between black and white women; and the economics of higher education. She has a Ph.D. from the University of California, Berkeley.

ANNE STEVENS is chair, president, and CEO, Carpenter Technology Corporation. Until October 1, 2006, she was group vice president, Canada, Mexico, and South America, Ford Motor Company, a position to which she was named in October 2003. At Ford, she was responsible for all operations in each country, including product development, manufacturing, purchasing, finance, and sales and marketing. Ford Motor Company's first female group vice president, Stevens joined the company in 1990 as a marketing specialist in the Plastic Products Division, Vehicle Exterior Systems. In 1992, she was named manager of the Quality Services Department at the Saline (Michigan) plant. She is a member of NAE.

GEORGE TAMARO is a partner at Mueser Rutledge Consulting Engineers. His technical interests are primarily in structural and geotechnical engineering. His work involves a broad range of analytical, design, and construction problems related to deep foundations and underground structures. He is also involved in the design and construction of containment facilities and the control of dam seepage using special barrier systems. Mr. Tamaro also has an interest in the preparation and training of young engineers who will someday be consultant engineers. He is particularly concerned with the development of engineers capable of analyzing, designing, and installing safe, economically constructed facilities. He is a member of NAE.

MARIE C. THURSBY is a member of the strategic management faculty and holds the Hal and John Smith Chair in Entrepreneurship at Georgia Institute of Technology. Before joining Georgia Tech in 2002, she was a member of the economics faculty at Purdue University, where she held the Burton D. Morgan Chair of International Policy and Management. Dr. Thursby has developed and directed three major multidisciplinary programs for research and curriculum development, including Purdue's Center for International Business Education and Research; the Technology Transfer Initiative; and the Innovation Realization Lab, which teams Ph.D. students in science and engineering with M.B.A. students to focus on the interface between technical, management, and economic issues involved in moving fundamental research into the marketplace.